Coastal Environments of India

P. Thambidurai · Anil Kumar Dikshit
Editors

Coastal Environments of India

A Comprehensive Approach

Springer

Editors
P. Thambidurai
Department of Coastal Disaster Management
School of Physical, Chemical and Applied Sciences
Pondicherry University—Port Blair Campus
Sri Vijaya Puram, Andaman and Nicobar Islands, India

Anil Kumar Dikshit
Environmental Science and Engineering Department
Indian Institute of Technology Bombay
Mumbai, Maharashtra, India

ISBN 978-3-031-97340-6 ISBN 978-3-031-97341-3 (eBook)
https://doi.org/10.1007/978-3-031-97341-3

This Springer imprint is published by the registered company Springer Nature Switzerland AG
The registered company address is: Gewerbestrasse 11, 6330 Cham, Switzerland

Preface

Coastal environments are one of our planet's most essential and complex environments. It is the most productive and valued ecosystem. It is an integrated system of terrestrial, marine, and atmospheric processes that all interact in shaping it. The dynamic drivers and functions on the continental margin include the estuarine, nearshore areas and the continental shelf. Around 60% of major cities are located in coastal zones of the world, and 40% of the population on the planet lives within 100 km of a coastal zone. Since the majority of humans have intertwined relationships with coastal resources, it is amplifying the questions of sustainability and development in our world today. The present book envisages serving as a comprehensive collection of chapters highlighting various challenges in the coastal environment in the present times. This could expand the knowledge resources and may collaboratively transform the coastal ecosystems to achieve a more sustainable future while ensuring balanced developmental activities. The scope of the book was intended to include the following themes:

Marine Coastal Ecosystem: Estuaries include mangrove forests, salt marshes, seagrass meadows, coral reefs, littoral forests, mudflats, tidal pools, and an intertidal ecosystem.

Coastal Biogeochemistry: Ocean Biogeochemistry, Weathering, Leaching, Nutrient cycle, Coastal upwelling, Bioturbation, Putrid system, Gas hydrates, and Bioremediation.

Coastal Hydrology: Hydrology and hydro-geochemistry of the coastal aquifer system, Groundwater and tidal dynamics, Seawater intrusion, Submarine groundwater discharge (SGD), Coastal palaeohydrology, Impact of land reclamation on coastal groundwater systems, Coastal aquifer management and seawater intrusion control, Sea level change, and Coastal aquifers.

Coastal Hazards: Storms, Flooding and erosion, Pollution, Trawling & human development, Coastal hazards, and Climate change.

Coastal Management and Policies: Coastal management policies, Coastal zone management, Coastal area management policies, Integrated coastal zone management and mitigation.

We plan to run out various volumes of this book series, each specific to other coastal zones of the globe, with the present volume being dedicated specifically to the Indian coasts.

We expect this book to serve the interests of Researchers, Scientists, Policymakers, NGOs, Social Scientists, Graduate and Post-Graduate Students, Research scholars, and Professors, having a keen involvement in coastal environments.

Sri Vijaya Puram, India P. Thambidurai
Mumbai, India Anil Kumar Dikshit

Acknowledgements The editors are thankful to all the authors for contributing their chapters to this book and for their unfailing commitment and cooperation.

Introduction

This book has fourteen chapters in all. First Chapter is about the "Vertical Distribution of Meiobenthose from the Mangrove Sediments of Port Blair, South Andaman". The study results highlighted how sites affect the response of a given genus of the nematode to the sedimentary environment, and it is the first of its kind in the Andaman Islands. Second Chapter is "Study on Mangrove Ecosystem in Northern Kerala by Using Remote Sensing and GIS Techniques". The study showed most of the mangrove vegetation in the Kannur district of Kerala is located along the backwater channels and on the banks of estuarine water bodies in the form of patches or narrow continuous belts. The vegetation of mangrove forests in this area was estimated using remote sensing and GIS techniques, which provide quantitative information on understanding the spatial distribution of mangrove forests. Third Chapter presents "Biogeochemical Dynamics of Pulicat Lagoon, Bay of Bengal in India". This study examined the ecological implications of the hydrodynamic conditions, such as tidal actions and current speeds, revealing that salinity, dissolved oxygen, and water transparency were significantly influenced by chlorophyll-a concentrations. Moreover, modelling results showed that rapid tidal mixing constrained nutrient availability and reduced water transparency in certain areas. The title of Fourth Chapter is "Geochemical and Microfaunal Studies of Sediments from Off-Shore Krishna-Godavari Basin: Implications for Palaeo-Environment and Hydrocarbon Prospecting". Pyritized nature and less abundance of the faunal populations coupled with high organic carbon values (0.9–2.9%) and clarification of organic matter as evidenced by the presence of lignite suggested anoxic conditions and plausible source rock for hydrocarbon accumulation. The interpretations of this study were consistent with the exploration results, which proved this well to be one of the few productive oil wells in the KG Basin. "High-Energy Deposits in the Shadow Zone of Rameswaram Island, Southeast Coast of India", are discussed in Fifth Chapter. The work was an attempt to understand the dynamics of the 2004 Indian Ocean tsunami (IOT) waves in the Rameswaram shadow zone (East coast of India) formed by virtue of the geographical position of Sri Lanka. The present study concluded that the swash-zone sediments were transported and overlain by tsunami sediments. Simultaneously, the dune sediments were liquified and deformed but were not significantly covered by the tsunami deposits.

Sixth Chapter presents a "Recent Trend Analysis of Groundwater Water Level Fluctuations in Coastal Aquifers of Lower Palar Basin, Tamil Nadu, India". In the present study attempt, the trends in groundwater levels from 2005 to 2018 were analyzed using Box-Jenkins and ARIMA models in the coastal plains of Palar River Basin, Tamil Nadu. The study results showed that about 12.82% of the total area falls under the 'Excellent Fit', 34.62% of the total area falls under the 'Good Fit', 18.59% of the total area falls under the 'Poor Fit' and 33.97% of the total area falls under the 'No Fit' groundwater recharge zones. "Simulation of Groundwater Flow in Coastal Aquifer of Ramanathapuram, Tamil Nadu, India", has been discussed in Seventh Chapter. This study demonstrated the effectiveness of numerical modelling in managing coastal groundwater resources and provided a robust tool for predicting future water dynamics and informing sustainable water management strategies. The findings underscore the importance of detailed hydrological models in addressing groundwater challenges in coastal regions, offering valuable insights to manage water resources sustainably. Eighth Chapter is about "Estimation of Submarine Groundwater Discharge in Coastal Aquifer of South Andaman, India", which is the first effort utilizing an integrated hydrological modelling technique to evaluate SGD along the South Andaman shoreline. The results, obtained under steady-state conditions, indicated that the current daily average SGD was approximately 0.0034 m^3/day per meter of shoreline. Additionally, the study reveals a strong exponential relationship between SGD and groundwater levels in the coastal unconfined aquifer, emphasizing the dynamic nature of groundwater-sea interactions. "Heavy Metals Eminence in Coastal Groundwater of Manali Industrial Region in Chennai, India", has been presented in Ninth Chapter, Groundwater representing shallow aquifers was collected and analysed for trace elements. The concentration of Al, Cr, Mn, Fe, Co, Ni, Cu, Zn, Pb, As, Se, Mo, Cd and Ba ions were determined using Inductively Coupled Plasma Mass Spectrometry (ICP-MS Agilent 7700X). The calculated heavy metal pollution index (HPI) values in the study area ranged from 32.53 to 1423.75, with an average index of 156.79. The values of HPI based on mean concentration were found to be 156.79, which is very high to the critical pollution index value of 100. The spatial distribution maps of heavy metals and HPI values suggested that the southwest and central-eastern parts of the region, especially the Manali and Ennore areas, are highly contaminated groundwater with the metals.

Tenth Chapter discusses the "Shoreline Dynamics in Thiruvananthapuram District of Kerala, India", where the study aimed to detect and understand the village-wise dynamics of Shoreline change (SLC) in Thiruvananthapuram district, Kerala, India, between 2003–2014 and 2014–2023. A significant increase in the erosional patterns was found in the second decade (2014–2023). It intensely demands constant monitoring, and urgent coastal management solutions are the need of the hour. Furthermore, immediate mitigative measures could be beneficial to prevent coastal erosion by the stakeholders, policymakers, and government bodies. Similarly, Eleventh Chapter, titled "Analysing Shoreline Changes: A Case Study of Ernakulam Coast in Kerala, India" is also on SLC. Here the study attempted to comprehensively assess shoreline dynamics on the Ernakulam coast of Kerala using remote sensing images and geographical information systems (GIS) over the last 35 years, from 1988 to 2023. The result revealed significant changes involving erosion accretion on the entire coast of Ernakulam. Additionally, the study observed that no region remained unchanged during the study period. "Assessment and Prediction of Shoreline Change Using Geospatial Techniques: A Case Study of North Goa Coast of India" was the topic of focus of the Tewlfth Chapter. The study identified areas of vulnerability that need prompt intervention along the North Goa coast of India using several multi-temporal satellite data and geo-statistical techniques for the period 1992 to 2022. The authors estimated the shoreline change rates and also forecasted future shoreline positions from 2022 to 2030 and 2050. They predicted an average shoreline positional change of 7.10 m from 2022–2030, with an anticipated rise to 24.84 m by 2050. While from 2022 to 2030, the coastline will shift by –26.13 m.

Thirteenth Chapter is on the "Abundance of Microplastic Pollution in Sediment of Puducherry Coast". The present study focuses on the availability, distribution, and impact of season on microplastic distribution in sediments from 2021 to 2023 on the continental shelf off the Puducherry coast, a Union territory of India. The study result reveals that the abundance of microplastics in the marine bottom sediments ranged between 5.25 ± 4.774 and 12 ± 10.469 particles/kg^{-1}. The size of the microplastics isolated from the bottom sediments from the study areas was found to have ranged between 60 μm and 800 μm. A total of 734 particles were observed from bottom sediments out of which fiber was observed to be the most dominant in the study. The availability of polymers was specific to the coastal zone. The presence of Polyvinyl alcohol (PVA), Polyethylene terephthalate glycol (PET-G) and Polyvinyl alcohol (PVA) in the Puducherry marine environment aligns with the region's status as a tourist hotspot, introducing the possibility of these polymers originating could be tourist-related activities. The "Occurrence of Microplastics in Coastal Waters of Kerala" has been presented in Fourteenth Chapter. The presence of microplastics in the coastal water samples collected from the southwestern coast/Malabar coast of Kerala at 40 sites covering five districts, including Kollam, Alappuzha, Ernakulam, Thrissur, Malappuram, and Kozhikode, was ascertained. Finally, the study confirmed the presence of polypropylene (PP), polyamide, nylon, and polyvinylpyrrolidone and that these had the influences of physico-chemical parameters of water.

Having the typical scope of each chapter, now the reader should be free to go through the book in a way he/she likes.

P. Thambidurai
Anil Kumar Dikshit

Contents

About the Editors

Dr. P. Thambidurai is an Assistant Professor in the Department of Coastal Disaster Management at Pondicherry University—Port Blair Campus, Sri Vijaya Puram, Andaman and Nicobar Islands, India. He received his Ph.D. in Hydrogeology from the Department of Earth Sciences, Indian Institute of Technology Bombay, India, and the *Gian Maria Zuppi* scholarship for high-achievement in research in the international student category, awarded by the University of Venice during his doctoral degree. He completed his graduate and post-graduate degrees in Applied Geology from Periyar University, Salem, India. He joined Amrita Vishwa Vidyapeetham, Kerala, as a Research Associate after finishing his Ph.D. in 2015 and then served there as an *Assistant Professor* till 2018. He has intensive field and consultancy experience in Geo Technical Mining Solutions, Dharmapuri, Tamil Nadu, India.

He has vast field experience in hydrogeology, where he contributed to building the national-level policy called *Revitalization of Rivers in India Draft Recommendation—The Fundamentals* to improve the groundwater and surface water resources. This River policy has been adopted by the Government of India—Ministry of Environment, Forest and Climate Change and Jal Sakthi to Rejuvenate 13 major rivers in India. He has published about 30 scientific articles in international journals, conference proceedings, and book chapters.

Dr. Anil Kumar Dikshit is Higher Administrative Grade (HAG) Professor in the Environmental Infrastructure and Clean Tech (EICT) Group and In-charge of GIS Laboratory for Environmental Management and Sustainability (GEMS) of the Environmental Science and Engineering (ESE) Department at Indian Institute of Technology, Bombay, India.

He did his graduation and masters' in Civil Engineering in 1985 and 1987, respectively, and completed Ph.D. in Civil and Environmental Engineering from Cornell University, USA, in the year 1994.

He has more than 35 years of experience in Teaching, Research and Consultancy at Cornell University, USA; IIT Kanpur, IIT Kharagpur and IIT Bombay, India; Kigali Institute of Science and Technology, Rwanda; University of Kwazulu-Natal, South Africa; Nanyang Technological University, and National University of Singapore, Singapore; Karlsruhe University, Germany; INSA de Lyon, France; Mälardalen University, and Blekinge Institute of Technology, Sweden; Royal Bhutan Institute of Technology, Bhutan; and Asian Institute of Technology, Bangkok. He has guided 40 masters' and 26 doctoral scholars while 6 research scholars are working on their PhDs. He has also worked on more than 150 national and international research and consultancy projects. The results of academic research work have provided 2 patents, 10 books, 11 chapters in edited books, 200 reports, 150 publications in international and national journals, 140 in international and national conferences, while many more papers are under review. In addition, more than 20 prototypes/equipment related to various environmental pollution management topics; academic versions of more than 10 environmental modeling and simulation software(s); and about 7 GIS applications on environmental management problems have been developed under his guidance.

Vertical Distribution of Meiobenthose from the Mangrove Sediments of Port Blair, South Andaman

N. Mohammed Afnan, C. K. Shahal Miraj, and P. M. Mohan

1 Introduction

Mangroves are extreme halophiles, and these woody plants are mainly seen in the intertidal regions of the tropics and subtropics, i.e., 30° N to 37° S (Kathiresan & Bingham, 2001). They are also regarded as one of the most carbon-dense ecosystems in the world, with a very high productivity rate. Given its abundance, India ranks fourth globally for mangroves (Ragavan et al., 2021). Mangroves and their role as shoreline protectors also support many organisms. These include macrofauna, meiofauna, and microfauna (Chatterjee et al., 2021). Meiofauna plays a major role in the nutrient cycling of the mangrove sediment. There have been significant studies on the meiofauna composition of Indian mangroves (Kondalarao, 1983; Abdullah & Lee, 2017). This group of organisms is the most abundant and, at the same time, diversified fauna in the mangroves. They benefit mangroves, including sediment bioturbation and organic matter recycling (Michelet et al., 2021). The meiofauna exhibits patchy distribution in vertical and horizontal zones of sediments (Sahoo et al., 2013). Meiofauna is considered an essential part of the coastal ecosystem's function. They formed a vital link between the macro-benthic and primary producers. Mangrove meiofauna are critical and integral components of the mangrove ecosystem that determine its structure and function. They are effectively described as the essential connection links to the higher trophic levels as they devour the organic matter in the mangrove sediments and are, in turn, eaten by the fishes and other scavengers present in the marine environment (Rosado-Solorzano & Guzman del Proo, 1998; Savurirajan et al., 2012). They are also considered tools for detecting environmental changes and health because of the high species diversity, short life cycle, abundance, and wide distribution (Hu et al., 2023). Meiofauna is crucial to the litter decomposition of mangrove

N. Mohammed Afnan · C. K. Shahal Miraj · P. M. Mohan (✉)
Department of Ocean Studies and Marine Biology, Pondicherry University - Port Blair Campus, Sri Vijaya Puram, Andaman and Nicobar Islands, India
e-mail: pmmnpu@gmail.com

P. Thambidurai and A. K. Dikshit (eds.), *Coastal Environments of India*,
https://doi.org/10.1007/978-3-031-97341-3_1

sediment (Nagelkerken et al., 2008). Furthermore, large numbers and short lifespans made this organism the most important environmental indicator by way of its proliferation or decrease in its diversity based on the ecological benefit or stress. However, their studies are very limited, and their understanding is primarily neglected by biologists as well as environmentalists. Mangrove in the Andaman Islands is one of the unique locations in India, and the meiofauna studies have been documented in the past four decades in these coastal regions (Rao, 1980; Ansari & Ingole, 1983; Mohan et al., 2012; ; Savurirajan et al., 2012; Mohan & Dhivya, 2013; Arunima et al., 2023). No pertinent literature mentions the vertical distribution patterns of meiofauna from the Andaman Islands. Hence, the present study attempted to identify the vertical variation of meiofauna in the mangrove sediment in sites specific to Port Blair coastal environment patches.

2 Materials and Methods

2.1 Study Area

A total of 8249 km^2 area is exhibited for the Andaman and Nicobar group of islands, of which 6408 km^2 shows the Andaman Islands and 1841 km^2 covers the Nicobar group of Islands. There are 872 islands located at 193 km from Cape Negrais in Myanmar. The islands are situated 1255 km away from Kolkata and 1190 km from Chennai. There is a total of 872 islands, out of which humans inhabit 38 islands. This group of islands comes under the zone of tropical rainforests, and the tropical hot and humid climate with abundant rain has resulted in a rich and varied flora and fauna diversity in the region. Sample collections were undertaken from mangrove habitats in and around the Port Blair coastal environment in three different stations, viz., Burmanala (BU), Carbyns Cove (CC), and Chidiyatapu (CT), were selected and surveyed for the presence of meiofauna in mangrove sediments (Fig. 1). One-time sampling was carried out in all the locations and nine samples were collected from each location, with three vertical samples in each station. The study location, Burmanala, is situated in the east of the South Andaman. Burmanala is an intertidal area spread over 2.4 km in length (Fig. 2a). The area has mangrove vegetation on the northern and southern tips that serve as nursery grounds for several species of fish and invertebrates. It has rich faunal diversity belonging to major phyla Porifera, Coelenterata, Annelida, Arthropoda, Mollusca, and Echinodermata, and includes many of the minor phyla and several vertebrate classes. This meiofauna composition serves as a prime feeding site for several species. *Rhizophora* sp. was the dominant species of mangroves in this area. The station Carbyns Cove had muddy sediment (Fig. 2b) and was relatively polluted due to human interference, which provides a very characteristic environment. This creek is a highly dynamic mangrove-dominated creek due to the seawater influx during high tide twice a day. *Avicennia marina* is dominated by this mangrove creek, followed by *Rhizophora* sp. The study site,

Chidiyatapu, is located at the southeastern end of the Sri Vijaya Puram (Port Blair). This pebbly coastal area has coarse sand and detritus mixed with muddy sediments (Fig. 2c). The mangrove in this area is in a creek close to the sea and covers a large stretch of intertidal environment parallel to the coast. *Rhizophora* sp. was the dominant species in this area. Species such as *Avicennia marina*, *Bruguiera* sp., and *Sonneratia* sp. were also present in this environment.

2.2 Sample Collection

The samples were collected in each location where the dominant mangrove species were presented during low tide from the mangrove forest floor with the help of a plastic hand-held corer, which specified 15 cm length and 5 cm diameter contained with plastic tube as a corer, it able to penetrate at a depth of 9 cm from the surface. The collected samples were made into three slices at a depth of 0–3 cm, 3–6 cm, and 6–9 cm, and each slice was kept separately in polythene bags, and further $MgCl_2$ was added as a narcotizing agent. The in-situ parameters, such as salinity, pH, and temperature of pore water, were measured during the sampling activity. Before the samples were preserved in the laboratory, *Rose Bengal* was used as a stain at 1% concentration and in a 4% buffered formalin solution for a day. Further, the sediment samples were washed thoroughly through 0.5 mm and 0.063 mm size sieves for meiofauna separation. The sieved samples retained on 0.063 sieves were decanted 7–8 times to remove sediment, and finally, the organisms were preserved in 4% formalin for long-term storage. The meiofauna were sorted into different groups and were enumerated under a stereomicroscope (Leica M205C). The species identification was carried out based on the key proposed by Higgins and Thiel (1988), Giere (2009), and Rao (2021). Meiofauna was mounted in glass slides based on the (using the formalin-ethanol-glycerol) technique suggested by Seinhorst (1959), and later, it was identified as possible as to the level of taxa under a DIC compound microscope (Olympus BX 51). The *in-situ* soil temperature was measured by a Mextech ST9264 digital thermometer. The salinity was measured from sediment pore waters using an ATAGO hand-held refractometer. The dissolved oxygen was measured from the surface sediment waters using the Winkler method.

2.3 Statistical Analysis

The populations of different organisms identified in the sediment were compared across the different study sites. Bray–Curtis similarity index analysis was performed to quantify the similarity of the species composition between three different sites based on the relative abundance of each species at each site. The studies were used to evaluate the results, which ensured the quality of output with respect to different

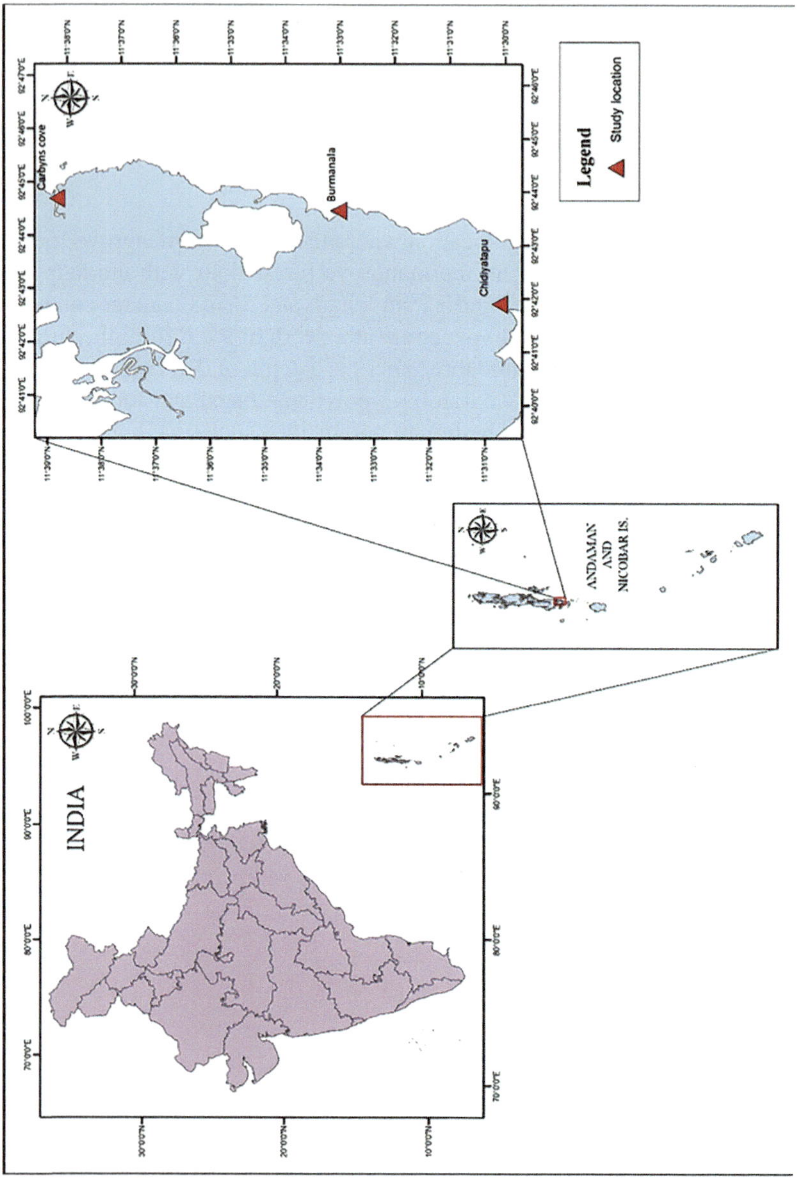

Fig. 1 Location map of sampling sites

Fig. 2 Study sites, **a** Burmanala, **b** Carbyns Cove, **c** Chidiyatapu

locations between the species' similarities. The result arrived through the standard Primer 7 software using biological samples.

3 Results and Discussion

A total of 3109 organisms were identified under vertical variation of meiofauna distribution from different stations in South Andaman. These organisms belonged to 9 groups, namely—nematoda, polychaeta, copepoda, foraminifera, diatoms, gastrotricha, kinorhyncha, oligochaeta, and halacaroidea, along with a very few numbers of organisms categorized as "others," which include some identified (Decapod larva, Pteropoda, Ostracoda, Platyhelminthes) and unidentified taxa. The study mainly focused on Chidiyatapu (CT), Carbyns Cove (CC), and Burmanala (BU), and among the three stations, half the number of meiofauna was found (Fig. 3) from the sediments of Carbyns Cove (1558 individuals), followed by Burmanala and Chidiyatapu with 824 individuals (27%) and 727 organisms (23%), respectively.

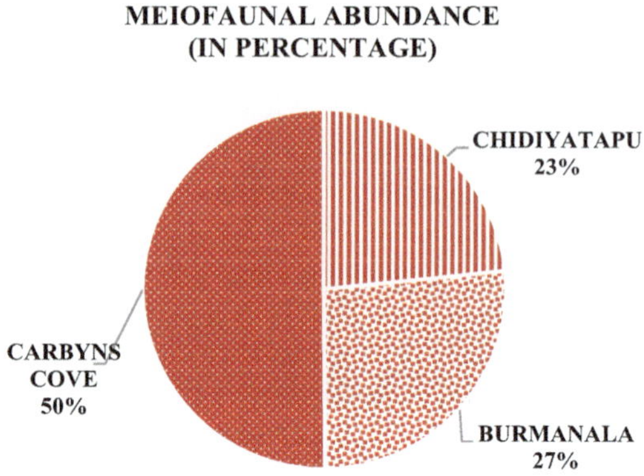

Fig. 3 Percentage of meiofauna observed from the three sampling sites

3.1 Meiofauna Composition

Nematodes were the most abundant organisms recorded in all three sampling sites, accounting for 73% of the total meiofauna population (Fig. 4). The overall abundance was followed by diatoms (9%), copepoda (7%), and foraminifera (5%).

Polychaeta was found in only about 2% of the population, while the remaining groups (gastrotricha, kinorhyncha, Oligochaeta, halacaroidea, and others) were at just 1% each. Among these groups, polychaeta and gastrotricha were not recorded from the Carbyns Cove sediments, while kinorhyncha and oligochaeta were found only from the Carbyns Cove site. A total of 54 species of meiofauna were encountered in total during this study is presented in Table 1. Out of which 19 species were from nematoda, eight species from diatoms, seven species from foraminifera, six species from harpacticoid copepoda, followed by three species each from polychaete larvae and oligochaeta, two species from gastrotricha, and one species from kinorhyncha. Further, the species were identified till the genus level (Fig. 5). The identified nematoda belong to 13 different families and the species are *Anticomopsis*

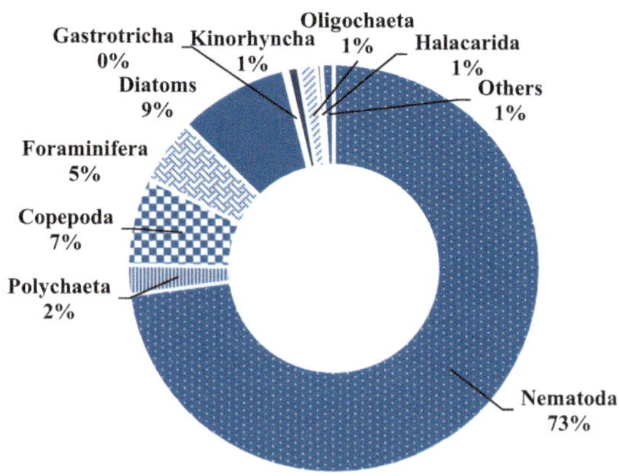

Fig. 4 Overall meiofauna composition

sp., *Halaimus* sp., *Thoracostoma* sp., *Anoplostoma* sp., *Cobbia* sp., *Elzalia* sp., *Echinotheristus* sp., *Daptomena* sp., *Linhomoeus* sp., *Terschellingia* sp., *bathyespsilonema* sp., *Desmodora* sp., *Microlaimodes* sp., *Desmoscolex* sp., *Tricoma* sp., *Paracomesoma* sp., *Sabatieria* sp., *Pselionema* sp. and *Leptolaimus* sp. Eight species of diatoms were also identified, namely *Biddulphia* sp., *Gyrosigma* sp., *Trachyneis* sp., *Actinoptychus* sp., *Surirella* sp., *Grammatophora* sp., *Fragilaria* sp. *and Chaetoceros* sp., belonging to eight families. The Foraminifera identified belonged to 7 families and three orders, and the species are *Ammonia* sp., *Bolivina* sp., *Elphidium* sp., *Quinqueloculina* sp., *Spiroloculina* sp., *Spirillina* sp. *and Ammodiscus* sp. Among the Phylum Arthropoda, six species of Copepoda belonging to five families and a crab larva were recorded. The Copepod species identified were *Ameira* sp., *Cletodyes* sp., *Apodospyllus* sp. *Ectinosoma* sp., *Microsetella* sp., and *Euterpina* sp. Among the phylum annelida, three species were identified from class polychaeta. *Nereis* sp., *Pisione* sp., *Syllides* sp. and three were from class oligochaeta, *Aktedrilus* sp., *Olavius* sp. *and Randiella* sp. Two species of gastroticha, *Macrodasya* sp. and *Pseudostomella* sp., and one species of kinorhyncha (*Echinoderes*) were also recorded.

3.2 *Vertical Distribution of Meiofauna*

The observations on the vertical distribution of meiofauna from mangrove sediments were carried out separately for each depth range from 3 to 9 cm as follows: top 3 cm, 3–6 cm, and 6–9 cm. The general trend of meiofauna groups was found to decrease in number with increasing depth, with only a few exceptions (Table 2). The top layer sediment showed the maximum number of individuals (1491), followed by 3–6 cm and 6–9 cm with 1060 and 558 individuals, respectively (Fig. 6). The samples from CC showed a significant reduction in numbers regarding the increase in depth (Fig. 7). The CT sample also showed highest abundance in 0–3 cm, while 3–6 cm and 6–9 cm had the same number of individuals. This was due to more organisms other than nematodes in 6–9 cm sediment than 3–6 cm. In the case of Burmanala, depth 3–6 cm samples had the most organisms, followed by 0–3 cm and 6–9 cm. This could be due to coarse sediment in the site, through which water percolates faster, causing organisms to migrate to deeper areas.

Nematoda was the most abundant group in all the segments listed above, except for the deeper sediments (6–9 cm) of Burmanala, where benthic diatoms dominated. The diatoms were observed in considerably high numbers in all three layers of the BU sample, showing a distribution pattern opposite to that of Nematoda. In Burmanala, the deepest layer (6–9 cm) had the highest number of diatoms, followed by the middle layer (3–6 cm), and the least at the top layer (0–3 cm). They were considerably less numbered in other stations and were absent in the CT between 3 and 6 cm sediment sample (Table 2). Copepoda and foraminifera were shown from all the sediment sections, similar to nematodes, although in significantly lesser numbers, making them the third and fourth most abundant groups, respectively (Fig. 8).

Table 1 Classification of meiofauna taxa collected during the study

Group	Phylum	Class	Order	Family	Genus/species
NEMATODA	Nematoda	Enoplea	Enoplida	Anticomidae	Anticomopsis sp.
				Oxystominidea	Halalaimus sp.
				Leptosomatidae	Thoracostoma sp.
				Anoplostomatidae	Anoplostoma sp.
		Chromadorea	Monhysterida	Xyalidae	Cobbia sp.
					Elzalia sp.
					Echinotheristus sp.
					Daptonema sp.
				Linhomoeidae	Linhomoeus sp.
					Terschellingia sp.
			Desmodorida	Epsilonemstidse	Bathyespsilonema sp.
				Desmoscolecidae	Desmodora sp.
				Microlaimoidae	Microlaimodes sp.
			Desmoscolecida	Desmoscolecidae	Desmoscolex sp.
					Tricoma sp.
			Araeolaimida	Comesomatidae	Paracomesoma sp.
					Sabatieria sp.
			Plectida	Ceramonematidae	Pselionema sp.
				Leptolaimidae	Leptolaimus sp.
FORAMINIFERA	Foraminifera	Globothalamea	Rotaliida	Ammoniidae	Ammonia sp.
				Bolivinitinae	Bolivina sp.
				Elphidium	Elaphidium sp.

(continued)

Table 1 (continued)

Group	Phylum	Class	Order	Family	Genus/species
		Tubothalamea	Milliolida	Hauerinidae	Quinqueloculina sp.
				Spiroloculinidae	Spiroloculina sp.
			Spirillinida	Spirillinidae	Spirillina sp.
				Ammodiscidae	Ammodiscus sp.
DIATOMS	Bacillariophyta	Bacillariophyceae	Biddulphiales	Biddulphiaceae	Biddulphia sp.
			Naviculales	Plleurosigmataceae	Gyrosigma sp.
				Naviculaceae	Trachyneis sp.
			Coscinodiscales	Helipltaceae	Actinoptychus sp.
			Surirellales	Surirellaceae	Surirella sp.
			Striatellales	Striatellaceae	Grammatophora sp.
			Fragilariales	Fragilariaceae	Fragilaria sp.
			Chaetocerotanae	Chaetocerotaceae	Chaetoceros sp.
COPEPODA	Arthropoda	Copepoda	Harpacticoida	Ameiriae	Ameira sp.
				Cletoidae	Cletodes sp.
				Paramesochridae	Apodospyllus sp.
				Ectinosomatidae	Ectinosoma sp.
					Microsetella sp.
				Tachidiidae	Euterpina sp.
POLY- CHAETA	Annelida	Polychaeta	Phyllodocida	Nereididae	Nereis sp.
				Sigalionidae	Pisione sp.
				Syllidae	Syllides sp.
				Polychaete larva	

(continued)

Table 1 (continued)

Group	Phylum	Class	Order	Family	Genus/species
OLIGO- CHAETA		Oligochaeta	Tubificida	Naididae	Aktedrilus sp.
					Olavius sp.
			Randiellida	Randielldae	Randiella sp.
GASTROTRICHA	Gastrotricha		Macrodasyida	Macrodasyidae	Macrodasya sp.
				Thaumastodermatidea	Pseudostomella sp.
KINORHYNCHA	Kinorhyncha	Cyclorhagida	Echinorhagata	Echinoderidae	Echinoderes sp.
OTHERS	Mollusa	Gastropoda	Pteropoda		Unidentified sp
	Platyhelminthes				Unidentified sp
	Arthropoda	Malacostraca	Decapoda	Crab Larva	
		Ostracoda			Unidentified sp

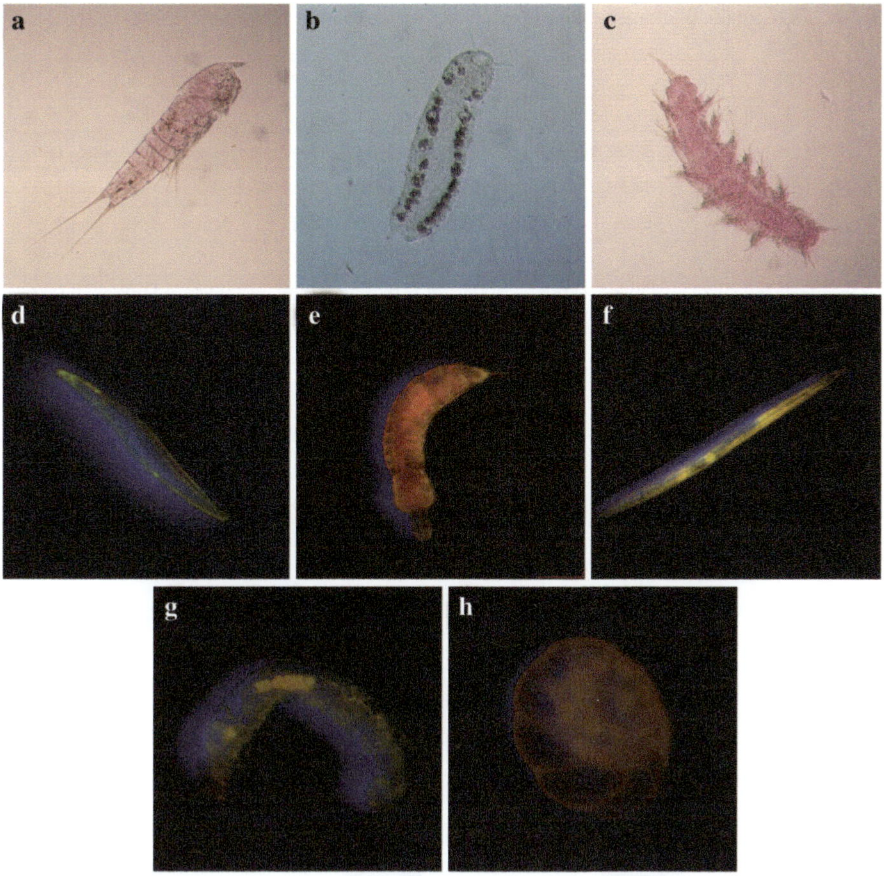

Fig. 5 Identified Meiofauna groups in the study. **a** Harpacticoida, **b** Gastrotricha, **c** Polychaete Larvae, **d** Diatom, **e** Kinorhyncha, **f** Nematoda, **g** Oligochaeta, **h** Foraminifera

Meanwhile, polychaeta were absent in all the samples collected from the CC site. They were observed in all segments of CT, with more numbers found in 0–3 cm, followed by 6–9 cm and 3–6 cm, respectively. In Burmanala, polychaeta were seen only from the 3–6 cm layer. kinorhyncha and Oligochaeta were found only in the CC site samples. While the top layer had the highest number of kinorhyncha, oligochaeta was seen mainly in the middle layer (3–6 cm) and was very scarce on the other two layers. Halacarid prefers the middle layer (3–6 cm), as observed in all stations, but only one specimen was collected from the top layer (BU 0–3 cm). Gastrotricha was utterly absent in the CC site sediments, but it was present only in the middle layer of the BU site. The top and deep layers of the CT site also recorded gastrotricha in a limited number. There were 29 individuals from different identified (decapod larva, pteropoda, ostracoda, platyhelminthes), and unidentified groups were also recorded.

Table 2 Distribution of meiofauna taxa across various depths in all three stations (NE—Not Encountered)

Name of the Group	Stations								
	Chidiyatapu			Burmanala			Carbyns Cove		
	0–3 cm	3–6 cm	6–9 cm	0–3 cm	3–6 cm	6–9 cm	0–3 cm	3–6 cm	6–9 cm
Nematoda	281	141	124	094	217	075	742	435	157
Polychaeta	050	002	010	NE	012	NE	NE	NE	NE
Copepoda	042	011	017	024	057	002	035	013	015
Foraminifera	012	004	005	046	032	026	027	006	003
Diatoms	004	NE	001	058	065	102	029	009	002
Gastrotricha	005	NE	002	NE	002	NE	NE	NE	NE
Kinorhyncha	NE	NE	NE	NE	NE	NE	021	004	003
Oligochaeta	000	000	000	000	NE	NE	004	036	003
Halacaroidea	NE	002	001	001	003	NE	NE	002	004
Others	007	003	003	004	003	001	005	001	002
TOTAL	**401**	**163**	**163**	**227**	**391**	**206**	**863**	**506**	**189**

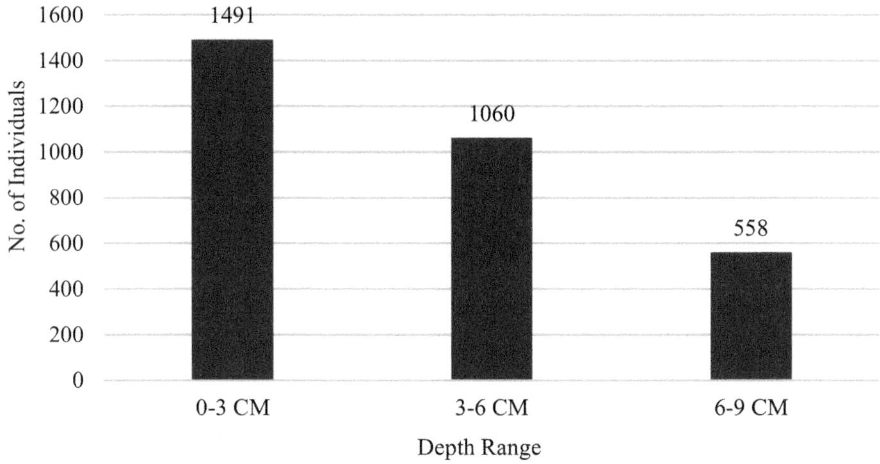

Fig. 6 Total number of individuals across different vertical sections

3.3 Diversity Indices

A Bray - Curtis Similarity Index analysis was performed on the identified organisms in dissimilar species with respect to different locations, as presented in Fig. 9. The cluster was calculated for all samples from the same stations into the same clade, except Burmanala. While the samples BU 3 (0–3 cm) and BU 9 (6–9 cm) were kept in the same clade, BU 6 (3–6 cm) showed a different pattern and clustered with the

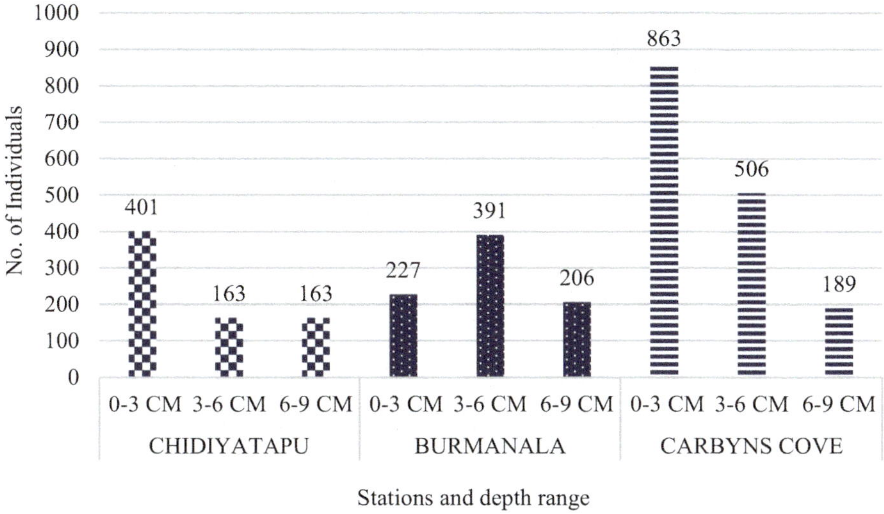

Fig. 7 Number of individuals from each vertical section of different locations

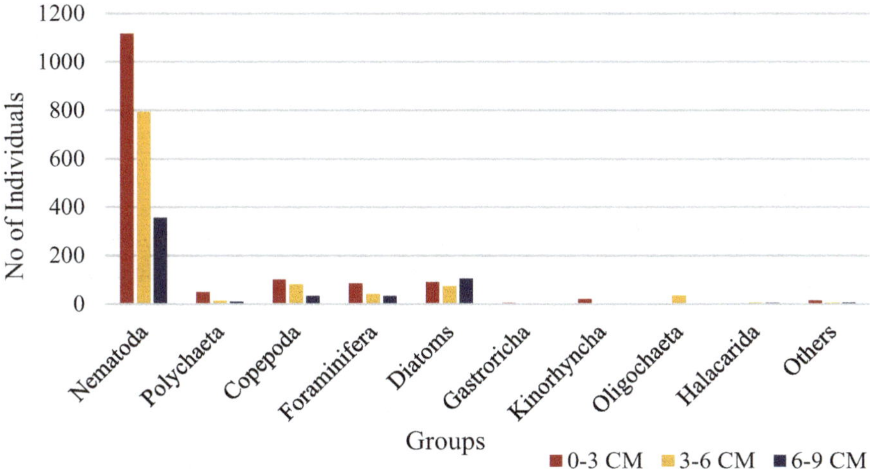

Fig. 8 Distribution of meiofauna groups in different depth ranges

CT samples and had more similarity to that of CT 9 (6–9 cm). As the index tries to determine the similarity in species abundance across different environments, the BU 6 showed more similarity towards the species abundance of the CT site samples. This dissimilarity in the BU 6 compared to the top and bottom layers may be because of the significantly higher number of organisms in this layer, which favours more affinity towards the pattern of species abundance in CT samples. Various indices were used to better understand the organism groups and their composition on the different layers of

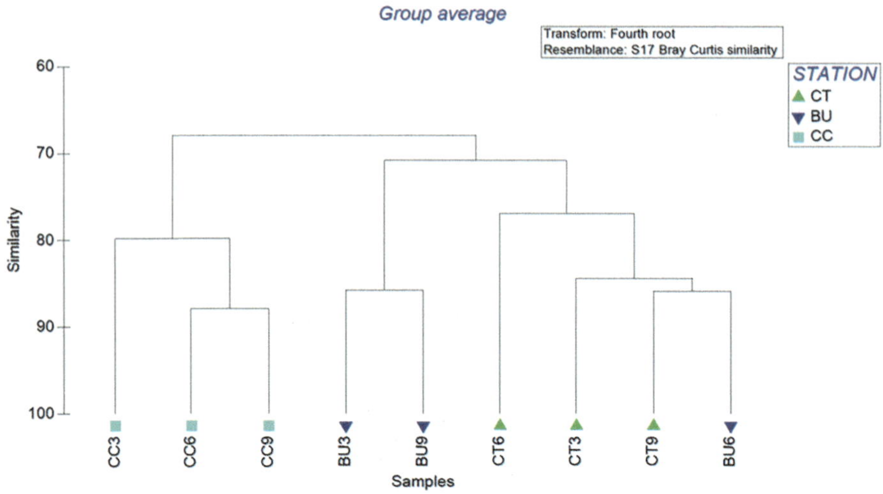

Fig. 9 Statistical summary of cluster organisms in the study

all three sampling sites. Margalef's species richness index (Margalef, 1958) measures the species richness by comparing the total number of species and individuals present in the sample. The index was worked out for the individual vertical sections of all three stations, and the average was calculated for the whole station. The CT and CC have recorded a species richness of 1.1, while BU gave a record of a slightly lesser value than 1 (0.9). The higher species richness value in CT and CC indicates the higher number of taxa in these areas, thus making it more prosperous than the other stations. Pielou's evenness index (Pielou, 1966) was highest in BU (0.7), indicating a higher degree of even distribution of species compared to CT and CC, which had 0.4 and 0.3 values, respectively. Though the number of individuals was higher in CC, an uneven distribution with higher densities of limited taxa was noted. The Shannon-Weiner and Simpson diversity indices showed the highest values in BU ($H' = 1.2$ and $\lambda = 0.7$), followed by CT ($H' = 0.8$ and $\lambda = 0.4$), and lowest at CC ($H' = 0.7$ and $\lambda = 0.3$). The diversity indices also indicate that CC is less diverse, although they have the maximum number of individuals. The absence of several taxa from the station was attributed to the negative impact on the indices.

3.4 *Physicochemical Analysis*

The physicochemical parameters of the three sampling stations were also collected, which can potentially serve as an essential parameter in the species distribution of meiofauna. Salinity was found to be the same, with a value of 34 PSU at BU and CT and slightly less (33.5 PSU) at the CC station. The temperature conditions in CT and BU showed almost the same values (26.5 and 26.8 °C), while that of CC was slightly higher (29.3 °C). In the case of the Dissolved Oxygen (DO), all three stations had similar values. i.e., the DO value recorded from CT and BU was 4.3 while that in CC was 4.2, which was not a significant difference. The pH value of CC was found to be 7.8, while CT and BU recorded 8.3 and 8.4 pH, respectively. All the parameters showed a similar trend between BU and CT, while a slight variation was observed in CC. Another main difference between the three stations was the sediment texture. The CT and BU were both sandy environments, but the former had silty sand while the latter had coarser sediment. The CC site had an entirely different texture with muddy sediment. This sediment texture might be one of the main factors that affect the meiofauna vertical distribution and abundance.

3.5 *Discussion*

Meiofauna is an integral part of the mangrove ecosystem, associated with other flora and fauna. A quantitative approach was carried out through the vertical profile to understand the meiofauna distribution in the study areas of South Andaman. The earlier studies of meiofauna in India mainly focused on the surface and a few on the vertical distributions. The present study on their vertical distribution is the first of its kind in the Andaman Islands, providing a scope for comparing the distribution patterns seen in various other marine environments. Nematodes are the most abundant group of organisms recorded from all three locations studied. The study supports the typical structures of mangrove intertidal meiofauna distribution reported previously in the literature from Andaman (Rao & Rao, 1988; Mohan et al., 2012; Ghosh & Mandal, 2019). Arunima et al. (2023) recorded foraminiferans as the dominant group and nematodes as the second most abundant, but from sublittoral sediments. Further, previous studies reported that the occurrence of meiofaunawas more abundant in the uppermost surface of sediment (Dye, 1983; Soto et al., 2015). The present study also finds the same ambience, except for the case of Burmanala, which has a coarse sand sediment habitat. One of the significant differences observed between the three sampling locations is the sediment texture. In addition, Martins et al. (2015) have reported that the sandy sediments used have fewer meiofauna because of the highly hydrodynamic environment. They added that the sandy sediments have the fewest individuals in the topmost stratum under consideration. This variation was also observed in this study, as the muddy sediments of the CC site, which hold more water, had a significantly higher number of organisms. In contrast, the top layer of the BU sample, with coarse sand, had a lesser abundance. Apart from the sediment characteristics, another factor to consider was the dominant mangrove species present in the area. *Avicennia marina* was the dominant mangrove species at CC, which has more robust and thicker roots than other mangrove species, thus holding the soil tighter to prevent erosion. The lesser the erosion, the more the chances of survival for the meiofauna (Asokan et al., 2023). The coarse sandy environment of BU supported the second most abundant population, while the silty sand environment of CT was the least abundant. The BU station also exhibited a good amount of diversity and evenness.

Nematodes were found in higher numbers, even in the deepest layers, in all three locations. The primary trend is that as the depth increases, the oxygen penetration is limited, making a redox potential discontinuity, and further, this can cause a reduction in the groups in deeper layers. The abundance of nematodes even here is attributed to the specialized adaptations shown by nematodes in these environments. A few of the nine groups identified were unavailable in all stations, possibly because they preferred certain sediment types or mangrove species. The present study concluded that the vertical distribution of meiofauna availability is based on individual group characteristics, their survival based on oxic or suboxic conditions, and the amount of detrital availability at vertical depth. Since no earlier study was carried out for vertical profiling of meiofauna in mangrove sediments of South Andaman, it is essential to extend the work on this aspect to understand individual groups' or species' behaviour regarding the depth to be evaluated. However, the surface distribution regarding the station-wise distribution pattern of meiofauna was similar to that of the distribution patterns previously reported from intertidal regions, with minor variations. Future studies, with reference to expected climatic change situations, need to be understood in deeper aspects of vertical profiling and individual meiofauna role in the mangrove environment.

Acknowledgements The authors thank the Head of the Department, Department of Ocean Studies and Marine Biology, Pondicherry University Port Blair Campus, Sri Vijaya Puram, and Pondicherry University for providing the laboratory facilities to complete the research work.

The authors declared no conflict of interest.

References

Abdullah, M, M., & Lee, S, Y. (2017). Structure of mangrove meiofaunal assemblages associated with local sediment conditions in subtropical eastern Australia. *Estuarine, Coastal and Shelf Science, 198*, 438–449.

Ansari, Z. A., & Ingole, B. S. (1983). Meiofauna of some sandy beaches of Andaman Islands. *Indian Journal of Geo-Marine Sciences, 12*, 245–246.

Arunima, T. K. A., Afnan, N. M., Swathi, V., Heera, N. S., Shah, K., & Mohan, P. M. (2023). Distribution of meiofauna in the sublittoral sediments of the tropical island ecosystem, off Port Blair, Andaman Sea. *Indian Journal of Geo-Marine Sciences, 52*, 130–144.

Asokan, H. N., Jaisankar, I., Lele, N., Jerard, B. A., & Jnu-Eiacp. (2023). Geodiversity and impact on environment. *An Overview of Andaman and Nicobar Mangroves -Existing Day State of Affairs, 29*, 3–25.

Chatterjee, T., Ingole, B., & Guru, B. C. (2021). Trends in Mangrove meiobenthic studies in India: An overview. *Acta Biologica, 28*, 67–80. https://doi.org/10.18276/ab.2021.28-08

Dye, A. H. (1983). Vertical and horizontal distribution of meiofauna in the mangrove sediments in Transkei, Southern Africa. *Estuar Coast Shelf S, 16*, 591–598.

Ghosh, M., & Mandal, S. (2019). Does vertical distribution of meiobenthic community structure differ among various mangrove habitats of Sundarban Estuarine System? *Regional Studies in Marine Science, 31*, Article 100788. https://doi.org/10.1016/j.rsma.2019.100778

Giere, O. (2009). *Meiobenthology. The microscopic motile fauna of aquatic sediments* (2nd edn., 527p). University of Hamburg, Springer-Verlag.

Higgins, R. P., & Thiel, H. (1988). *Introduction to the study of meiofauna* (488p). Smithsonian Institution Press.

Hu, M. C., Guo, Y. Q., Shih, Y. J., Liu, K., Li, C. X., Ji, F. F., & Chu, T. J. (2023). Comparison of the meiofauna and marine nematode communities before and after removal of *Spartina alterniflora* in the mangrove wetland of Quanzhou bay. *Fujian Province. Water, 15*, 3829. https://doi.org/10.3390/w15213829

Kathiresan, K., & Bingham, B. L. (2001). Biology of mangroves and mangrove ecosystems. *Advances in Marine Biology, 40*, 81–251.

Kondalarao, B. (1983). Distribution of meiofauna in Gautami Godavari estuarine system. *Mahasagar Bulletin of the National Institute of Oceanography, 16*, 453–457.

Margalef, D. R. (1958). Information theory in ecology. *Genetics and Systematics, 3*, 36–71.

Martins, M. O., Almeida, T. C. M., & Domenico, M. D. (2015). Vertical distribution of meiofauna on reflective sandy beaches. *Brazilian Journal of Oceanography, 63*, 469–480.

Michelet, C., Zeppilli, D., Hubas, C., Baldrighi, E., Cuny, P., & Dirberg, G. (2021). First assessment of the benthic meiofauna sensitivity to low human-impacted mangroves in French Guiana. *Forests, 12*, 338.

Mohan, P. M., & Dhivya, P. (2013). A review on meiofaunal study in India. *Journal of the Andaman Science Association, 18*, 1–24.

Mohan, P. M., Dhivya, P., Sachithanandam, V., & Ragavan, P. (2012). Distribution of mangrove meiofaunal composition in relation to organic carbon and carbonate in Port Blair, South Andaman, India. In B. Nagarajan, C. Kunhikannan, K. R. Sasidharan and N. Krishnakumar (Eds,). *Institute of forest genetics and tree breeding* (pp. 47–54). Coimbatore. ISBN. 978-81-900346-4-7.

Nagelkerken, I., Blader, S. J. M., Bouillon, S., Green, P., Haywood, M., Kirton, L. G., Meynecke, J. O., Pawlik, J., Penrose, H. M., Sasekumar, A., & Somerfield, P. J. (2008). The habitat function of mangroves for terrestrial and marinefauna: A review. *Aquatic Botany, 89*, 155–185.

Pielou, E. C. (1966). The measurement of diversity in different types of biological collections. *Journal of Theoretical Biology, 13*, 131–144.

Ragavan, P., Kathiresan, K., Rana, T. S., Saxena, A., Mohan, P. M., Jayaraj, R. S. C., Ravichandran, R., & Mageswaran, T. (2021). *Indian mangroves: A photographic field identification guide* (p. 208). New India Publishing Company.

Rao, G. C. (1980). On the zoogeography of the interstitial meiofauna of the Andaman and Nicobar Islands, Indian Ocean. *Records of Zoological Survey of India, 77*, 153–178.

Rao, G. C. (2021). *Meiofauna of marine sediments—A fascinating world of animal life* (p. 528). Nature Books India.

Rao, M., & Rao, G. M. N. (1988). Mangrove populations of the Godavari Delta complex. *Indian J Mar Sci, 17*, 326–329.

Rosado-Solorzano, R., & Guzman Del Proo, S. A. (1998). Preliminary trophic structure model for Tampamachoco Lagoon, Veracruz, Mexico. *Ecological Modelling, 109*, 141–145.

Sahoo, G., Suchiang, S. R., & Ansari, Z. A. (2013). Meiofauna-Mangrove interaction: a pilot study from a tropical mangrove habitat. *Cah. Biol. Mar, 54*, 349–358.

Savurirajan, M., Jayabarathi, R., Padmavati, G., & Ganesh, T. (2012). Species composition, abundance and diversity of meiobenthiv fauna in mangrove sediments of South Andaman Islands. In D. R. Singh, et al. (Eds.), *Souvenir, national seminar on innovative technologies for conservation and sustainable utilization of island biodiversity* (pp. 83–89). Central Agricultural Research Institute.

Seinhorst, J. W. (1959). A rapid method for the transfer of nematodes from fixative to anhydrous glycerin. *Nematologica, 4*, 67–69.

Soto, E., Cabellero, W., & Quiroga, E. (2015). Composition and vertical distribution of metazoan meiofauna assemblages on the continental shelf off central Chile. *Latin American Journal of Aquatic Research, 43*, 922–935.

Study on Mangrove Ecosystem in Northern Kerala by Using Remote Sensing and GIS Techniques

P. Ashila Babu, M. A. Mohammed-Aslam, and Pawan Kumar Gautam

1 Introduction

Remote sensing and GIS are effective tools for monitoring and analyzing wetland environments (Berlanga-Robles & Ruiz-Luna, 2002). Coarse-resolution imagery, such as the Landsat multispectral scanner (MSS), has proven valuable for large-scale mangrove forest monitoring and mapping (Howari et al., 2009). However, limited research has explored the use of satellite remote sensing beyond aerial extent mapping and change detection in mangrove studies (Heumann, 2011a, 2011b). Remote sensing provides a fast and efficient approach for mapping and studying mangrove vegetation. The accuracy of vegetation delineation and classification significantly impacts analytical outcomes (Ibrahim-Bathis, 2021). The quality of newly generated maps depends on sensor specifications, analytical methods, and training datasets. Proper data selection and processing are crucial for achieving research objectives related to remote sensing data analysis (Heumann, 2011a, 2011b; Kuenzer et al., 2011a, 2011b; Mumby, 2000; Pavithra et al., 2019; Wang et al., 2003; Kumar et al., 2018; Singh et al., 2018; Gautam et al., 2020; Gautam, 2023; Gautam et al., 2022; Gautam, 2024; Gautam & Kumar, 2025). Accuracy assessment of classified images of mangrove areas is highly complicated (Blasco et al., 1998; Pham et al., 2019). A variety of airborne and satellite sensor systems, along with multiple image processing methods, are available for the study of mangrove ecosystems. Owing to the availability of a huge repository, the multi-temporal data of Landsat

P. Ashila Babu (✉)
Department of Geology, Government College, Kasaragod, Kerala, India
e-mail: ashilababu03@gmail.com

M. A. Mohammed-Aslam
Department of Geology, Central University of Karnataka, Kalaburagi, Karnataka, India

P. K. Gautam
Department of Geology, School of Earth and Environmental Sciences, Babasaheb Bhimrao Ambedkar University, Lucknow, India

P. Thambidurai and A. K. Dikshit (eds.), *Coastal Environments of India*,
https://doi.org/10.1007/978-3-031-97341-3_2

is highly useful in the study of mangrove vegetation. Recently, Sentinel satellite data offers another useful database in coastal wetland studies. Satellite-derived data can be used to carry out the change detection studies. The coastal region is one of the unique landscapes in Kerala. This region covers the most densely populated region. The protection of this region, which is experiencing rapid and severe environmental impact, is of paramount importance. Major problems facing the coastal region coastal erosion are uncontrollable water scarcity, runoff, increased salinity in soil and water, destruction of inland fisheries resources, and nature becoming unsuitable for indigenous farming practices. The reason for these environmental issues is the damage caused to the natural environment of the coastal areas by the irrational interventions of human beings. The present study focuses on the characterization of mangrove vegetation in northern parts of Kerala, India by 'remote sensing' and GIS ('geographic information system') techniques. Approximately 59 species within 41 genera, comprising 29 families, are reported among the Indian mangroves. The west coast of India alone has 34 species in 25 genera, with 21 families. Nearly 25 species of mangrove are confined to the coast in the eastern part of India, which have not been reported on the western coast. Similarly, only 8 types of mangroves are seen on the western coast of India, and they are not found on the eastern coast.

2 Study Area

Kerala is gifted with 44 major rivers, and that maximum number is located in Northern Kerala, including Kasaragod and Kannur districts of the state. The area of coverage of untouched mangroves is mostly scattered in Ernakulam, Kannur (known as the capital of mangroves in Kerala), and Kozhikode Districts. The study is focused on the major parts of the Kannur and Kasaragod districts, which have not been studied much. These districts are bordered by the Arabian Sea to the west and the Western Ghats to the east. The Kannur and Kasaragod districts receive rainfall about 70% during the southwest monsoon, with the rest in the northeast monsoon. The climate of these districts is classified as hot or tropical, warm, and humid. The average maximum temperature reaches 31.2 °C, while the minimum averages 23.6 °C. The southwest monsoon, accompanied by thunderstorms, begins in late May or early June and lasts until September. The northeast monsoon follows in October and November, while the dry season occurs from December to February. The hottest period extends from March to May, with January and February being the coolest months of the year. The study area is geographically positioned between N 11.8689°–E 75.35546° and N 12.5°–E 75.0°. Figure 1 illustrates the sampling locations. The study area map was developed using Survey of India (SOI) toposheets at a 1:50,000 scale. A total of 18 accessible station sites were selected for sampling in this remote sensing study of mangrove forests.

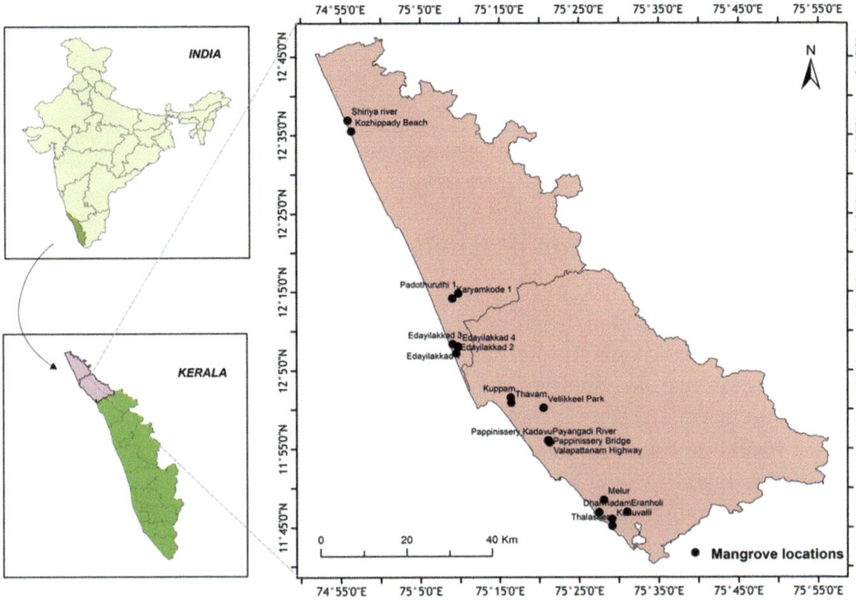

Fig. 1 Location of study area

3 Method and Material

3.1 Data Used (Sentinel-2 Data)

The mission of Copernicus Sentinel-2 contains a group of two polar-orbiting satellites, which is aimed at the study of land surface Fig. 1. The swath width of this satellite is 290 km. This satellite mission provides optical imagery data having a high spatial resolution, which bridges the data continuity for the moderate resolution data of other previous satellite missions. This mission has plentiful potential in the application of land-based studies (Baloloy et al., 2020; Immordino et al., 2019; Manna & Raychaudhuri, 2020; Prakash Hati et al., 2020). The mission consists of two satellites, Sentinel-2A and Sentinel-2B. Sentinel-2A was launched on June 23, 2015, while Sentinel-2B was launched on March 7, 2017. These satellites (Fig. 2) are equipped with the Multispectral Imager (MSI) sensor, which provides high-resolution optical imagery. For this study, two Sentinel-2A images, dated January 12, 2020, were obtained from the ESA Copernicus Open Access Hub (https://scihub.copernicus.eu/dhus/#/home). These images cover the Kannur and Kasaragod districts. The downloaded images underwent pre- and post-processing to ensure accurate analysis and interpretation. Sentinel-2's optical instrumentation includes 13 spectral bands with varying spatial resolutions: four bands at 10 m, six bands at 20 m, and three bands at 60 m (Manna & Raychaudhuri, 2020; Wang et al., 2020; Gautam et al., 2024a, b, c; Chandra et al., 2024). Detailed information on Sentinel-2's multispectral bands is provided in Table 1.

4 Result

4.1 Visual Interpretation of Satellite Data

The mangrove vegetation in the coastal region has a unique appearance that has been studied on the basis of photo interpretation for visual interpretation, such as tone, texture, associations, etc., are important in the analysis (Piovan, 2020; van Genderen, 2016) as follow, (a) The tone of mangrove features varies with the band combination. The best band combination for the discrimination of mangroves is infrared regions with bands 8, 4, and 3 (NIR, RED, and GREEN) in the Sentinel-2 images, (b) Texture is described as having "smoothness" or "roughness" of image features, resulting due tonal variations. The mangrove trees are characterized by bushy plants that appear very rough in the satellite images, (c) Association is one of the best keys to identifying the mangrove in satellite images because mangrove trees are always associated along the estuaries, brackish river banks, sea coast, and near the river delta. Five mangrove localities in the Kasaragod and Kannur districts of Kerala state were selected for the current mangrove vegetation mapping. These five mangrove locations were Shiriya and Pallam Lake in the Kasaragod District and Kunhimangalam, Velichangool, and Vellikkeel in the Kannur districts (Fig. 2). Satellite images of each location in true colour composite and false colour composite images are also shown (Figs. 3, 4, 5, 6 and 7).

4.2 NDVI Images of Mangrove Vegetation

The Normalized Difference Vegetation Index (NDVI) is a measure of vegetation reflectance based on the red (RED) and near-infrared (NIR) spectral bands, with values ranging from -1 to $+1$. NDVI values help assess the relative abundance and health of vegetation (Carlson & Ripley, 1997; Chen et al., 2004; Jiang et al., 2006). Sentinel-2 satellites are equipped with the necessary spectral bands for NDVI calculation: Band 8 (NIR) and Band 4 (RED) (Veloso et al., 2017; Vrieling et al., 2018; Zhang et al., 2017). The NDVI for Sentinel-2 data is computed using Eq. 1.

$$NDVI = (NIR - RED)/(NIR + RED) \tag{1}$$

By calculating the NDVI for the present study area of satellite image, the value was found to range from -0.47 to $+0.80$. Since the mangrove vegetation is an evergreen tropical forest type, the NDVI value of this vegetation has always been higher value,

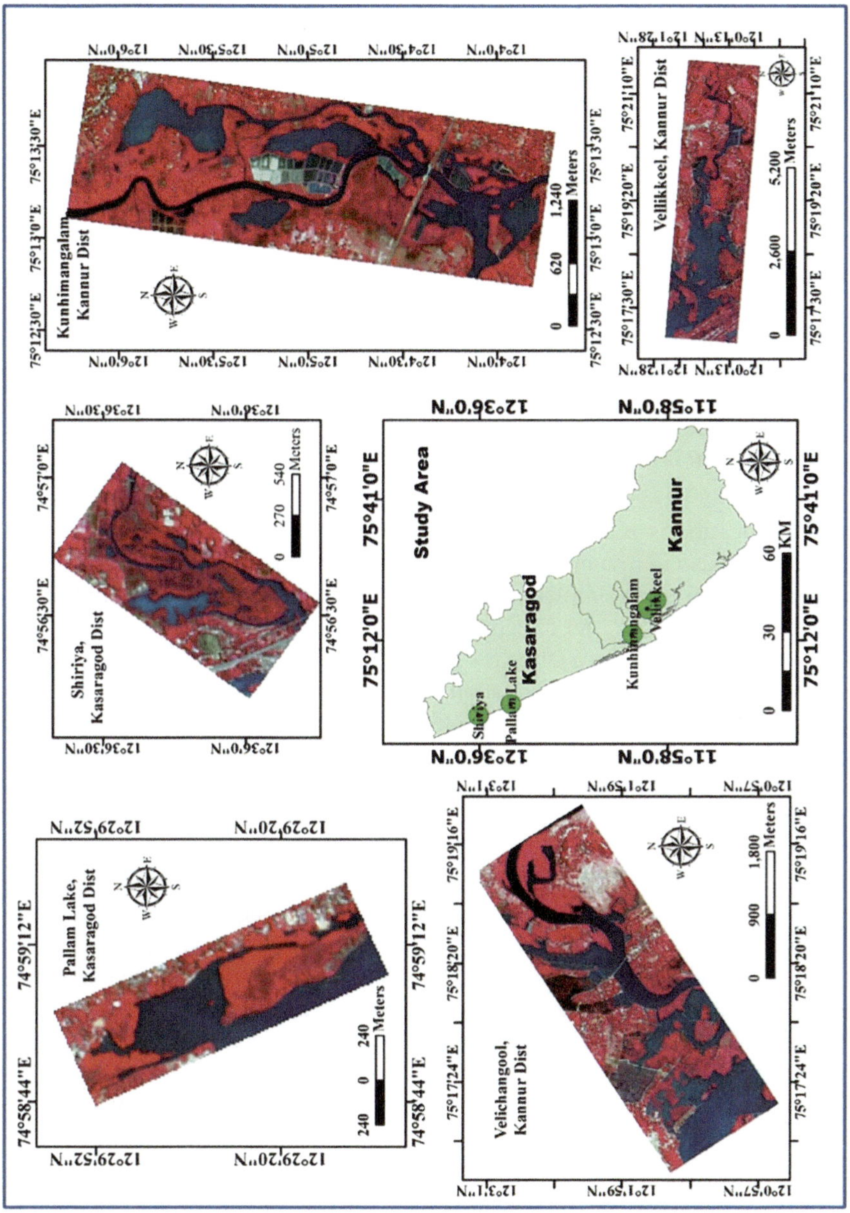

Fig. 2 Map of selected mangrove localities in the northern districts of Kerala Sentinel-2A image 2020

Table 1 Imaging sensor characteristics of Sentinel-2A

Bands	Sentinel-2A		
	Bandwidth (nm)	Central wavelength (nm)	Spatial resolution (m)
Band 2—Blue	66	492.4	10
Band 3—Green	36	559.8	10
Band 4—Red	31	664.6	10
Band 8—NIR	106	832.8	10

at more than + 0.7, thereby showing a higher density of green leaves (Chen et al., 2017; Kawamuna et al., 2017; Wachid et al., 2017). In contrast, areas with little to no vegetation show NDVI values close to zero, while water bodies typically display negative NDVI values (Chen, 2020; Wachid et al., 2017). The NDVI maps for the five selected mangrove locations are shown in Figs. 3d, 4d, 5d, 6d and 7d.

4.3 Image Classification of Satellite Data

Image classification is a fundamental digital image processing technique used to categorize all pixels in a satellite image into specific land cover or land use classes for a given study area. In simpler terms, it involves extracting meaningful information from a multispectral raster image by assigning each pixel to a particular class or theme based on satellite data. Here, the objective of the classification of Sentinel data is to categorize the mangrove vegetation from the other classes. For this purpose, a supervised classification system was selected, where the true mangrove pixel had been identified, and the computer system was classified as a mangrove (Abburu & Babu Golla, 2015; Kettig & Landgrebe, 1976). The supervised classification method requires a training dataset for each class, consisting of distinct and representative samples from the target classes in the image. In this study, the sample sets were identified using Sentinel-2A satellite images to ensure accurate classification. The major objective of the classification in the present was to identify the true mangrove vegetation using the sentinel-2A images. Hence fore, five classes were selected. The class 'deep water' indicates water bodies with maximum depths, having large volumes of water. The class 'shallow water' indicates the water bodies with shallow depth. This class is inclusive of the water body with less depth, such as the peripheral of a lake or river. In 'built-up land' classes, all classes of other green vegetation are included, such as roads, settlements, urban areas, open soil and sand exposure in the river side or lake side etc. 'Vegetation other than mangrove' classes include all the green vegetation other than true mangrove pixels, which include such as coconut trees, plantation trees, scrubs and grasses, trees, etc. 'Mangrove vegetation' class include only the true mangrove pixel including all the species. Trial and error methods were followed to improve the classification accuracy. The processes were repeated until the final map met the classification quality. From the classified image,

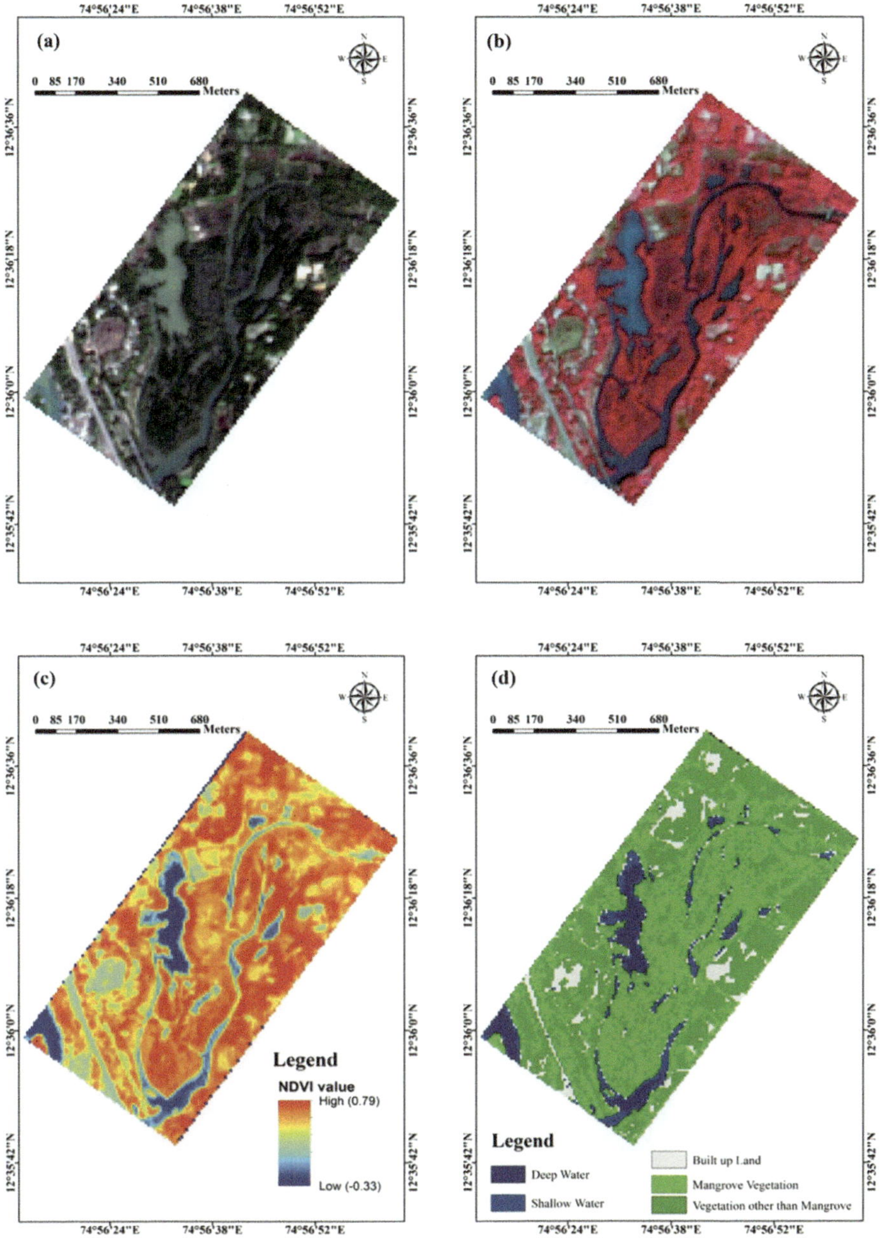

Fig. 3 **a** True colour composite image (RED, BLUE, GREEN), **b** false-colour composite image (NIR, RED, BLUE), **c** NDVI images, and **d** classified images derived from Sentinel-2A of Shiriya region in the Kasaragod district

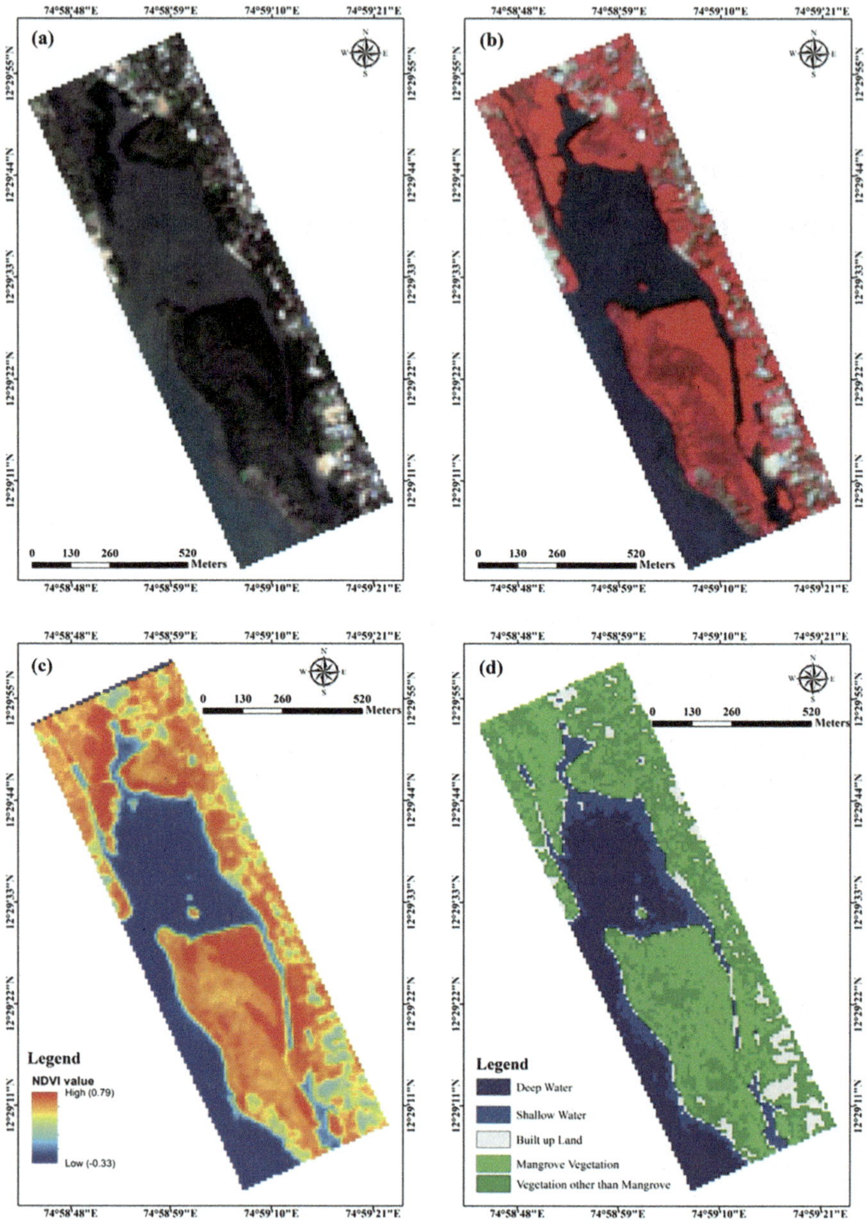

Fig. 4 **a** True colour composite image (RED, BLUE, GREEN), **b** false-colour composite image (NIR, RED, BLUE), **c** true colour composite image (RED, BLUE, GREEN), and **d** false-colour composite image (NIR, RED, BLUE) of Sentinel-2A Pallam Lake region in the Kasaragod district

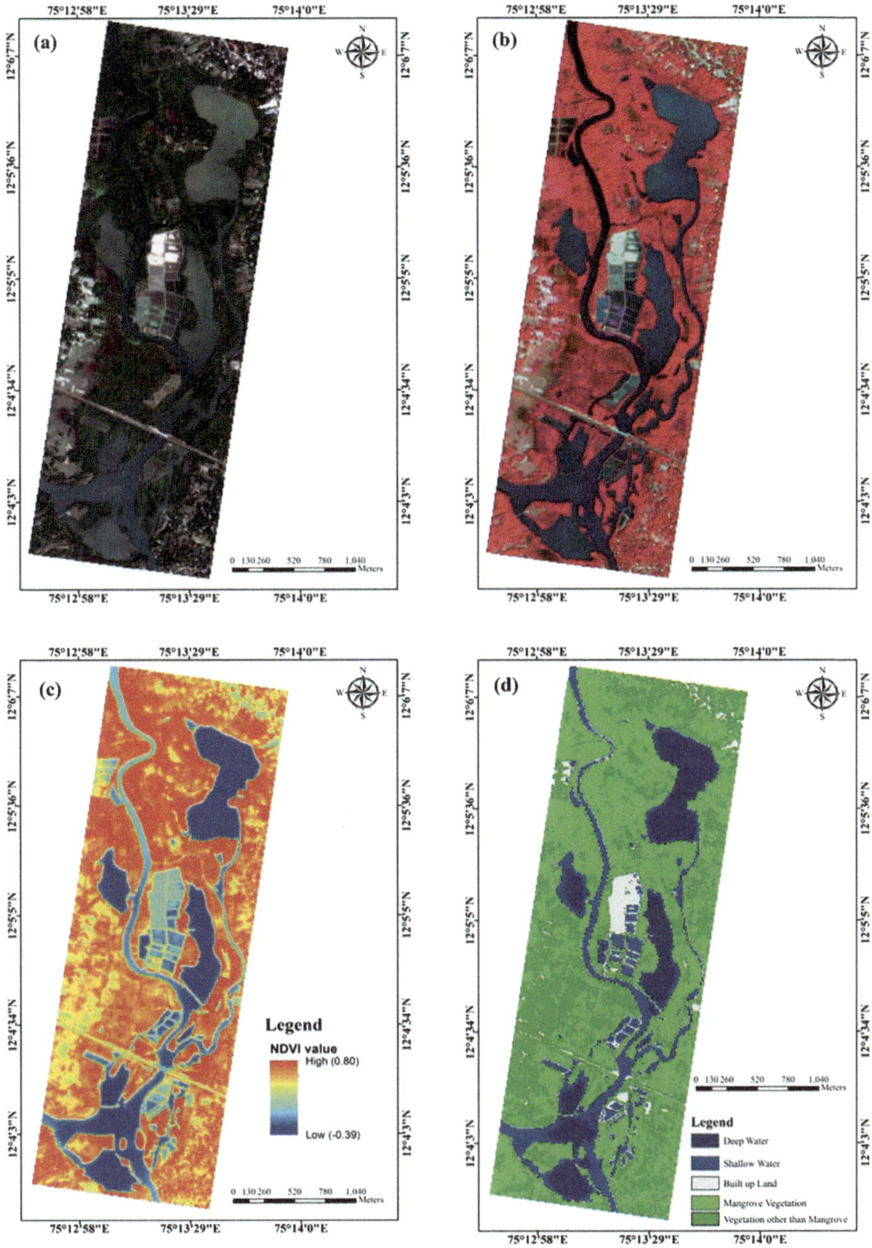

Fig. 5 **a** True colour composite image (RED, BLUE, GREEN), **b** false colour composite image (NIR, RED, BLUE), **c** true colour composite image (RED, BLUE, GREEN) and **d** false colour composite image (NIR, RED, BLUE) of Sentinel-2A Kunhimangalam region in the Kannur district

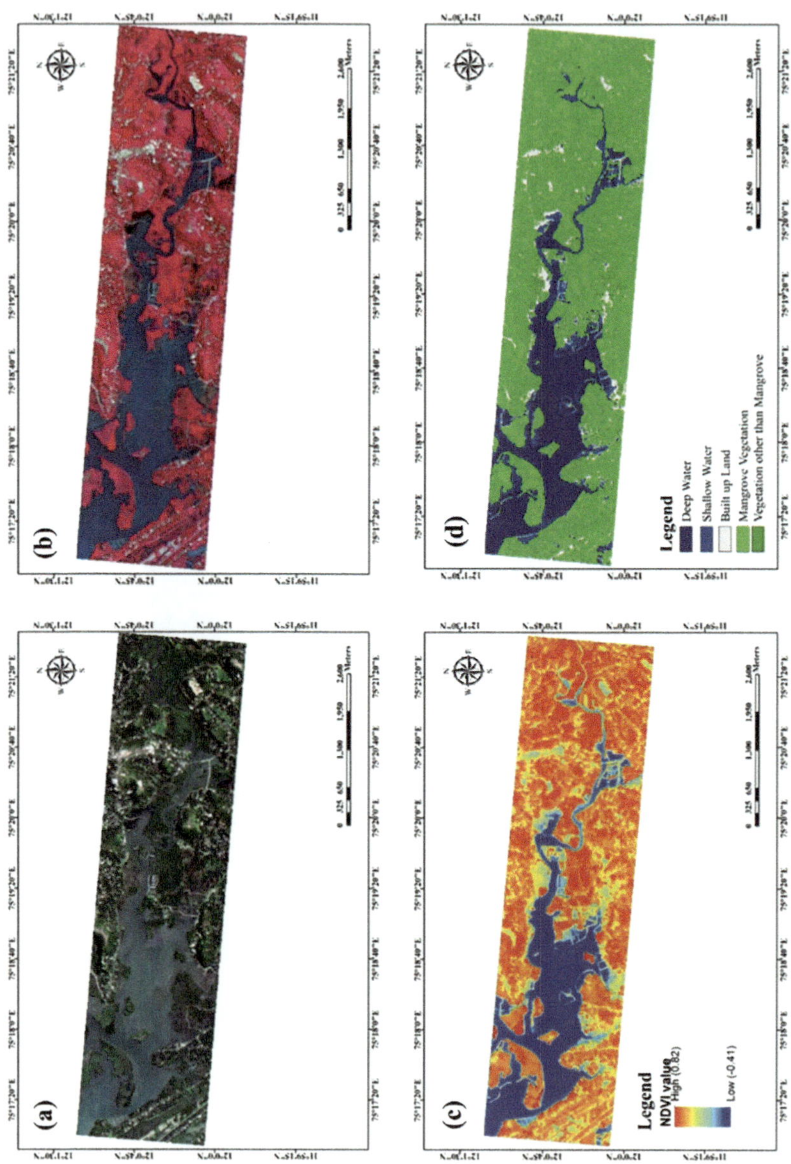

Fig. 6 **a** True colour composite image (RED, BLUE, GREEN), **b** false-colour composite image (NIR, RED, BLUE), **c** true colour composite image (RED, BLUE, GREEN), and **d** false-colour composite image (NIR, RED, BLUE) of Sentinel-2A Vellikkeel region in the Kannur district

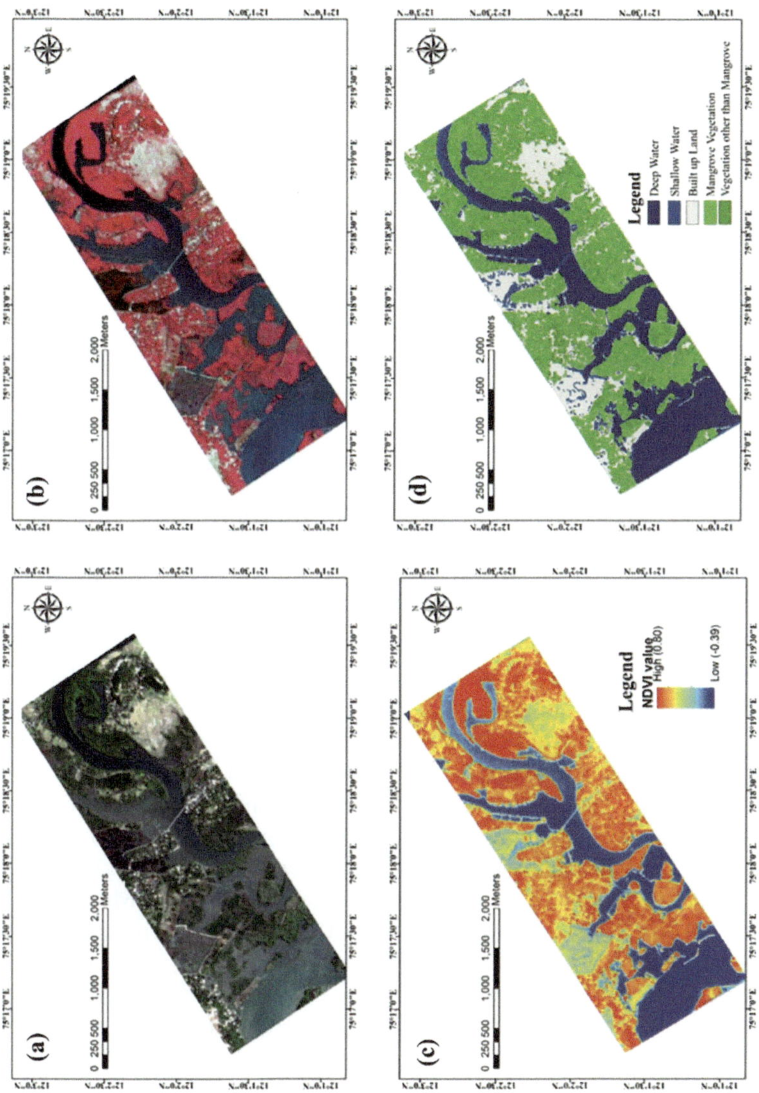

Fig. 7 **a** True colour composite image (RED, BLUE, GREEN), **b** false colour composite image (NIR, RED, BLUE), **c** true colour composite image (RED, BLUE, GREEN) and **d** false colour composite image (NIR, RED, BLUE) of Sentinel-2A Velichangool region in the Kannur district

the mangroves from the other green vegetation were clearly separated. The classified maps for each mangrove location are shown in (Fig. 8). Google Earth images have very good use in the mapping of mangrove vegetations (Hossain et al., 2016; Mohammed-Aslam, 2015; Tieng et al., 2019). Google Earth Image data was used to extract mangrove vegetation. The Mangrove vegetation on the coast of Kerala is seen in the form of patchy and thin formations. These small diversities of mangrove trees are very difficult to identify from satellite images. The highest resolution, which is publicly available, is Sentinel data, with 10 m of spatial resolution. Even these patchy mangroves are still lesser in dimeson such that less than the 10-m area in the field. Such thin formations of mangroves can be identified by the high spatial resolution images which are tagged in the Google Earth images. Though these images are not available for digital image extraction, they can be made available for visual identification. The following examples demonstrate the importance of these Google Earth mages in identifying and monitoring the mangrove vegetation in the study location (Table 2; Fig. 9).

5 Discussion

Remote sensing techniques have been utilized to distinguish mangrove vegetation from non-mangrove areas, aiding in the precise characterization of mangrove ecosystems. Optical remote sensing is widely recognized as a crucial method for studying mangrove distributions. The selection of suitable satellite sensor data plays a vital role in enhancing the understanding and classification of mangrove vegetation, particularly with the advancement of emerging technologies. In this study, the current spatial distribution of mangroves has been assessed in detail across five locations using Sentinel-2A satellite imagery. Additionally, Google Earth images have been visually analyzed to identify thin and fragmented mangrove formations, along with their temporal variations. The integration of remotely sensed data has yielded robust results, reinforcing the effectiveness of this approach in monitoring mangrove ecosystems. To comprehensively assess patchy mangrove regions, high-resolution satellite data is essential. Several studies have successfully employed remote sensing for mangrove evaluation and classification using satellite data (Green & Mumby, 2000; Kovacs et al., 2004; Chauhan & Dwivedi, 2008). Geographic Information Systems (GIS) and remote sensing have become indispensable tools for monitoring, planning, and managing fragile wetland ecosystems (Everitt et al., 2008; Roby et al., 2014). These techniques provide quantitative insights into the spatial distribution of mangrove ecosystems (Giri et al., 2007). The density of mangrove forests has a direct correlation with vegetation thickness (Kuenzer et al., 2011a, 2011b), and the Normalized Difference Vegetation Index (NDVI) is widely applied to differentiate mangroves from other vegetation types (Green et al., 1998). Anthropogenic activities such as aquaculture expansion have significantly impacted wetland ecosystems by replacing mangrove forests (Seto & Fragkias, 2007). In India, mangroves span approximately 4827 km^2, with 57% located along the east coast, 23% along the west

Fig. 8 Temporal Mangrove degradation can be viewed from Google Earth images at Location Pappinissery, Kannur district

Table 2 Summary of mangrove vegetation in northern Kerala

Sl. No	Places/locations	Area in hectare	District
1	Kumbla and Shiriya backwaters	50.98	Kasaragod
2	Chandragiri backwater and Pallam Lake	18.98	Kasaragod
3	Kunhimangalam	213.03	Kannur
4	Velichangool	199.87	Kannur
5	Vellikkeel	309.7	Kannur
Total		792.56	

Fig. 9 Mangrove location at Kunhimangalam—**a** and **b** temporal variations. Image date December 16, 2019

coast, and the remaining 20% in the Andaman and Nicobar Islands (Venkataraman & Wafar, 2005). Given the ecological significance of these coastal wetlands, mangrove conservation has become a critical topic of contemporary wetland research. Remote sensing applications have substantially contributed to assessing mangrove resources (Fei et al., 2011). Mangroves, alongside coral reefs and coastal upwelling systems, rank among the world's most ecologically and economically valuable natural ecosystems (Wang, 1998). Monitoring and mapping mangrove vegetation have traditionally relied on moderate to coarse-resolution satellite imagery, including Landsat Multispectral Scanner (MSS) (Howari et al., 2009), Landsat Thematic Mapper (TM) and Enhanced Thematic Mapper Plus (ETM+) (Howari et al., 2009; Andriamparany & François, 2010; Ferreira et al., 2009; Lee & Yeh, 2009), SPOT (Prasad et al., 2009), and Advanced Spaceborne Thermal Emission and Reflection Radiometer (ASTER) (Saito et al., 2003; Giri et al., 2007; Vaiphasa et al., 2006a, 2006b). In addition, visual interpretation techniques have been effectively applied in mangrove studies (Prasad et al., 2009; Selvam et al., 2003). Various classification approaches have been employed for mangrove mapping, including supervised classification (Paling et al., 2008; Howari et al., 2009; Lee & Yeh, 2009; Andriamparany & François, 2010), unsupervised classification (Ferreira et al., 2009; Giri et al., 2007; Paling et al., 2008), neural network classification (Mas, 2004; Seto & Fragkias, 2007), and object-based methods (Berlanga-Robles & Ruiz-Luna, 2002; Myint et al., 2008; Wang et al., 2004). These advanced techniques provide valuable insights into mangrove health, structure, and spatial dynamics, ensuring more effective conservation and management strategies.

The Findings of Study Area

A remote sensing method was employed to delineate the mangrove vegetation from non-mangrove regions for effectively characterizing mangrove belts. Optical remote sensing detection is employed as an important method in the investigation of mangroves. The selection of satellite data from appropriate sensors helps in the better understanding and characterization of mangrove vegetation owing to the emerging technologies. Areal coverage of the present status of mangroves has been carried out at 5 places in detail. Sentinel-2A satellite data has been used in the present study. Google Earth images have also been interpreted visually to obtain information about the thin patchy formations of mangroves and their temporal changes. The remotely sensed analysis provided strong results for the integrated approach in monitoring mangrove vegetation in the study area. High-resolution data are required for a detailed understanding of the characteristics of patchy mangrove regions.

6 Conclusion

The delineation of mangrove vegetation from non-mangrove areas plays a crucial role in effectively characterizing mangrove belts. The analysis of remotely sensed data provided a strong foundation for an integrated and scientific approach to the conservation and monitoring of mangrove ecosystems in the study area. High-resolution imagery is essential for a detailed assessment of the processes influencing fragmented mangrove formations. The findings of this study highlight the current distribution and status of mangrove ecosystems along the coastal regions of Kasaragod and Kannur Districts. Furthermore, intensive anthropogenic activities in certain areas have contributed significantly to the degradation of mangrove forests. Agricultural land conversion and settlements are a few of these causes. In addition, uncontrolled sand mining and irrational disposal of wastes have caused severe threats to the mangrove vegetation. Progressive increments in the area of mangrove vegetation from past decades in some places have been noticed. It was mainly because of the mangrove plantation by the Forest Department. The present situation demands greater attention in conserving the remaining mangrove vegetation. Otherwise, the threat to the coastal ecology will be alarming in extent. Improving the vegetal cover of mangrove vegetation minimizes the effects of disturbance to the stable ecosystem, salinity of soils of coasts, degrading quality of water and coastal erosion.

References

Abburu, S., & Babu Golla, S. (2015). Satellite image classification methods and techniques: A review. *International Journal of Computer Applications, 119*, 20–25. https://doi.org/10.5120/21088-3779

Andriamparany, R., & Francois, F. (2010). Dynamics of mangrove forests in the Mangoky River Delta, Madagascar, under the influence of natural and human factors. *Forest Ecology and Management, 259*, 1161–1169.

Baloloy, A. B., Blanco, A. C., Raymund Rhommel, R. C., & Nadaoka, K. (2020). Development and application of a new mangrove vegetation index (MVI) for rapid and accurate mangrove mapping. *ISPRS Journal of Photogrammetry and Remote Sensing, 95*, 95–117. https://doi.org/10.1016/j.isprsjprs.2020.06.001

Berlanga-Robles, C. A., & Ruiz-Luna, A. (2002). Land use mapping and change detection in the coastal zone of northwest Mexico using remote sensing techniques. *Journal of Coastal Research, 18*, 514–522.

Blasco, F., Gauquelin, T., Rasolofoharinoro, M., Denis, J., Aizpuru, M., & Caldairou, V. (1998). Recent advances in mangrove studies using remote sensing data. *Marine and Freshwater Research, 49*, 287–296.

Carlson, T. N., & Ripley, D. A. (1997). On the relation between NDVI, fractional vegetation cover, and leaf area index. *Remote Sensing of Environment, 62*, 241–252. https://doi.org/10.1016/S0034-4257(97)00104-1

Chandra, S., Gautam, P. K., & Singh, A. P., et al. (2024). Site selection for suitability of dam construction using analytic hierarchy process (AHP): A review study on Rihand Dam, Uttar Pradesh, India. *Arabian Journal of Geosciences, 17*, 293. https://doi.org/10.1007/s12517-024-12097-x

Chauhan, H. B., & Dwivedi, R. M. (2008). Inter-sensor comparison between RESOURCESAT LISS III, LISS IV and AWiFS with reference to coastal land use/land cover studies. *International Journal of Applied Earth Observation and Geoinformation, 10*, 181–185.

Chen, B., Xiao, X., Li, X., Pan, L., Doughty, R., Ma, J., Dong, J., Qin, Y., Zhao, B., Wu, Z., Sun, R., Lan, G., Xie, G., Clinton, N., & Giri, C. (2017). A mangrove forest map of China in 2015: Analysis of time series Landsat 7/8 and Sentinel-1A imagery in Google Earth Engine cloud computing platform. *ISPRS Journal of Photogrammetry and Remote Sensing, 131*, 104–120. https://doi.org/10.1016/j.isprsjprs.2017.07.011

Chen, J., Jönsson, P., Tamura, M., Gu, Z., Matsushita, B., & Eklundh, L. (2004). A simple method for reconstructing a high-quality NDVI time-series data set based on the Savitzky-Golay filter. *Remote Sensing of Environment, 91*, 332–344. https://doi.org/10.1016/j.rse.2004.03.014

Chen, N. (2020). Mapping mangrove in Dongzhaigang, China using Sentinel-2 imagery. *Journal of Applied Remote Sensing, 14*(01), Article 014508. https://doi.org/10.1117/1.jrs.14.014508

Everitt, J. H., Yang, C., Sriharan, S., & Judd, F. W. (2008). Using high resolution satellite imagery to map black mangrove on the Texas Gulf Coast. *Journal of Coastal Research, 24*, 1582–1586.

Fei, S. X., Shan, C. H., & Hua, G. Z. (2011). Remote sensing of mangrove wetland identification. *Procedia, Environmental Sciences, 10*, 2287–2293.

Ferreira, M. A., Andrade, F., Banderiba, S. O., Cardoso, P., Nogueira Mendes, R., & Paula, J. (2009). Analysis of cover change (1995–2005) of Tanzania/Mozambique transboundary mangroves using Landsat imagery. *Aquatic Conservation, 19*, 38–45.

Gautam, P. K., Singh, D. S., Kumar, D., & Singh, A. K. (2020). A GIS-based approach in drainage morphometric analysis of Sai River Basin, Uttar Pradesh, India. *Journal of the Geological Society of India, 95*, 366–376. https://doi.org/10.1007/s12594-020-1445-9

Gautam, P. K., Singh, D. S., Singh, A. K., & Kumar, D. (2022). Geomorphic analysis of Baghain River, Yamuna Basin, and its implication for drainage characteristic and tectonics using remote sensing and GIS techniques. *Journal of the Geological Society of India, 98*(11), 1573–1584. https://doi.org/10.1007/s12594-022-2214-8

Gautam, P. K. (2023). The significance of morphometric analysis of Shimsha River, Karnataka, India to understand the hydrological and morphological characteristics. *River, 2*(4), 490–505. https://doi.org/10.1002/rvr2.66

Gautam, P. K. (2024). Drainage analysis of the Karanja River basin, Karnataka, India using Geo-informatics. *Applied Geomatics 16*, 1025–1046. https://doi.org/10.1007/s12518-024-00584-5

Gautam, P. K., Singh, D. S., Kumar, D., & Singh, A. K. (2024a). Grain size and sediment load control on channel migration patterns of the Ghaghara River from Faizabad to Deoria, Ganga Plain, India. *Arabian Journal of Geosciences, 17*, 295. https://doi.org/10.1007/s12517-024-121 04-1

Gautam, P. K., Chandra, S., & Henry, P. K. (2024b). Estimation of flood inundation in river basins of Uttar Pradesh using Sentinel 1A-SAR data on Sentinel Application Platform (SNAP). *Arabian Journal of Geosciences, 17*, 107. https://doi.org/10.1007/s12517-024-11910-x

Gautam, P. K., Singh, D. S., & Singh, A. K. (2024c). Quantitative assessment of channel migration in the Ghaghara River, Ganga Plain, India. *Journal of the Geological Society of India, 100*(7), 919–935. https://doi.org/10.17491/jgsi/2024/173935

Gautam, P. K., & Kumar, N. (2025). Geo-hydrological analysis and watershed delineation of the Sarayan River, Ganga Plain, India: Integrating geospatial data for water resource management. *Arabian Journal of Geosciences, 18*, 111. https://doi.org/10.1007/s12517-025-12255-9

Giri, C., Pengra, B., Zhu, Z., Singh, A., & Tieszen, L. (2007). Monitoring mangrove forest dynamics of the Sundarbans in Bangladesh and India using multi-temporal satellite data from 1973 to 2000. *Estuarine, Coastal and Shelf Science, 73*, 91–100.

Green, E., & Mumby, P. (2000). Mapping mangroves. In A. J. Edwards (Ed.), *Remote sensing handbook for tropical coastal management* (pp. 183–198). UNESCO.

Green, E. P., Mumby, P., Clark, C. D., Mumby, P. J., Edwards, A. J., & Ellis, A. C. (1998). Remote sensing techniques for mangrove mapping. *International Journal of Remote Sensing, 19*, 935–956.

Heumann, B. W. (2011a). An object-based classification of mangroves using a hybrid decision tree–support vector machine approach. *Remote Sensing, 3*(11), 2440–2460. https://doi.org/10.3390/rs3112440

Heumann, B. W. (2011b). Satellite remote sensing of mangrove forests: Recent advances and future opportunities. *Progress in Physical Geography, 35*(1), 87–108.

Hossain, M. D., & Nuruddin, A. A. (2016). Soil and mangrove: A review. *Journal of Environmental Science and Technology, 9*(2), 198–207. https://doi.org/10.3923/jest.2016.198.207

Hossain, M. S., Bujang, J. S., Zakaria, M. H., & Hashim, M. (2016). Marine and human habitat mapping for the Coral Triangle Initiative region of Sabah using Landsat and Google Earth imagery. *Marine Policy, 72*, 176–191. https://doi.org/10.1016/j.marpol.2016.07.003

Howari, F. M., Jordan, B. R., Bouhouche, N., & Wyllie-Echeverria, S. (2009). Remote-sensing assessment of mangrove forests and seagrass beds in the northwestern part of the United Arab Emirates. *Journal of Coastal Research, 25*, 48–56.

Ibrahim-Bathis, K. (2021). Insight into the spatial-temporal extent of mangrove forests in the northern coast of Kerala. In *Remote sensing of ocean and coastal environments* (pp. 375–392). https://doi.org/10.1016/B978-0-12-819604-5.00021-4

Immordino, F., Barsanti, M., Candigliota, E., Cocito, S., Delbono, I., & Peirano, A. (2019). Application of Sentinel-2 multispectral data for habitat mapping of Pacific Islands: Palau Republic (Micronesia, Pacific Ocean). *Journal of Marine Science and Engineering, 7*(9), 316. https://doi.org/10.3390/jmse7090316

Jiang, Z., Huete, A. R., Chen, J., Chen, Y., Li, J., Yan, G., & Zhang, X. (2006). Analysis of NDVI and scaled difference vegetation index retrievals of vegetation fraction. *Remote Sensing of Environment, 101*(3), 366–378. https://doi.org/10.1016/j.rse.2006.01.003

Kawamuna, A., Suprayogi, A., & Wijaya, A. (2017). Analisis kesehatan hutan mangrove berdasarkan metode klasifikasi NDVI pada citra Sentinel-2 (studi kasus: Teluk Pangpang Kabupaten Banyuwangi). *Jurnal Geodesi Undip, 6*(1), 277–284.

Kettig, R. L., & Landgrebe, D. A. (1976). Classification of multispectral image data by extraction and classification of homogeneous objects. *IEEE Transactions on Geoscience Electronics, 14*(1), 19–26. https://doi.org/10.1109/TGE.1976.294460

Kovacs, J. M., Flores-Verdugo, F., Wang, J., & Aspden, L. P. (2004). Estimating leaf area index of a degraded mangrove forest using high spatial resolution satellite data. *Aquatic Botany, 80*(1), 13–22. https://doi.org/10.1016/j.aquabot.2004.06.001

Kuenzer, C., Bluemel, A., Bluemel, A., Quoc, T. V., & Dech, S. (2011a). Remote sensing of Mangrove ecosystems: A review. *Remote Sensing, 3*, 878–928.

Kuenzer, C., Bluemel, A., Bluemel, A., Vo Quoc, T., & Dech, S. (2011b). Remote sensing of mangrove ecosystems: A review. *Remote Sensing, 3*, 878–928.

Kumar, D., Singh, D. S., Prajapati, S. K., Khan, I., Gautam, P. K., & Vishawakarma, B. (2018). Morphometric parameters and neotectonics of Kalyani River Basin, Ganga Plain: A remote sensing and GIS approach. *Journal of the Geological Society of India, 91*, 679–686. https://doi.org/10.1007/s12594-018-0923-9

Lee, T., & Yeh, H. (2009). Applying remote sensing techniques to monitor shifting wetland vegetation: A case study of Danshui River estuary mangrove communities. *Taiwan. Ecol Eng, 35*, 487–496.

Manna, S., & Raychaudhuri, B. (2020). Retrieval of leaf area index and stress conditions for Sundarban mangroves using Sentinel-2 data. *International Journal of Remote Sensing, 41*(3), 1019–1039.

Mas, J. F. (2004). Mapping land use/cover in a tropical coastal area using satellite sensor data, GIS, and artificial neural networks. *Estuarine, Coastal and Shelf Science, 59*, 219–230.

Mohammed-Aslam, M. A. (2015). Satellite based spectral signatures of coastal wetlands and mangrove vegetation of Kasaragod, India. In *Biodiversity conservation—Challenges for future* (pp. 211–214). Bentham Science Publishers.

Mumby, P. J. (2000). Mapping Mangroves. In *Remote sensing handbook for tropical coastal management* (pp. 183–198).

Myint, S. W., Giri, C. P., Wang, L., Zhu, Z., & Gillette, S. C. (2008). Identifying mangrove species and their surrounding land use and land cover classes using an object-oriented approach with a lacunarity spatial measure. *GIS Sci Remote Sensing, 45*, 188–208.

Paling, E. I., Kobryn, H. T., & Humphreys, G. (2008). Assessing the extent of mangrove change caused by cyclone Vance in the eastern Exmouth Gulf, northwestern Australia. *Estuarine, Coastal and Shelf Science, 77*, 603–613.

Pavithra, B., Kalaivani, K., & Ulagapriya, K. (2019). Remote sensing techniques for mangrove mapping. *International Journal of Engineering and Advanced Technology, 8*(3S), 27–30.

Pham, T. D., Yokoya, N., Bui, D. T., Yoshino, K., & Friess, D. A. (2019). Remote sensing approaches for monitoring mangrove species, structure, and biomass: Opportunities and challenges. *Remote Sensing, 11*(3), 230.

Piovan, S.E. (2020). Remote sensing. In *Geospatial approach* (pp. 171–197). Springer. https://doi.org/10.1007/978-3-030-42439-8_7

Prakash Hati, J., Samanta, S., Rani Chaube, N., Misra, A., Giri, S., Pramanick, N., Gupta, K., Datta Majumdar, S., Chanda, A., Mukhopadhyay, A., & Hazra, S. (2020). Mangrove classification using airborne hyperspectral AVIRIS-NG and comparing with other spaceborne hyperspectral and multispectral data. *Egyptian Journal of Remote Sensing and Space Science, 24*, 273–281.https://doi.org/10.1016/j.ejrs.2020.10.002

Prasad, P. R., Reddy, C. S., Rajan, K. S., Raza, S. H., & Dutt, C. B. (2009). Assessment of tsunami and anthropogenic impacts on the forest of the North Andaman Islands, India. *Int J Remote Sensing, 30*, 1235–1249.

Roby, T. J., Nair, P. V., & Joyce, J. (2014). GIS techniques for mapping highly fragmented ecosystems: A case study on the Myristica swamp forests of Southern Kerala, India. *Research Journal of Recent Sciences, 3*(ISC-2013), 110–119.

Saito, H., Bellan, M. F., Al-Habshi, A., Aizpuru, M., & Blasco, F. (2003). Mangrove research and coastal ecosystem studies with SPOT-4HRVIR and TERRA ASTER in the Arabian Gulf. *Int J Remote Sensing, 24*, 4073–4092.

Selvam, V., Ravichandran, K. K., Gnanappazham, L., & Navamuniyammal, M. (2003). Assessment of community-based restoration of Pichavaram mangrove wetland using remote sensing data. *Current Science, 85*, 794–798.

Seto, K. C., & Fragkias, M. (2007). Mangrove conversion and aquaculture development in Vietnam: A remote sensing-based approach for evaluating the Ramsar Convention on Wetlands. *Global Environ Change, 17*, 486–500.

Singh, D. S., Tiwari, A. K., & Gautam, P. K. (2018). The Burhi Gandak: Most sinuous river. In D. Singh (Ed.), *The Indian Rivers* (Vol. 2, pp. 209–219). Springer Hydrogeology, Singapore. https://doi.org/10.1007/978-981-10-2984-4_17

Tieng, T., Sharma, S., Mackenzie, R. A., Venkattappa, M., Sasaki, N. K., & Collin, A. (2019). Mapping mangrove forest cover using Landsat-8 imagery, Sentinel-2, very high resolution images and google earth engine algorithm for entire Cambodia. *IOP Conference Series: Earth and Environmental Science, 266*, Article 012010. https://doi.org/10.1088/1755-1315/266/1/012010

Vaiphasa, C., Andrew, S. K., & Willem, F. B. (2006a). A post-classifier for mangrove mapping using ecological data. *ISPRS Journal of Photogrammetry and Remote Sensing, 61*, 1–10.

Vaiphasa, C., Andrew, S. K., & Willem, F. B. (2006b). A post-classifier for mangrove mapping using ecological data. *ISPRS J Photogr Remote Sensing, 61*, 1–10.

Van Genderen J. (2016). Fundamentals of satellite remote sensing: an environmental approach. *International Journal of Digital Earth*, 1275–1276. https://doi.org/10.1080/17538947.2016.1209938

Veloso, A., Mermoz, S., Bouvet, A., Le Toan, T., Planells, M., Dejoux, J. F., & Ceschia, E. (2017). Understanding the temporal behavior of crops using Sentinel-1 and Sentinel-2-like data for agricultural applications. *Remote Sensing of Environment, 199*, 415–426. https://doi.org/10.1016/j.rse.2017.07.015

Venkataraman, K., & Wafar, M. (2005). Coastal and marine biodiversity of India. *Journal of Marine Sciences, 34*(1), 57–75.

Vrieling, A., Meroni, M., Darvishzadeh, R., Skidmore, A. K., Wang, T., Zurita-Milla, R., Oosterbeek, K., O'Connor, B., & Paganini, M. (2018). Vegetation phenology from Sentinel-2 and field cameras for a Dutch barrier island. *Remote Sensing of Environment, 215*, 517–529. https://doi.org/10.1016/j.rse.2018.03.014

Wachid, M. N., Hapsara, R. P., Cahyo, R. D., Wahyu, G. N., Syarif, A. M., Umarhadi, D. A., Fitriani, A. N., Ramadhanningrum, D. P., & Widyatmanti, W. (2017). Mangrove canopy density analysis using Sentinel-2A imagery satellite data. *IOP Conference Series: Earth and Environmental Science, 70*, Article 012020. https://doi.org/10.1088/1755-1315/70/1/012020

Wang, S. (1998). Status of coastal wetlands and their protection in Guangdong province. *Chongqing Environmental Science, 20*(1), 4–11.

Wang, D., Wan, B., Liu, J., Su, Y., Guo, Q., Qiu, P., & Wu, X. (2020). Estimating aboveground biomass of the mangrove forests on northeast Hainan Island in China using an upscaling method from field plots, UAV-LiDAR data, and Sentinel-2 imagery. *International Journal of Applied Earth Observation and Geoinformation, 85*, Article 101986. https://doi.org/10.1016/j.jag.2019.101986

Wang, L., Sousa, W., & Gong, P. (2004). Integration of object-based and pixel-based classification for mangrove mapping with IKONOS imagery. *Int J Remote Sensing, 25*, 5655–5668.

Wang, Y., Bonynge, G., Nugranad, J., Traber, M., Ngusaru, A., Tobey, J., Hale, L., Bowen, R., & Makota, V. (2003). Remote sensing of mangrove change along the Tanzania coast. *Marine Geodesy, 26*(1–2), 35–48. https://doi.org/10.1080/01490410306708

Zhang, T., Su, J., Liu, C., Chen, W. H., Liu, H., & Liu, G. (2017). Band selection in Sentinel-2 satellite for agriculture applications. In *ICAC 2017—2017 23rd International conference on automation and computing (ICAC)* (pp. 1–6), Huddersfield, UK. https://doi.org/10.23919/IConAC.2017.8081990

Biogeochemical Dynamics of Pulicat Lagoon, Bay of Bengal in India

G. Edwin Jeba Kumar and Usha Natesan

1 Introduction

Lagoons and estuaries define the intermediate zone between coastal regions and watersheds. These areas are vital for capturing and converting organic and inorganic nutrients from river discharges (Torregroza-Espinosa et al., 2020). Assessing nutrient changes alongside water, salt, and nutrient stabilities is essential for understanding the biological dynamics of lagoon systems (Menció et al., 2023). The biological dynamics are closely interconnected and significantly influenced by natural factors and human activities (Cepic et al., 2021). However, the lack of data on nutrient concentrations and seasonal fluctuations, particularly concerning land runoff and atmospheric inputs, continues to pose challenges for thorough assessments of lagoon metabolism (Khan & Mohammad, 2014; Devlin & Brodie, 2023). Even without direct measurements, these challenges can be addressed with the help of Comprehensive Biogeochemical Models (CAM), which can be utilized regionally to provide integral estimates over large areas (Wong et al., 2024; Hauck et al., 2023; Upton et al., 2024). In the context of Pulicat Lagoon, the biogeochemical dynamics are primarily influenced by its hydrology, sediment composition, and anthropogenic activities. The interaction between freshwater inflow from nearby rivers, saline water from the Bay of Bengal, and tidal movements creates a complex environment where various chemical processes occur. Nutrient inputs, particularly nitrogen and phosphorus, are critical drivers of primary productivity; however, their sources and concentrations exhibit significant temporal and spatial variability (Reddy et al., 2012). Understanding these dynamics is crucial, especially given the lagoon's role in maintaining ecological

G. Edwin Jeba Kumar (✉)
Department of Geology, Alagappa Government Arts College, Karaikudi, India
e-mail: rsedwin@gmail.com

U. Natesan
National Institute of Technical Teachers Training and Research, Chennai, India

P. Thambidurai and A. K. Dikshit (eds.), *Coastal Environments of India*,
https://doi.org/10.1007/978-3-031-97341-3_3

balance and supporting its biotic community (Chacón Abarca et al., 2021). Therefore, improvement and supervision are vital to ensure the health of this exceptional ecosystem. Notably, the modification of flow patterns in lagoons has exacerbated tidal inlet dynamics, thereby impacting the ecological functioning of these systems (Behja et al., 2024; Srichandan et al., 2015; Tosi et al., 2024). As a result, managing these limited resources has become imperative (Patten., 2010). This study will establish a foundation for determining the biogeochemical variations and their influence on Pulicat Lagoon, a necessary step toward managing the coastal lagoon efficiently to guarantee its future health.

2 Material and Methods

2.1 Study Area

Pulicat Lagoon, the second-largest brackish water lagoon in India, is located north of Chennai along the eastern seaboard. It was formed from the backwaters of the Bay of Bengal. (Krishnan & Sampath, 1973; Farooqui & Vaz, 2000; Nanda Kumar et al., 2008; Nagarjuna et al., 2010; Jeba Kumar & Natesan, 2015) and is also known as the "Lagoon of the Palar Basin" (Jeba Kumar & Natesan, 2015; Prabhakar & Zutshi, 1993). The lagoon originated approximately 6650 years ago during the Holocene geological period (Jeba Kumar & Natesan, 2015; Ramteke et al., 2023). It extends about 60 km in length, with a width ranging from 0.2 km to 17.5 km (Jeba Kumar & Natesan, 2015). The water surface area varies between a maximum of 450 km^2 and a minimum of 250 km^2 (Krishnan & Sampath, 1973) (Fig. 1). The mouth of Pulicat Lagoon is notably shallow, with a mean tidal amplitude of about 0.25 m, compared to 1 m along the adjacent coast (Jeba Kumar & Natesan, 2015). The periodic closure of the lagoon's inlet due to sandbar formation significantly affects water quality impedes navigation, and obstructs the natural flushing mechanisms of the lagoon (Jeba Kumar & Natesan, 2015; Nagarjuna et al., 2010). The dimensions of the inlet are crucial for regulating water exchange, including the movement of dissolved and suspended materials, which influence flushing rates and residence times (Jeba Kumar & Natesan, 2015; Krishnan & Sampath, 1973; Prabhakar & Zutshi, 1993).

2.2 Sampling Strategy

The primary data collection included surface water samples, current speed, and direction. Sampling stations were strategically selected to study the changes in biogeochemical and physical processes resulting from interactions with the sea, Buckingham Canal, and the Kalangi, Swarnamukhi, and Araniyar Rivers. A total of twenty

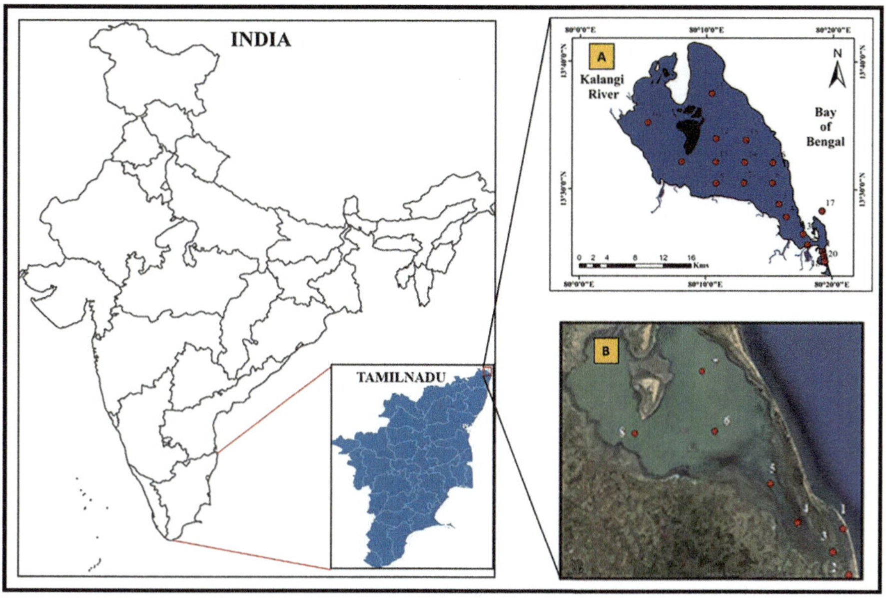

Fig. 1 Study area map, **a** water sampling, **b** current observation

stations for water sample collection and eight stations for current observations were identified from the mouth of the lagoon (Fig. 1). At each station, current details, specifically current speed and direction were measured in situ using floats at the surface during high tide on full moon or new moon days across different seasons such as southwest (SW), northeast (NE), post-monsoon (PM), and summer (S). The floats were released at each of the eight stations at known distances, and data regarding float velocity and movement direction were collected. Surface water samples were gathered from the sampling stations using 1-L polyethene plastic containers and. The samples were collected using a fishing boat, starting from Pulicat village (Station 1) and ending at Station 20 in Buckingham Canal. Samples were filtered and analyzed using established procedures by Grasshoff et al. (1999) (Table 1). Grasshoff et al. (1999) describe the Nitrate-Nitrite method as a process where nitrate is reduced to nitrite. The total nitrite concentration is then measured using diazotization and coupling reactions, and the nitrate concentration is determined by subtracting the initial nitrite content from the total. In the Total Phosphorus (TP) method, all phosphorus forms are converted into orthophosphate, which is then analyzed using a colourimetric approach.

This study uses a flexible mesh to model hydrodynamics, advection–dispersion patterns, and water quality in nearshore environments. A computational grid is employed for these simulations. The system includes a pre- and post-processing module that supports data analysis and enables the visualization of model outputs. These outputs are systematically stored in a structured database (Jinoj et al., 2024). The modelling approach is based on solving the incompressible Reynolds-averaged

Table 1 Description of parameters considered for analysis

Parameter	Abbreviation	Units	Technique/instrument used
Dissolved oxygen	DO	mg/L	Modified Winkler's method (de et al. 2024)
Biological oxygen demand	BOD	mg/L	Modified Winkler's method (de et al. 2024)
Ammonia–nitrogen	NH_3-N	µmol/L	Spectrophotometer method
Nitrite-nitrogen	NO_2-N	µmol/L	Spectrophotometer method Orellana et al. (2014)
Nitrate-nitrogen	NO_3-N	µmol/L	Spectrophotometer method Orellana et al. (2014)
Total nitrogen	TN	µmol/L	Spectrophotometer method
Inorganic phosphate	PO_4-P	µmol/L	Spectrophotometer method
Total phosphate	TP	µmol/L	Spectrophotometer method

Navier–Stokes equations in either two or three dimensions. It follows the assumptions of the Boussinesq approximation and hydrostatic pressure equations (Adandedji et al., 2017; Jinoj et al., 2024; Vardy & Tijsseling, 2022). The corresponding Eqs. (1) and (2) are provided below (Jinoj et al., 2024; Ala Amjadi & Mokhtari, 2019):

Local Continuity Equation

$$\frac{\partial h}{\partial t} + \frac{\partial hu}{\partial x} + \frac{\partial hv}{\partial y} = hS \quad (1)$$

where

h	Water depth
u and v	Water particle velocities in x and y directions
s	Energy-source dissipation term.

Momentum Equation for x and y components

$$\frac{\partial u}{\partial t} + \frac{\partial u^2}{\partial x} + \frac{\partial vu}{\partial y} + \frac{\partial wu}{\partial z} = fv - g\frac{\partial \eta}{\partial x} - 1/\rho_0 \frac{\partial Þ_a}{\partial x} - \frac{g}{\rho_0}\int_z^{\eta} \frac{\partial \rho}{\partial x}\partial z + F_u + \frac{\partial}{\partial z}\left[v_t \frac{\partial u}{\partial z}\right] + u_s S$$
$$\frac{\partial v}{\partial t} + \frac{\partial v^2}{\partial y} + \frac{\partial uv}{\partial x} + \frac{\partial wv}{\partial z} = fu - g\frac{\partial \eta}{\partial y} - \frac{1}{\rho_0}\frac{\partial Þ_a}{\partial y} - \frac{g}{\rho_0}\int_z^{\eta} \frac{\partial \rho}{\partial y}\partial z + Fv + \frac{\partial}{\partial z}\left[V_t \frac{\partial v}{\partial z}\right] + v_s S \quad (2)$$

where

t	Time
x and y	Cartesian Coordinates
η	Surface elevation
u and v	Velocity components in x and y direction
f	Coriolis parameter
g	Gravitational acceleration

ρ	Density of water
S	Energy-source dissipation term
F	Horizontal diffusion terms.

Computational Domain and Lagoons' Depth Profile

The computational domain was chosen with a southern and northern boundary, with a maximum depth of 70 m in the Bay of Bengal (Gangopadhyay, 2022). The western boundary was along the coastline, with zero depth. Bathymetry was prepared using data from the National Institute of Ocean Technology (NIOT) (Panda & Mohanty, 2008) and validated with Global Positioning System (GPS) observations for the study area. A mesh file was created using the 2D/3D grids tool in MIKE 21 software for the Pulicat lagoon, generating triangles with smooth boundaries and high resolutions in areas of particular attention (Panda & Mohanty, 2008) (Fig. 2). This data is then processed into the software, defining the spatial extent and topography of the model domain. Boundary conditions are established to accurately signify water flow dynamics and tidal interactions within the lagoon (Panda & Mohanty, 2008) (Fig. 3).

Calibration

Numerical models must be calibrated to be used effectively in real-world situations. Nevertheless, there isn't a widely recognized process for model calibration and validation in the literature (Hsu et al., 1999). Finding modifiable parameters about the model domain, bathymetry, bed roughness, and eddy momentum diffusivity is the major objective of calibration (Mardani et al., 2020; Williams & Esteves, 2017). Following model setup, these calibration parameters were adjusted and the sensitivity of the model was evaluated using field data (Dias & Lopes, 2006; Williams & Esteves, 2017). The model was validated without altering the calibration coefficients and was then employed to analyze the hydrodynamics of the study area across four seasons. This process typically involves quantitatively comparing model-generated data, such as water levels or velocities, with field data collected from the same location and period (Dias & Lopes, 2006). This iterative calibration and validation process enhances model accuracy and reliability. By ensuring that the model aligns with observed conditions, researchers can improve predictions of hydrodynamic behaviour, which is vital for effective water resource management and ecological assessments (Mardani et al., 2020; Williams & Esteves, 2017).

Particle Tracking

The Particle Tracking Module can simulate particle tracks and trajectories in water, including additional drift from the surface wind (Irfan et al., 2024). The technique

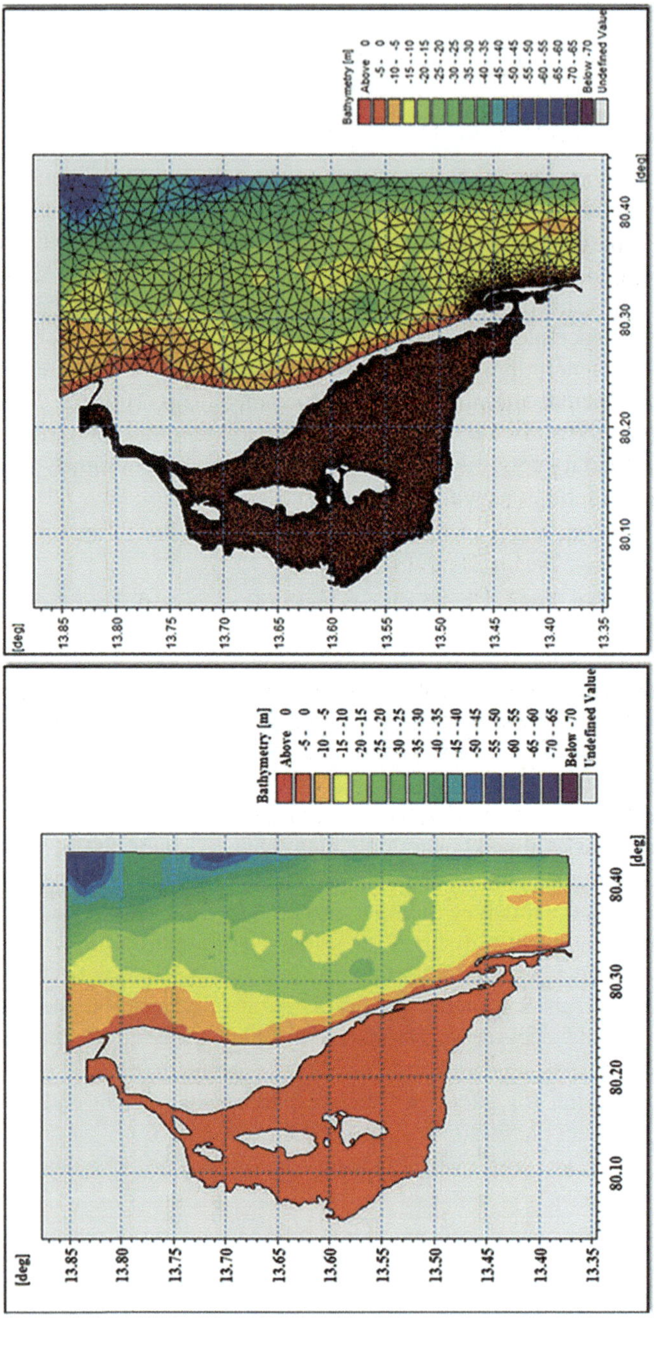

Fig. 2 Bathymetry and flexible mesh structure of model domain

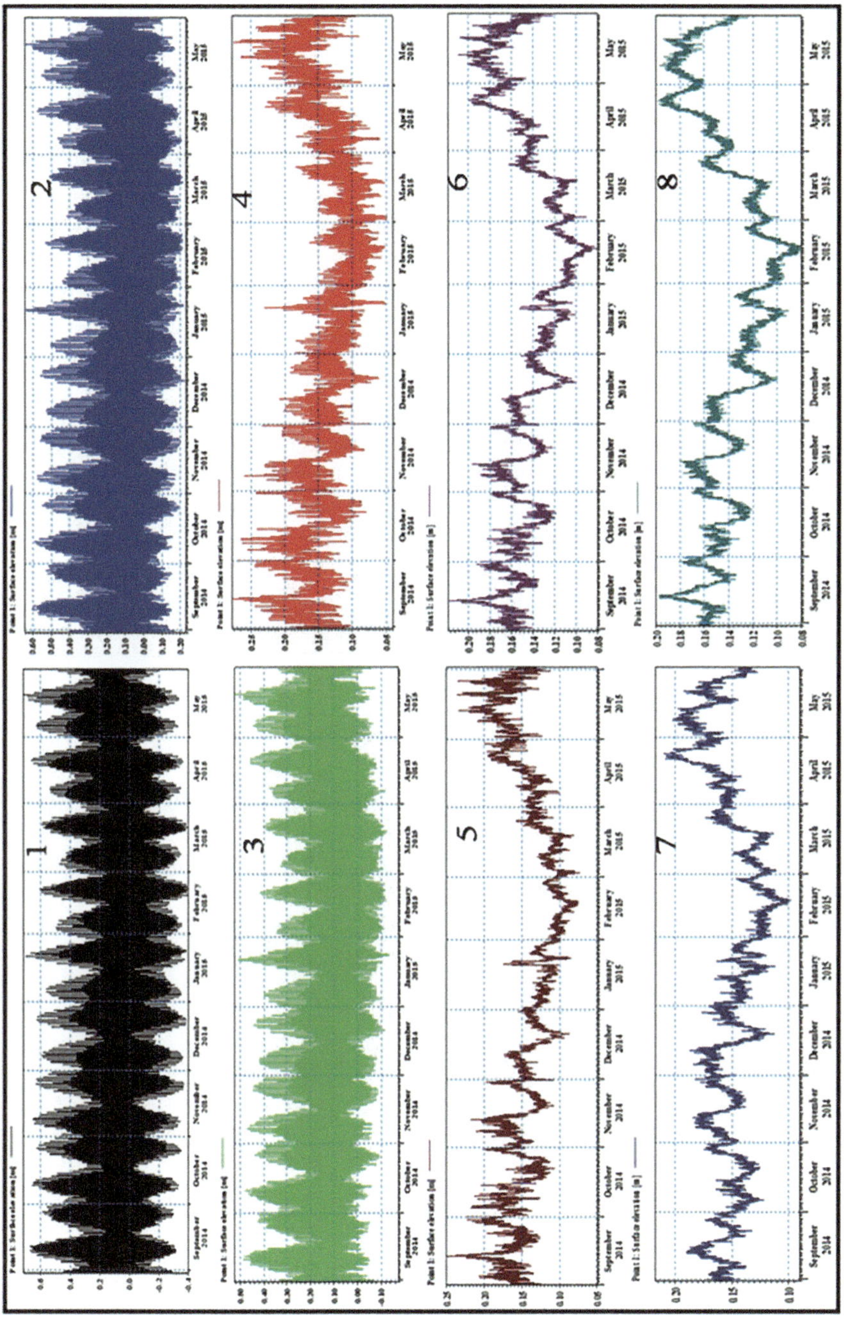

Fig. 3 Tidal variations at stations

follows the Langevin equation, which describes the transport and dispersion of particles based on stochastic differential equations (Spivakovskaya et al., 2005).

Model Run

Simulation of hydrodynamic conditions and tidal influences on the lagoon was carried out in the years 2014 (September and December) and 2015 (March and May), representing southwest monsoon, northeast monsoon, post-monsoon, and summer, respectively. Table 2 shows the input parameters for HD and PT simulations with the inputs listed below, the simulation was replicated for a model domain for different scenarios.

BOD—DO Dynamics

The ECO Lab is a customizable tool for aquatic ecosystem models, simulating water quality, eutrophication, heavy metals, and ecology (Nguyen et al., 2020). The module describes interactions between chemical and ecosystem variables, including sedimentation (Huggett et al., 2021). Advection–dispersion mechanisms are integrated for accurate simulations (Adandedji et al., 2017). The model offers transient solutions for shallow water bodies with irregular coastlines, complex bathymetry, and open boundaries (Nguyen et al., 2020), as presented in Table 3.

LOICZ Biogeochemical Modelling (LBM) Approach

Padedda et al. (2010) describe the LOICZ Biogeochemical Modelling approach as a comprehensive system to understand the flow of nutrients and other biogeochemical constituents within coastal ecosystems. This approach involves measuring inputs, outputs, and internal transformations of key elements, including carbon, nitrogen, and phosphorus (Ramesh et al., 2015). The interrelationships between the various components in the LBM approach are represented by Eq. (3).

$$\mathrm{dM/dt} = \Sigma \mathrm{Input} - \Sigma \mathrm{output} + \Sigma(\mathrm{source} - \mathrm{sink}) \tag{3}$$

where

dM/dt change in mass of the material of interest

Estimation of Water and Salt

The Land–Ocean Interactions in the Coastal Zone (LOICZ) budgeting approach is commonly used to examine the movement of water and salt in coastal areas (Ramesh et al., 2015). This method evaluates both the input and output of these elements, incorporating factors such as freshwater runoff (VQ), precipitation (VP), and groundwater

Table 2 HD and PT model setup

Hydrodynamic—model domain	
Parameters	Values
Mesh and bathymetry	Total nodes = 11,527 Minimum depth cutoff = 1 Map projection = Longitude /Latitude
Simulation period	347 days
Time step interval	60 s
No. of time steps	16,608
Simulation start date	07/01/2014
Simulation end date	06/12/2015
Solution technique Panda and Mohanty (2008)	Low-order, fast algorithm Minimum time step: 0.01 s Maximum time step: 15 s Critical CFL number: 0.8
Density	Barotropic
Eddy viscosity	Type—Smagorinsky Formulation (Constant ~ 0.28)
Resistance	Manning Number: Constant Value—32 m^ [1/3]/s Dahmani et al. (2021)
Boundary	Derived from SRTM Data and Insitu validation
HD result file	Type—2D Horizontal and Format—Area Series
PT model inputs	
Minimum particle mass	0 kg
Maximum particle age	1,584,000 s
Source type	Area source
Flux	500 kg/s
Reference	Position = 0.1 m
	Depth = 0.5 m
Horizontal dispersion	Dispersion coefficient formulation (0.1 m^2/s)
Vertical dispersion	No dispersion
Drift profile	Use raw data from hydrodynamics

discharge (VG) (Ramesh et al., 2015). The water budget equation is expressed as follows:

$$\frac{dVsys}{dt} = V_Q + V_P + V_G + V_O + V_E + V_R \tag{4}$$

where

V_{sys} Total system volume ($length^3$/time)

Table 3 ECOLAB model setup

Ecolab model inputs			
Solution technique	Time integration	Higher order	
	Space Integration	Higher order	
Constants Tri et al. (2018)	Description	Type	Value
	Latitude	Built-in	
	1storder decay rate at 20 o c	Constant	0.5/d
	Temperature coefficient	Constant	1.07
	Half-saturation oxygen concentration	Constant	2 mg/l
	Maximum oxygen production at noon	Constant	2 /d
	Secchi disc depths	Constant	0.4 m
	Time correction	Constant	0 h
	Respiration rate	Constant	0 /d
	Temperature coefficient, respiration	Constant	1.08
	Half saturation concentration for respiration	Constant	2 mg/l
	Sediment oxygen demand (SOD)	Constant	0.5/d
	Temperature coefficient for SOD	Constant	1.07
	Half saturation concentration for SOD	Constant	2 mg/l
Horizontal dispersion	BOD	Dispersion Coefficient formulation (0.01 m^2/s)	
	DO	Dispersion Coefficient formulation (0.01 m^2/s)	
Source	Specified concentration	BOD	1 mg/l
		DO	10 mg/l
Initial conditions	BOD	1.4 mg/l	
	DO	3 mg/l	
Boundary conditions Agostino (2023)	Code 2	BOD	2 mg/l
		DO	4.5 mg/l
	Code 3	BOD	2.7 mg/l
		DO	4.4 mg/l
	Code 4	BOD	2.6 mg/l
		DO	4.5 mg/l

V_Q Stream/river inflow ($length^3/time$)
V_P Direct precipitation inflow ($length^3/time$)
V_G Groundwater discharge ($length^3/time$)
V_O Additional inflows (e.g., sewage) ($length^3/time$)
V_E Evaporation rate (mm/day)
V_R Net residual flow ($length^3/time$)

When assuming $dV_{sys}/dt = 0$, the residual volume (V_R) is determined using the equation: Equation (5) (Newton et al., 2013)

$$V_R = V_Q - V_P - V_G - V_O - V_E \tag{5}$$

The exchange flow or mixing flux (V_x) can be determined using the following formula (Newton et al., 2013):

$$Vx = \frac{(V_Q S_Q + V_P S_P + V_G S_G + V_O S_O + V_R S_R)}{(S_{ocn} - S_{sys})} \tag{6}$$

where

S_{sys} Salinity within the system (psu)
S_{ocn} Salinity in the external environment (psu)
S_Q Salinity of river/stream inflow (psu)
S_P Salinity contribution from precipitation (psu)
S_G Salinity of groundwater discharge (psu)
S_O Salinity from additional sources (e.g., sewage) (psu)

Finally, the exchange time (τ) is calculated by dividing the system volume by the total of the absolute values of residual flow and exchange flow. Equation (7) (Newton et al., 2013):

$$\tau = \frac{V_{sys}}{|V_R| + |V_X|} \tag{7}$$

Estimation of Nutrient Balance

Assessing inputs, outputs, and the internal cycling of nutrients such as phosphorus and nitrogen is a key aspect of the estimation process (Swaney, 2011; Lønborg et al., 2021; Ramesh et al., 2015). Typical sources of inputs consist of groundwater inflows, atmospheric deposition, and nutrient masses from rivers (Swaney, 2011). The outputs comprise the nutrient export to the open ocean, sediment burial, and uptake by primary producers. The LOICZ plays a crucial role in measuring these fluxes and internal variations, which facilitates the assessment of the overall nutrient balance in coastal ecosystems (Gomez Castillo, 2022; Lønborg et al., 2021; Ramesh et al., 2015).

Assessment of Apparent Net Biogeochemical Processes

This assessment involves evaluating the net fluxes of various biogeochemical constituents, including carbon, nitrogen, and phosphorus, across different compartments of coastal environments (Lønborg et al., 2021; Ramesh et al., 2015). The study used simple system stoichiometry with a Redfield C: N: P ratio of 106:16:1 to calculate P-R and Nfix- Denit for unit time and surface area. (μmols/m^2/d) (Islam et al., 2019) (Eqs. 8 and 9).

$$P - R = -106 \times DIP_{obs} \tag{8}$$

$$N_{fix} - D_{enit} = DIN - (16 \times DIP_{obs}) \tag{9}$$

where

DIP_{obs} Measured Dissolved Inorganic Phosphorus (DIP) concentration (μmols/m^3)

DIN_{obs} Measured Dissolved Inorganic Nitrogen (DIN) concentration (μmols/m^3)

LBM—Input Details

The study computed the LBM budgets for the lagoon using the steady-state hypothesis for all seasons to determine seasonal variation and compare the annual budget with other systems. The required has been collected such as water circulation, salinity, and various forms of nutrients (N, P). Accurate measurements of water in and out of the system are crucial for nutrient flux measurements (Krasakopoulou & Pagou, 2011). The area and storage volume of the lagoon were measured at different reference elevations using a Geographical Information System (GIS). Precipitation data was obtained from the Indian Meteorological Department and was used for the budget calculation. Groundwater influx (V_G), is hard to measure and insignificant (Rengarajan & Sarma, 2015). Quantitative judgments of data are vital in constructing a reasonable budget, but this LBM approach has inherent limitations and restrictions specific to its application to coastal areas (Gazeau et al., 2004). Any exchange at the water–sediment interface is assumed to be negligible in this budget. The data collected from the bar mouth is considered an "oceanic" end member and used to prepare the budget for the inner boxes (Gazeau et al., 2004).

3 Results and Discussion

This study combines hydrodynamic modelling with biogeochemical simulations for the Pulicat Lagoon. It provides significant insights into the dynamics of this intricate lagoon ecosystem, supporting informed decisions for its protection and Ecosyst.

3.1 Water and Salt Budget

The Pulicat Lagoon is an estuarine ecosystem with a well-mixed zone of freshwater entering from rivers and tidally influencing seawater. The LBM was constructed for all seasons to understand seasonal variations (Fig. 4). Freshwater discharge varies throughout the seasons, with post-monsoon and summer discharging about 20.6% of freshwater, while southwest and northeast discharge 90.7%. Seasonal precipitation input to the lagoon ranges from 1.2×10^3 m^3 d^{-1} to 3.5×10^3 m^3 d^{-1}, with evaporation loss being less than 5% of the total water budget. A Strong seasonal variation in precipitation influences the lagoon water budget, with lean flows in post-monsoon and summer having a significant impact (Fig. 4). The residual flow within the lagoon was determined using the water balance equation, which illustrates changes in water levels and flows over a period. (Eq. 3) The estimated VR values are derived for the entire system moving toward the ocean; positive values signify inflow into the system, while negative values signify outflow from the system (Calvario-Martínez et al., 2025). (Fig. 4).

A tendency of decreased nutrient retention in the water column, increased nutrient export to the ocean, and decreased production and nitrogen fixation is evident in the water budgets calculated for the lagoon. The exchange flow of salinity was constructed after determining the salinities of the system, the rivers, and the adjacent ocean. The lagoon exhibits substantial seasonal variations of salinity, with upstream regions generally showing lower salinity values than mouth regions (Fig. 4). During the post-monsoon and summer seasons, seawater dominance was higher in the lagoon with high salinity values. The residual flux dominates the mixing flux during the northeast and southwest monsoon seasons. Mutual relations of both residual and mixing flux control the residence time of the system, which regulates and modifies biogeochemical reactions (Fig. 4). The total system residence time is 10.37 days, with seasonal variation in water exchange time largely controlled by riverine flow. The study found that biogeochemistry is highly dynamic during the wet season, while both ocean and freshwater properties control it during the dry season (Fig. 4).

3.2 Nitrogen Budget

Dissolved Inorganic Nitrogen (DIN) concentrations in river, lagoon, and ocean waters vary seasonally, providing significant information about nitrogen dynamics. (Liu et al., 2022) DIN levels in river water were consistently higher than in lagoon and ocean waters, particularly during the northeast monsoon season. Seasonal variations exhibited significant differences in DIN concentrations ($p < 0.05$), with the highest mean DIN values recorded during the northeast monsoon period (Fig. 4). Fluvial input of nitrogen to the lagoon was notably high during the monsoon, attributed to catchment flushing and industrial discharges. This resulted in the DIN inputs of approximately 0.96 times the fluvial flux in summer and 0.93 times during the

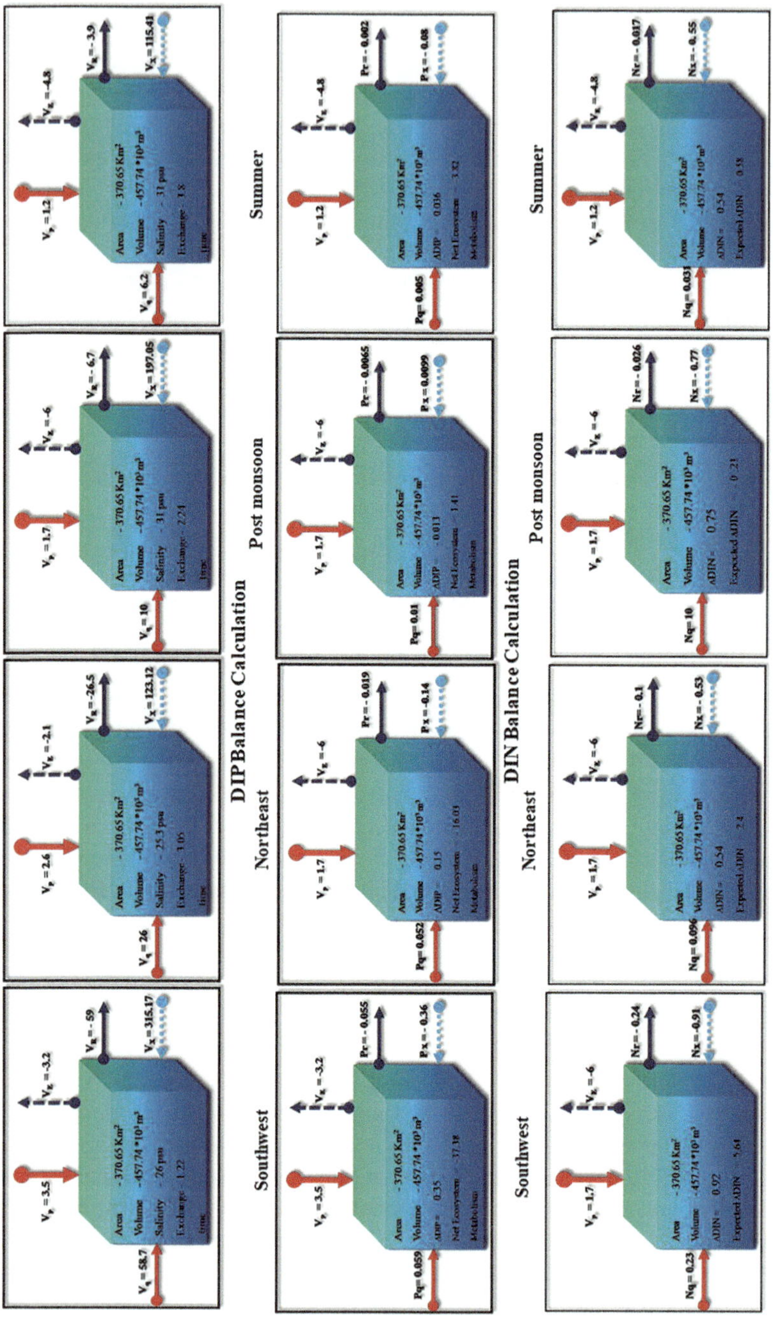

Fig. 4 Seasonal Variations in water, salt, and nutrient budgeting

northeast monsoon. The lagoon functioned as a net DIN sink across all seasons, with non-conservative DIN flux being positive in summer and negative during the southwest monsoon. Daily nitrogen imports into the lagoon exceeded daily exports, with ΔDIN values shifting from moderate removal (-16.8×10^3 mol N d^{-1}) in summer to high removal (-169.8×10^3 mol N d^{-1}) during the southwest monsoon. Moreover, about 3.9% of DIN input during the summer and 7.1% during the monsoon were either denitrified or buried rather than being remineralized. The annual net denitrification rate for the lagoon was measured at -1.45 mol N m^{-2} yr^{-1}, indicating the lagoon's significant role as a nitrogen sink influenced by anthropogenic (Rysgaard et al., 1996) activities (Fig. 4). Overall, these findings underscore the lagoon's function as a transitional nitrogen sink, primarily shaped by river transport rather than oceanic mixing, and highlight the considerable impact of human activities on nitrogen dynamics in coastal ecosystems.

3.3 Phosphorus Budget

The calculated riverine dissolved inorganic phosphate (DIP) values were comparatively lower than the system and ocean values during summer, whereas during the northeast season, the riverine DIP has almost similar values to the system (Fig. 4). Significant variation ($p < 0.05$) is observed between the DIP values in the system during all the seasons. Strong seasonal fluctuation is observed in the riverine import of DIP, which increases from 41.3×10^3 mol P d^{-1} in the dry season to 177.2 $\times 10^3$ mol P d^{-1} in the rainy season (Fig. 4). Throughout the season, $V_R DIP_R$ is negative, indicating consistent DIP output. The gradient in DIP concentration between the ocean and the lagoon determines how the $V_X DIP_X$ changes. During the summer and northeast monsoon, the inward mixing of DIP is between 6.2 and 1.2 ($\times 10^3$ mol P d^{-1}), which is less than river DIP inflow or residual outflow (Fig. 4). When DIP is positive, it means that the system is the source of it (Fig. 4), and when it is negative, it means that the system is the sink for DIP (Karl & Björkman, 2014). Non-conservative fluxes of DIP were lower during the post-monsoon and summer. A notable change over time was detected for ΔDIP, showing moderate levels of release in the summer and increased levels during the northeast monsoon season (Hung et al., 2007). This implies that 38% of phosphorus received in the system was released during summer and 35% during the northeast monsoon (Fig. 4). The lagoon consistently acted as a DIP output source, with inward mixing being much lower than river inflows or residual outflows. A positive ΔDIP indicates a net source of DIP, with annual totals of $+ 15.6 \times 10^6$ mol P yr^{-1}, attributed to organic matter oxidation.

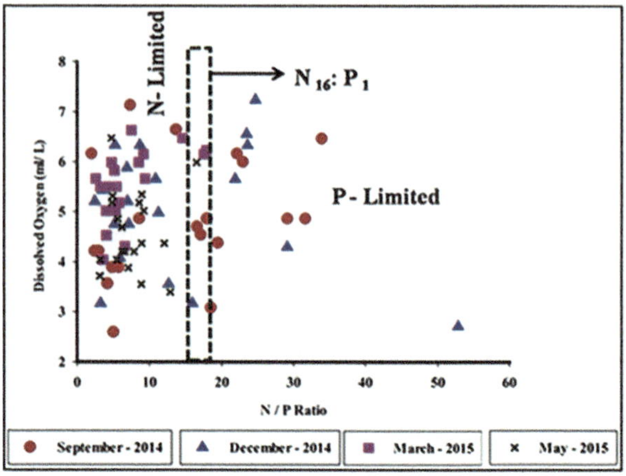

Fig. 5 Seasonal variations in nutrient limitation

3.4 Net Ecosystem Metabolism

The distinction between primary production and community respiration reflects a calculation of net carbon metabolism in balanced systems (Segura-Noguera et al., 2016). Net Ecosystem Metabolism (NEM) is determined stoichiometrically based on the change in dissolved inorganic phosphorus (ΔDIP) that is either generated or used (Fig. 5) within the lagoon (Feng et al., 2012). A net positive value of ΔDIP in the water column is equivalent to negative NEM values (Vybernaite-Lubiene et al., 2022). In the lagoon, the NEM showed a positive value throughout the monsoon season, suggesting net autotrophy, which means that an external supply of organic matter is unnecessary for decomposition (Muduli et al., 2013). Negative NEM values indicate that community respiration overpowers production in the lagoon (Feng et al., 2012). The highest net heterotrophy suggests a greater discrepancy between respiration and production at the southern entrance. (Delgadillo-Hinojosa et al., 2008). High waste loads from human activities in the estuary contribute to organic matter, prompting the system toward heterotrophic activity (Nixon, 1988). The NEM for the entire system during summer was about -8.33 mmol C m^{-2} d^{-1}, which is significant due to non-conservative sources of nutrients (DIN and DIP) (Jickells et al., 2000). Negative NEM was observed for all the seasons indicating that the system is net heterotrophic throughout the year (Weyhenmeyer, 2024). Thus, the entire system undergoes net mineralization of organic matter and is in a heterotrophic state during the summer, although lower in magnitude when compared to the monsoon heterotrophy.

3.5 Hydrodynamic Variations

The observed modelling values through the highest and lowest water 1evls usually occur at spring tides in all seasons. In summer, the high-water level was recorded as +

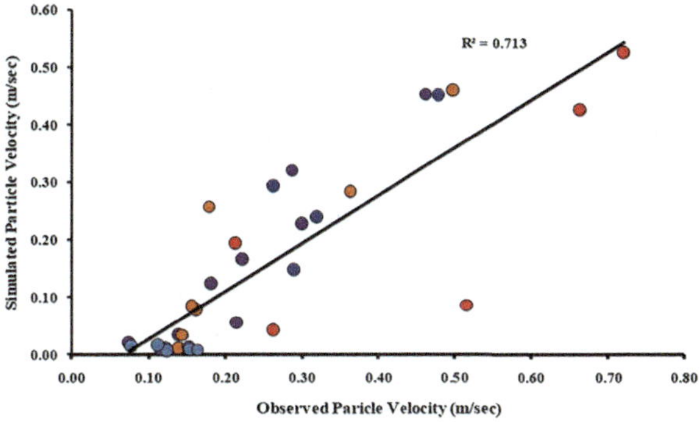

Fig. 6 Observed and simulated particle velocity

0.63 m at neap (13.5.2015) and spring (21.5.2015) tides (Fig. 6). This highest range associated with neap tide may be due to local weather disturbances such as barometric pressure fluctuations (Redzuan & Underwood, 2021). The results show water level variation as the maximum in post-monsoon followed by southwest monsoon, while the lowest water level variation with the minimum tidal range is observed during summer (Fig. 6). The interesting observation on water level variation is that during the latter part of summer, there is an increase in the amplitude, while during the southwest and northeast monsoon seasons, the sinusoidal nature of variation continues for the entire month (Fig. 6). Water level variation near Pulicat inlet was examined considering three important forcing factors namely, tide, freshwater, and wind. The latter two forces show significant seasonal variation (Srinivasan & Natesan, 2013).

3.6 *Tidal Influence*

After calibration and validation procedures (Sect. 2.3.3), the reproduction of the system model was considered (Mardani et al., 2020; Williams & Esteves, 2017) satisfactory (Fig. 5) and sufficient to simulate the hydrodynamic processes in the Pulicat Lagoon, and, therefore it could force the Lagrangian transport module for the water velocities and related water flows. The patterns cause the transport and distribution of pollutants in the lagoon. Water velocities at the mouth of the lagoon (up to 0.6 m/s) are higher than in other areas of the lagoon, which indicates a strong influence of inflow from the sea (Fig. 6). A Lagrangian particle tracking model was employed to investigate the advection of particles (water mass) and assess the influence of tides on the lagoon (Gu et al., 2024). Figure 12 shows the position of the particles at time step 2047 and corresponds to the tidal situation in the lagoon. Particle migration is due to residual tidal circulation in the lagoon. Particles near the mouth of Pulicat swift towards the lagoon, while particles at the back end of the lagoon (near Arambakkam) remained there during the count. It reflects that large tidal currents tend to affect particles in estuaries, and small tidal currents tend to affect particles in other areas. The saltwater mass stays near Annamalaicheri at the beginning of the simulation, but during the second phase, it is moved to the north (offshore) during low tides and to the south during high tides via the Buckingham Channel (Fig. 12). Particles in the centre of the lagoon (seawater) were trapped longer than expected. During the upwelling, weak currents carry them towards the Bay of Bengal, so that there is not enough time to move them into the entrance channel (Fig. 12). In the next flood water, most of these particles (seawater) move towards the lagoon and re-enter the interior of the lagoon. Thus, the particles move back and forth with each tidal cycle, and only a small fraction of these particles (seawater) leave the lagoon. The mass of water coming from different parts of the lagoon mixes in waves only in the entrance channel, from where they are quickly washed away under the influence of the strong currents prevailing here. In the simulation period, seawater reaches the centre of the lagoon near Sunnambukkulam in SW monsoon (~ 15.8 km), Kuruvithittu in NE Monsoon (~ 7.158 km), Karimanal in post-monsoon (~ 7.5 km) and upstream. Mangodu (~ 7.769 km) in summer (Fig. 12). The particles stay in the places closer to Sunnambukkulam and the far end of the Arambakkam, close to Irukkam and Venadu islands, for a longer period because of the slow dynamics there (Fig. 12). Due to the tidal currents in these areas cause the particles to be moved back and forth across the channel (Fig. 6). Since all different kinds of particles can be seen in the lagoon's centre, this area also serves as a mixing zone for other locations. During flood tide circumstances, the maximum current speed is recorded between stations 1 and 3, and it progressively declines towards the mid-off of the lagoon (Figs. 7, 8, 9, 10 and 11). The inlet has the greatest water current during ebb tide, reaching up to 0.6 m/s (Figs. 7, 8, 9 and 10). Along the X direction, especially at Pulicat mouth, predicts a maximum value of 0.6 m/s during flooding conditions, along the Y—direction, the current records a rise in the velocity due to tidal influence.

It is observed that the lagoon inlet is flood tide-dominated during the southwest monsoon, while during the northeast, it is ebb tide-dominated (Fig. 7, 8, 9, 10 and 11). Flood/ebb tide dominance critically depends on forcing functions such as tide and freshwater discharge (Smith & Anderson, 2018). The summer period is the driest in the area, and there is no freshwater discharge. The current at the inlet is mainly due to the tidal forcing. The particles (water mass) located near Pulicat's mouth rapidly progressed towards the lagoon, whereas the ones at the far end of the lagoon (near Arambakkam) remained there throughout the study period. It is observed that the lagoon inlet is flood tide-dominated during the southwest monsoon, while during the northeast, it is ebb tide-dominated (Figs. 7, 8, 9, 10 and 11). Flood/ebb tide dominance critically depends on forcing functions such as tide and freshwater discharge (Wang et al., 2021). The summer period was the driest in the area and had no freshwater discharge (Figs. 7, 8, 9 and 10). The current at the inlet is mainly due to the tidal forcing. The particles (water mass) located near Pulicat's mouth rapidly progressed towards the lagoon, whereas the ones at the far end of the lagoon (near Arambakkam) remained there throughout the study period (Fig. 12).

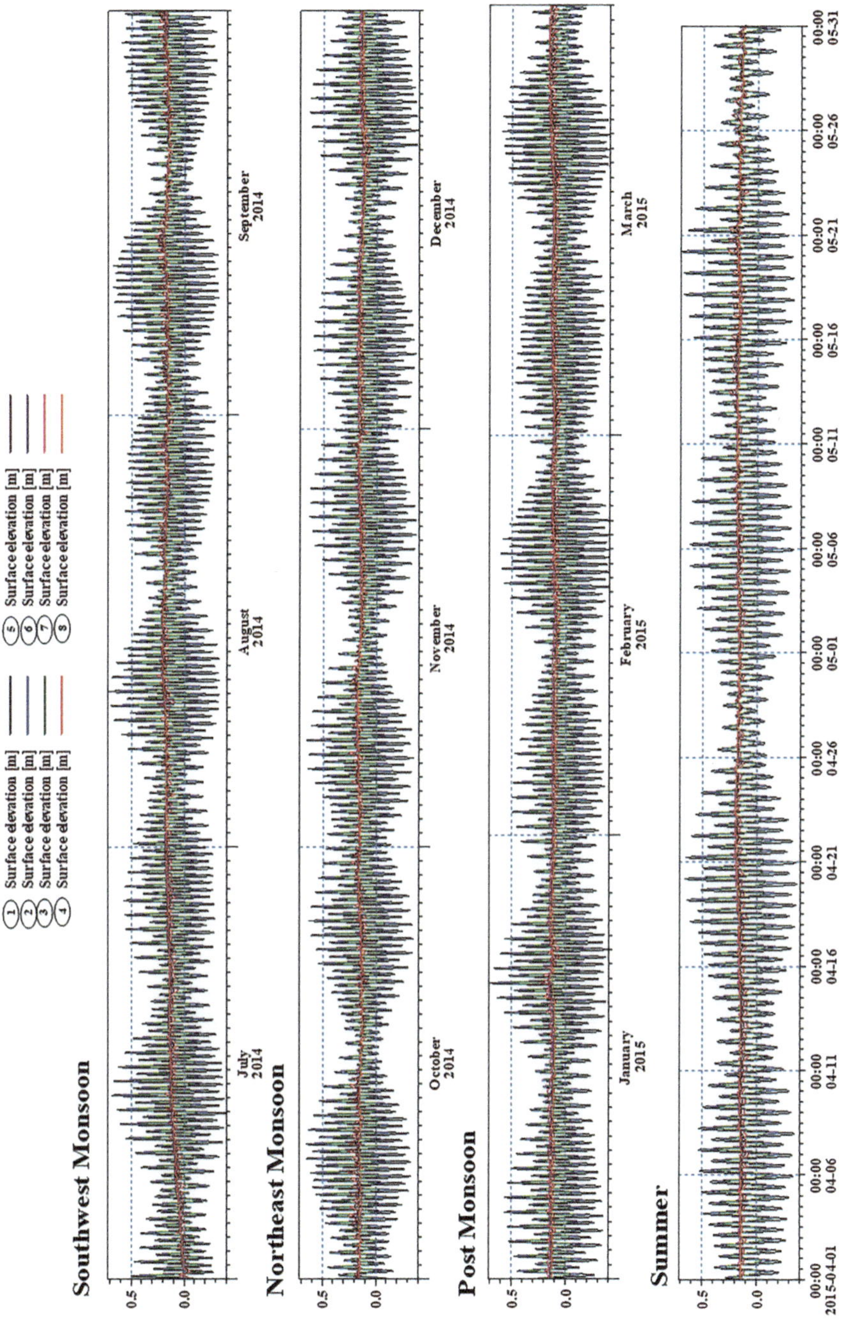

Fig. 7 Seasonal variations in water levels

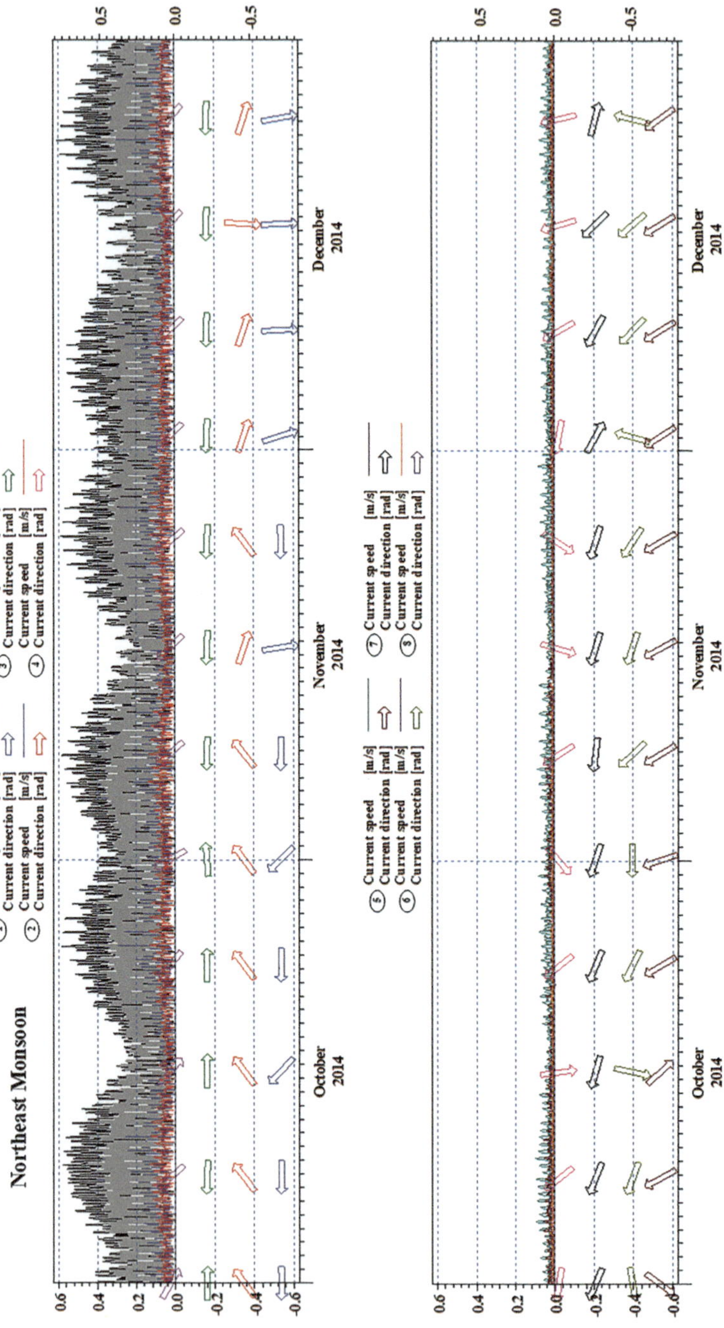

Fig. 8 Disparities in current speed and direction during southwest monsoon season

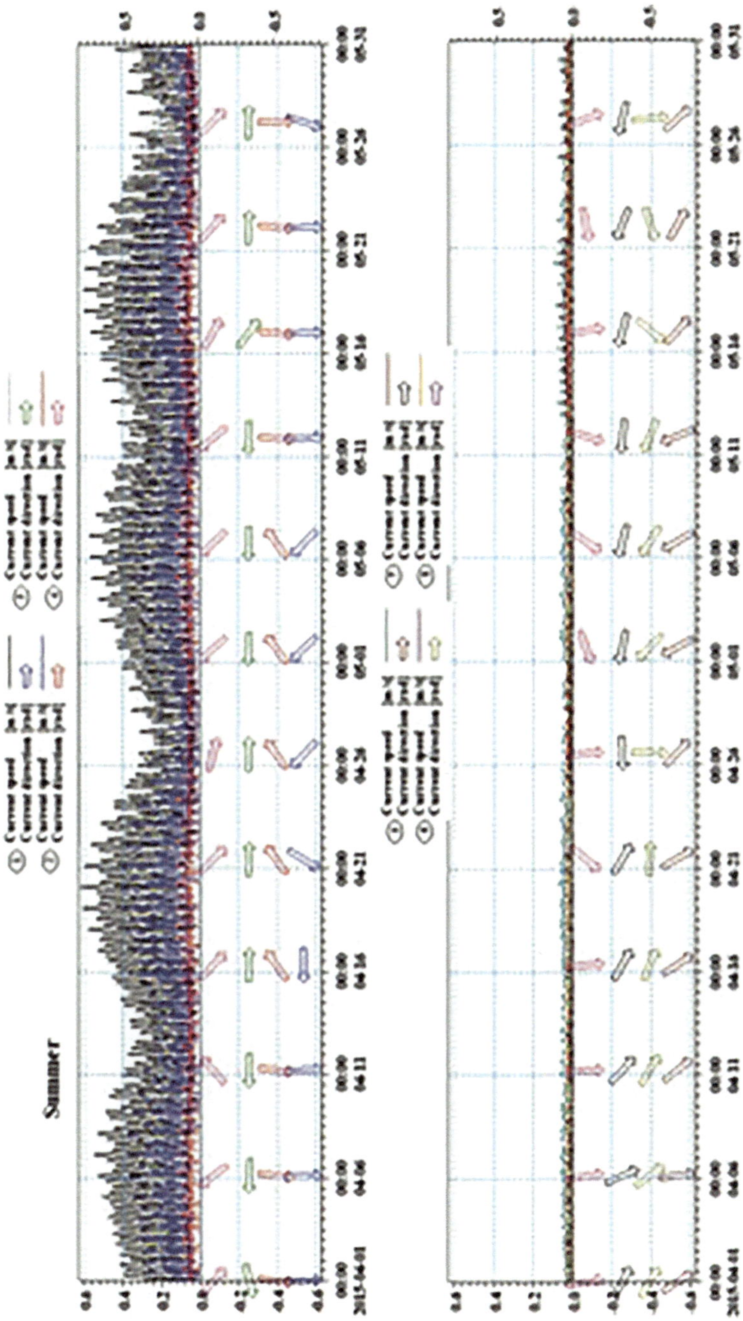

Fig. 9 Disparities in current speed and direction during northeast monsoon season

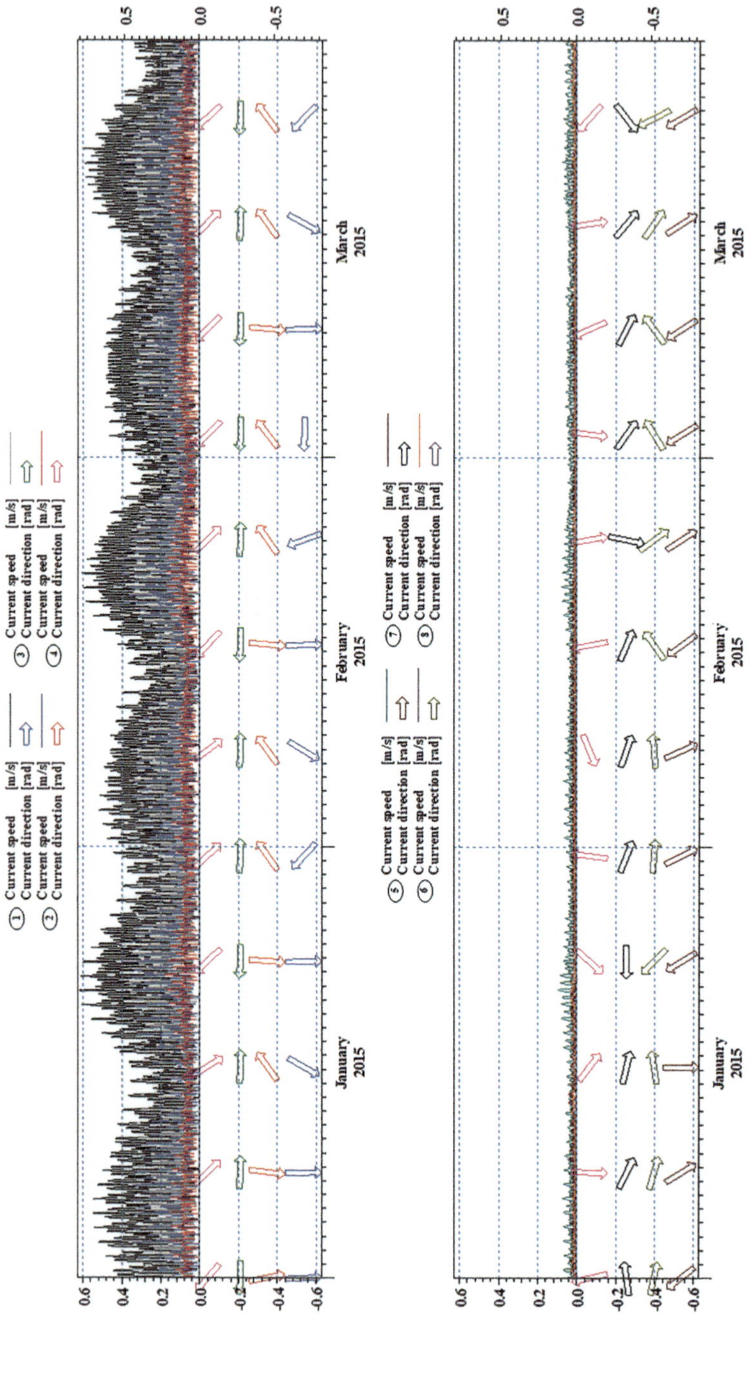

Fig. 10 Disparities in current speed and direction during post-monsoon season

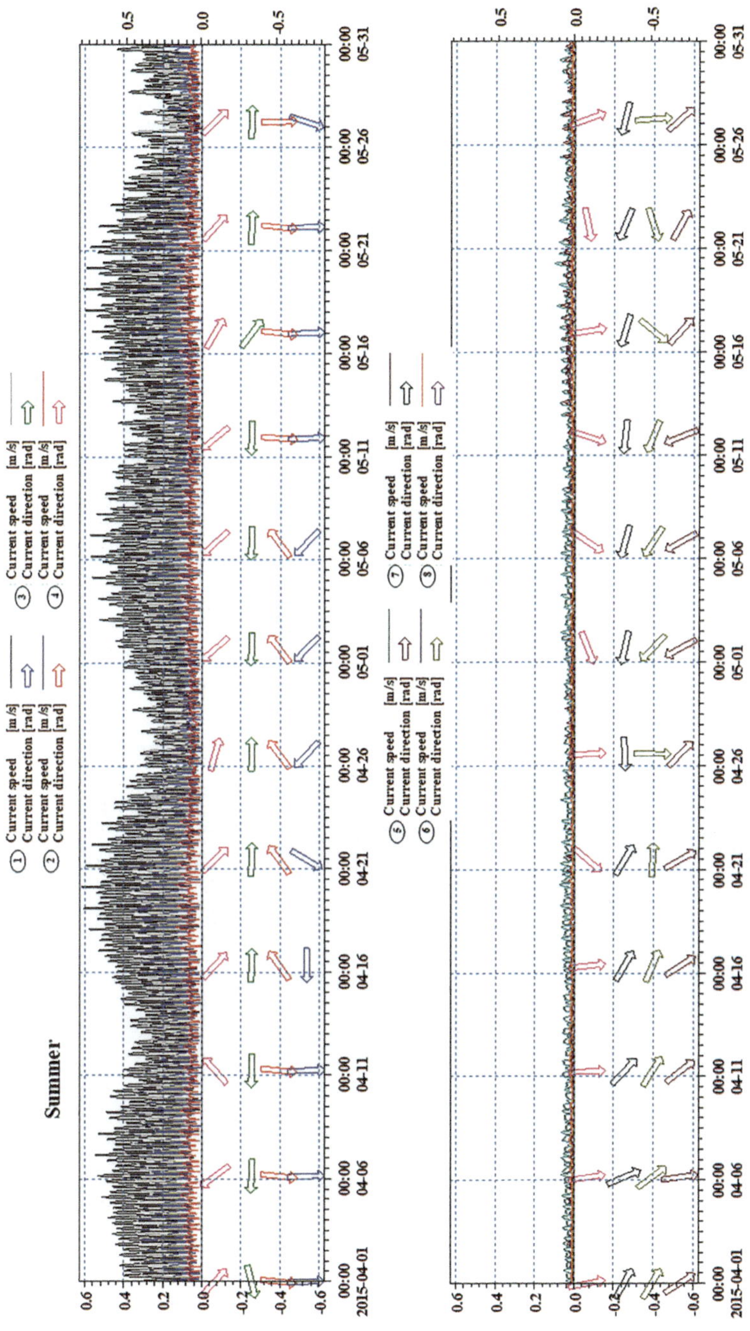

Fig. 11 Disparities in current speed and direction during the summer season

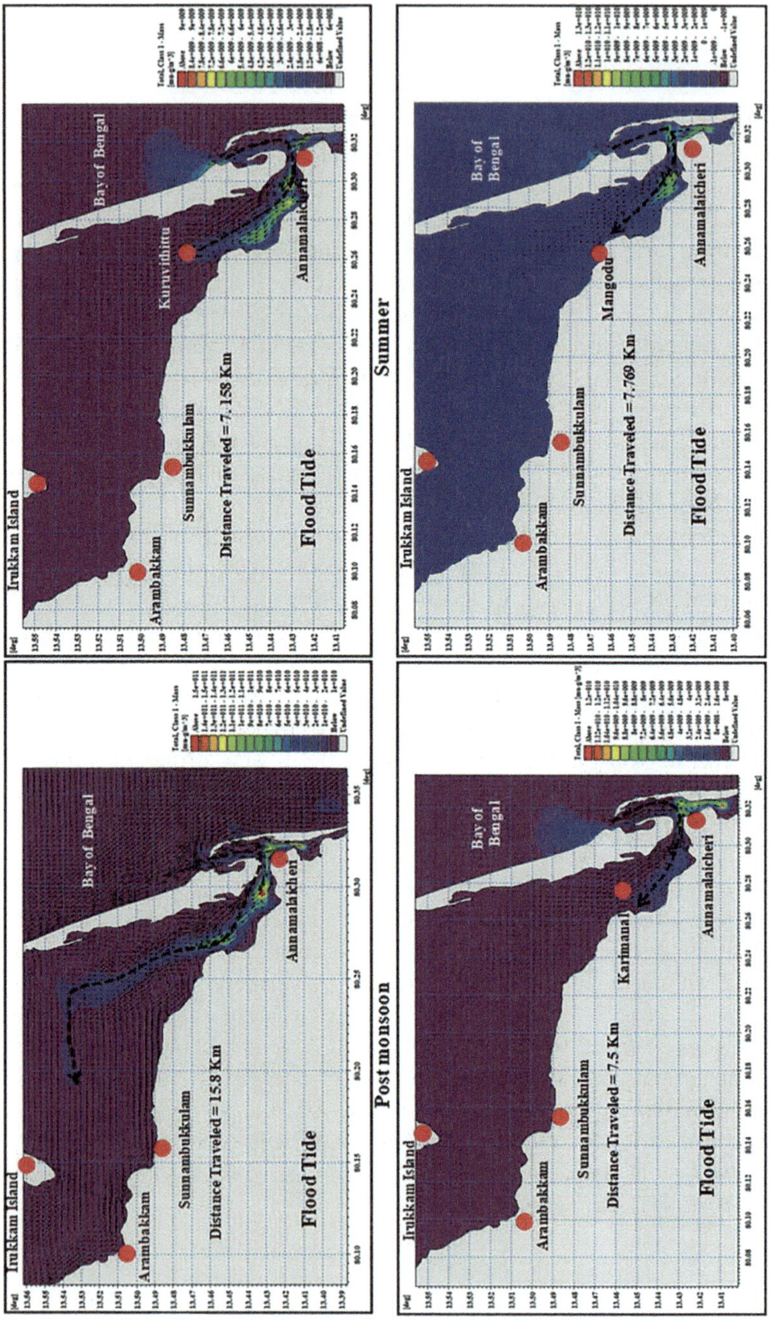

Fig. 12 Tide Influencing zones in different seasons

3.7 BOD - DO Dynamics

The levels of biochemical oxygen demand (BOD) and dissolved oxygen (DO) in Pulicat Lagoon vary during the day because of tidal changes. (Jeba Kumar & Natesan, 2015) (Fig. 13). Higher levels are typically observed during spring tides when wave action enhances the oxygenation of seawater. Sewage discharge characteristically exhibits low DO levels due to organic enrichment (Culhane et al., 2019). Along the Buckingham Canal, consistently high BOD-DO concentrations are observed when the sewage is discharged, indicating excessive organic matter in the effluents (Kumar & Reddy, 2009). This organic matter undergoes bacterial decomposition upon entering the lagoon, depleting dissolved oxygen and contributing to odour generation (Kumr & Natesan, 2015; Ortega-Arbulú et al., 2019) (Fig. 13).

4 Conclusion

The biogeochemical dynamics of Pulicat Lagoon, situated in the Bay of Bengal, are governed by a complex interplay of various environmental factors, anthropogenic influences, and natural processes. Significant variations in the lagoon ecosystem's nitrogen cycle, primary production, and water quality indices were observed during our study. Seasonal fluctuations, mainly influenced by monsoonal patterns, play a crucial role in shaping the biogeochemical dynamics of the lagoon. The post-monsoon period exhibits heightened water level variations, accompanied by increased nutrient inputs from terrestrial runoff, leading to elevated primary production and phytoplankton blooms. Increased tidal exchange and freshwater flow during the Southwest monsoon cause higher levels of biogeochemical activity and nutrient enrichment. Conversely, the summer season experiences minimal water level variations and reduced nutrient inputs, resulting in lower primary production rates and nutrient concentrations. Anthropogenic activities, including sewage discharge and agricultural runoff, further exacerbate nutrient loading and eutrophication processes leading to oxygen depletion and adverse impacts on aquatic ecosystems. The present study underscores the need for efficient management techniques to decrease nutrient pollution and promote the sustainable use of Pulicat Lagoon. Various Integrated Lagoon Management steps and plans are listed (Figs. 14 and 15).

- Data collection and monitoring are vital for managing Pulicat Lagoon, focusing on ecological and socio-economic aspects. This involves gathering baseline data on biodiversity, hydrology, and water quality, along with socio-economic surveys to understand community dependence on lagoon resources. Engaging local communities in this process fosters stewardship and ensures management reflects their needs.
- Implementation and capacity building emphasize training stakeholders—such as local fishermen and officials—in sustainable practices. Community engagement promotes a sense of ownership, while resource allocation and collaboration with

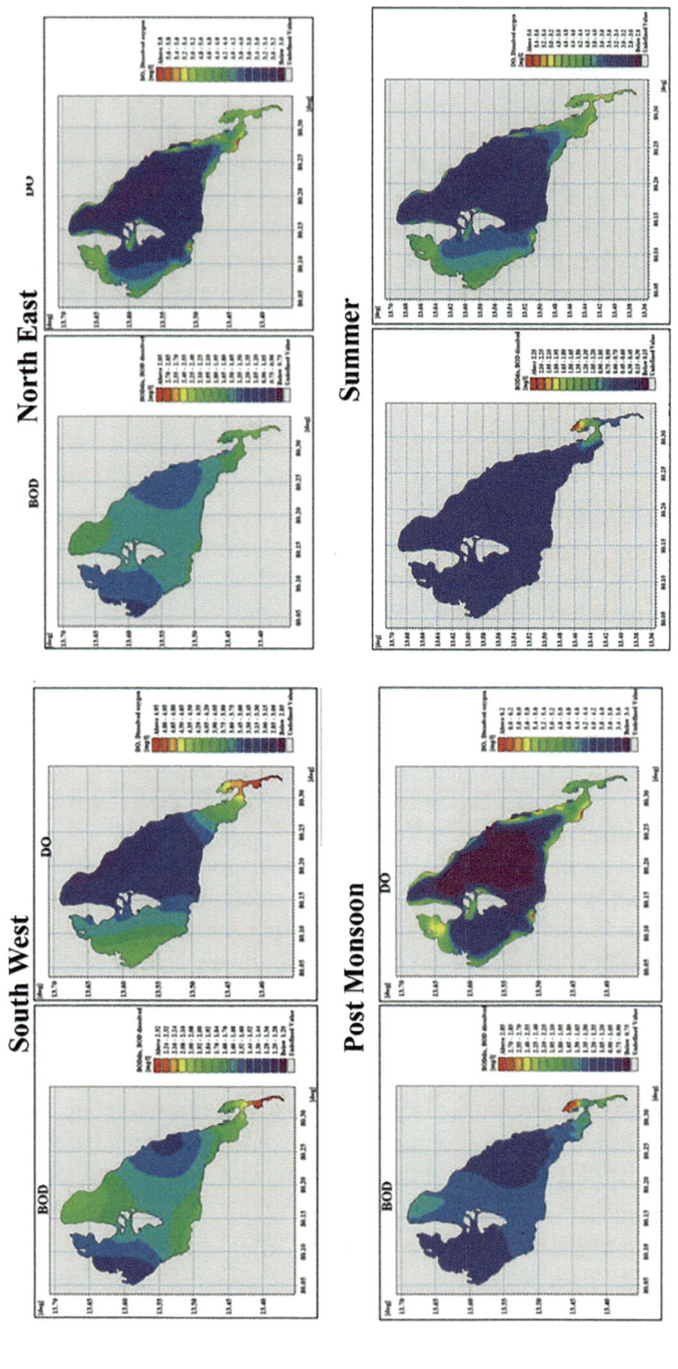

Fig. 13 Variations in BOD – DO dynamics in different seasons

Fig. 14 Integrated Pulicat lagoon management plans

local institutions enhance management efforts. Promoting eco-friendly practices and providing technical support are also crucial.

- Effective policy and conflict management are essential for success. This includes analyzing legal frameworks, proposing actionable policies, and establishing conflict resolution strategies. Integrating traditional knowledge enhances policy relevance, and encouraging public participation fosters transparency.
- Evaluation and adaptation are key for ongoing success, using performance indicators and feedback mechanisms to assess progress and make necessary adjustments. Documenting lessons learned and conducting risk assessments help anticipate challenges and refine management strategies.
- Assessment and planning involve conducting environmental impact assessments and SWOT analyses to identify critical factors. Setting clear goals and strategic plans, along with integrating scientific research and local knowledge, strengthens the overall approach, enhancing the resilience of Pulicat Lagoon and supporting the communities that depend on it.

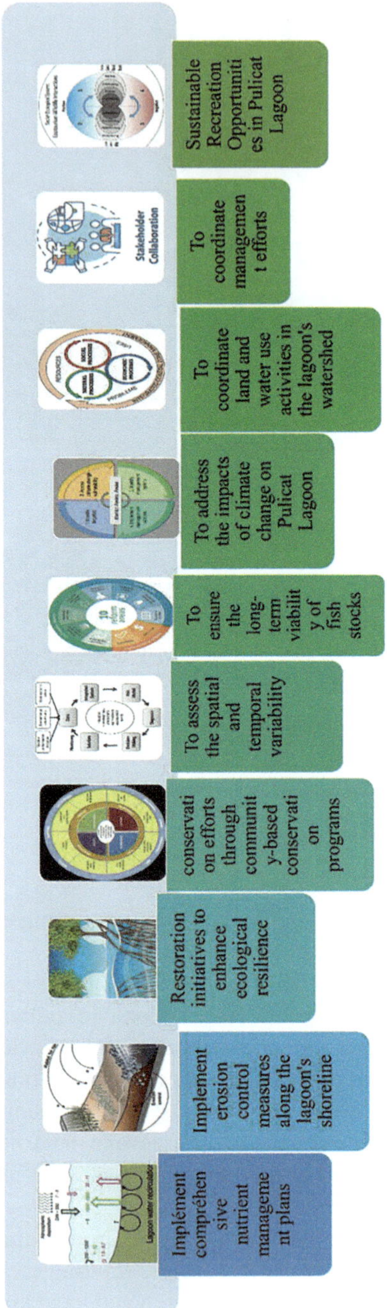

Fig. 15 Integrated lagoon management steps

Integrated Management Plans

Nutrient Management Strategies: Implement comprehensive nutrient management plans to minimize nutrient loading from agricultural runoff and sewage discharge. This includes promoting sustainable practices and utilizing advanced wastewater treatment technologies.

Erosion Control Measures: Implement erosion control measures along the lagoon's shoreline to mitigate sedimentation and maintain water quality, including the restoration of natural vegetation buffers and the construction of bioengineering structures.

Habitat Restoration Initiatives: Undertake habitat restoration initiatives to enhance the ecological resilience of Pulicat Lagoon, including the restoration of mangrove habitats, seagrass beds, and wetlands to provide critical habitats for fish, birds, and other wildlife.

Initiatives for Conservation in Local Communities: Involve local populations in conservation activities via community-driven conservation initiatives, environmental education programs, stakeholder workshops, and collaborative monitoring efforts to promote stewardship and understanding of the lagoon's ecological significance.

Integrated Water Quality Monitoring: Develop a comprehensive water quality monitoring program to assess spatial and temporal variations in key water quality parameters. This includes tracking nutrient concentrations, dissolved oxygen levels, and pollutant loads to inform management strategies and evaluate progress toward water quality improvement goals.

Sustainable Fisheries Management: Adopt sustainable fisheries management approaches to maintain healthy fish populations while supporting local fishing communities. Measures such as establishing marine protected areas, enforcing seasonal fishing bans, and regulating fishing gear usage can help prevent overfishing and minimize environmental harm.

Climate Change Adaptation Strategies: Design and implement climate adaptation policies to address the impacts of climate change on Pulicat Lagoon. These strategies should focus on mitigating risks associated with sea-level rise, increased storm frequency, and shifting rainfall patterns. Key actions include strengthening coastal defenses, maintaining habitat connectivity, and optimizing water resource management.

Integrated Coastal Management: Develop a holistic coastal zone management plan that integrates land and water resource planning within the lagoon's watershed. This approach should incorporate zoning regulations, land-use planning, and development controls to reduce pollution from land-based activities and safeguard critical habitats.

Stakeholder Collaboration and Partnerships: Encourage cooperation and alliances between government entities, non-profit organizations, research bodies, and local communities to orchestrate management initiatives, exchange resources and knowledge, and enhance capabilities for sustainable lagoon oversight.

Ecotourism Development and Sustainable Recreation: Promote ecotourism development and sustainable recreation opportunities in Pulicat Lagoon to generate alternative livelihoods for local communities, raise awareness of the lagoon's ecological value, and provide incentives for conservation and stewardship efforts. These integrated lagoon management plans aim to address the diverse ecological, social, and economic challenges facing Pulicat Lagoon while promoting its long-term sustainability and resilience.

References

Adandedji, F. M., Sintondji, L. O., Mama, D., & Vested, H. J. (2017). Pollutants dispersion in tropical coastal lake based on numerical modeling approach: a case study of Nokoué Lake in the Republic of Benin. *International Journal of Scientific and Engineering Research, 8*, 1369–1374.

Agostino, A. (2023). *Assessing the role of climate-driven extreme events on the character and flux of Australian natural organic matter* (Doctoral dissertation, University of New South Wales (Australia)).

Ala Amjadi, A., & Mokhtari, M. (2019). Effects of change in the parameters of initial water displacements on energy and amplitude amplification of tsunamis. *Journal of Seismology and Earthquake Engineering, 21*(2), 1–10.

El Behja, H., El M'rini, A., Nachite, D., Bouchkara, M., El Khalidi, K., Zourarah, B., Uddin, M.G., & Abioui, M. (2024). Evaluating coastal lagoon sustainability through the driver pressure state impact response (DPSIR) approach in Khenifiss Lagoon, Morocco. *Frontiers in Earth Science, 12*, 1322749.

Calvario-Martínez, O., Medina-Galvan, J., Domínguez-Jiménez, V. P., Alonso-Rodríguez, R., Sánchez-Rodríguez, M. A., Reyes-Velarde, P. M., Betancourt-Lozano, M., & Serrano-Hernández, D. (2025). A two-decade overview of the environmental carrying capacity in Bahía Santa Maria–La Reforma Coastal Lagoon system. *Journal of Marine Science and Engineering, 13*(2), 295.

Cepic, M., Bechtold, U., & Wilfing, H. (2021). Modelling human influences on biodiversity at a global scale–A human ecology perspective. *Ecological Modelling, 465*, Article 109854. https://doi.org/10.1016/j.ecolmodel.2021.109854

Chacón Abarca, S., Chávez, V., Silva, R., Martínez, M. L., & Anfuso, G. (2021). Understanding the dynamics of a coastal lagoon: Drivers, exchanges, state of the environment, consequences and responses. *Geosciences, 11*(8), 301.

Culhane, F. E., Briers, R. A., Tett, P., & Fernandes, T. F. (2019). Response of a marine benthic invertebrate community and biotic indices to organic enrichment from sewage disposal. *Journal of the Marine Biological Association of the United Kingdom, 99*(8), 1721–1734.

Dahmani, A. E. A., Mezouar, K., Cherif, Y. S., & Sallaye, M. (2021). Coastal processes and nearshore hydrodynamics under high contrast wave exposure, Bateau-cassé and Stamboul coasts, Algiers Bay. *Estuarine Coastal and Shelf Science, 250*, Article 107169. https://doi.org/10.1016/j.ecss.2021.107169

de Campos Júnior, E. O., Pereira, B. B., & Barros, N. O. (2024). Integrating real-time monitoring and ecotoxicology using a neotropical stream as a study case. *Hydrobiologia*, 1–14.

Delgadillo-Hinojosa, F., Zirino, A., Holm-Hansen, O., Hernández-Ayón, J. M., Boyd, T. J., Chadwick, B., & Rivera-Duarte, I. (2008). Dissolved nutrient balance and net ecosystem metabolism in a Mediterranean-climate coastal lagoon: San Diego Bay. *Estuarine, Coastal and Shelf Science, 76*(3), 594–607.

Devlin, M., & Brodie, J. (2023). Nutrients and eutrophication. In: Reichelt-Brushett, A. (Eds.), *Marine pollution—Monitoring, management and mitigation. Springer textbooks in earth sciences, geography and environment*. Springer. https://doi.org/10.1007/978-3-031-10127-4_4

Dias, J. M., & Lopes, J. F. (2006). Implementation and assessment of hydrodynamic, salt and heat transport models: The case of Ria de Aveiro Lagoon (Portugal). *Environmental Modelling & Software, 21*(1), 1–15.

Farooqui, A., & Vaz, G. G. (2000). Holocene sea-level and climatic fluctuations: Pulicat lagoon—a case study. *Current Science, 79*(10), 1484–1488. http://www.iisc.ernet.in/currsci/nov252000/1484.pdf

Feng, M. L., Sun, T., Zhang, L. X., & Shen, X. M. (2012). Net ecosystem metabolism simulation by dynamic dissolved oxygen model in Yellow River Estuary, China. *Procedia Environmental Sciences, 13*, 807–817.

Gangopadhyay, A. (2022). Introduction to ocean circulation and modeling (1st ed.). CRC Press. https://doi.org/10.1201/9780429347221

Gazeau, F., Smith, S. V., Gentili, B., Frankignoulle, M., & Gattuso, J. P. (2004). The European coastal zone: Characterization and first assessment of ecosystem metabolism. *Estuarine, Coastal and Shelf Science, 60*(4), 673–694.

Gomez Castillo, A. P. (2022). *Identification of seasonal and interannual drivers of primary production in two temperate estuaries using high-frequency environmental data* (Doctoral dissertation, University of Southampton).

Gordon, D. C., Boudreau, P. R., Mann, K. H., Ong, J. E., Silvert, W. L., Smith, S. V., Wattayakorn, G., Wulff, F., & Yanagi, T. (1996). *LOICZ biogeochemical modelling guidelines* (Vol. 5). LOICZ Core Project, Netherlands Institute for Sea Research.

Grasshoff, K., Kremling, K., & Ehrhardt, M. (1999). *Methods of seawater analysis* (3ed edn). Verlage Chemie.

Green, N. S., Li, S., Maul, J. D., & Overmyer, J. P. (2022). Natural and anthropogenic factors and their interactions drive stream community integrity in a North American river basin at a large spatial scale. *The Science of the Total Environment, 835*, Article 155344. https://doi.org/10.1016/j.scitotenv.2022.155344

Gu, B. H., Woo, S. B., Kwon, J. I., Park, S. H., & Kim, N. H. (2024). Case study of contaminant transport using Lagrangian particle tracking model in a macro-tidal estuary. *Water, 16*(4), 617.

Hauck, J., Gregor, L., Nissen, C., Patara, L., Hague, M., Mongwe, P., Bushinsky, S., Doney, S. C., Gruber, N., Quéré, C. L., Manizza, M., Mazloff, M., Monteiro, P. M. S., & Terhaar, J. (2023). The southern ocean carbon cycle 1985–2018: Mean, seasonal cycle, trends, and storage. *Global Biogeochemical Cycles, 37*(11). https://doi.org/10.1029/2023gb007848

Hsu, M. H., Kuo, A. Y., Kuo, J. T., & Liu, W. C. (1999). Procedure to calibrate and verify numerical models of estuarine hydrodynamics. *Journal of Hydraulic Engineering, 125*(2), 166–182.

Huggett, R. D., Purdie, D. A., & Haigh, I. D. (2021). Modelling the influence of riverine inputs on the circulation and flushing times of small shallow estuaries. *Estuaries and Coasts, 44*(1), 54–69.

Hung, J. J., Wang, S. M., & Chen, Y. L. (2007). Biogeochemical controls on distributions and fluxes of dissolved and particulate organic carbon in the Northern South China Sea. *Deep Sea Research Part II: Topical Studies in Oceanography, 54*(14–15), 1486–1503.

Irfan, T., Isobe, A., & Matsuura, H. (2024). A particle tracking model approach to determine the dispersal of riverine plastic debris released into the Indian Ocean. *Marine Pollution Bulletin, 199*, Article 115985. https://doi.org/10.1016/j.marpolbul.2023.115985

Islam, M. J., Jang, C., Eum, J., Jung, S.M., Shin, M.S., Lee, Y., Choi, Y. and Kim, B. (2019). C: N: P stoichiometry of particulate and dissolved organic matter in river waters and changes during decomposition. *Journal of Ecology and Environment, 43*, 4 (2019). https://doi.org/10.1186/s41610-018-0101-4

Jeba Kumar, E., & Natesan, U. (2015). Status and critical knowledge gaps for integrated lagoon management—Pulicat lagoon. *India. J Coast Conserv, 19*, 475–489. https://doi.org/10.1007/s11852-015-0401-0

Jickells, T., Andrews, J., Samways, G., Sanders, R., Malcolm, S., Sivyer, D., Parker, R., Nedwell, D., Trimmer, M., & Ridgway, J. (2000). Nutrient fluxes through the Humber Estuary—Past, present and future. *Ambio, 29*(3), 130–135. https://doi.org/10.1579/0044-7447-29.3.130

Jinoj, T. P. S., Panda, U., Vijay, A., Ezhilarasan, P., Kumaraswami, M., & Murthy, M. R. (2024). Hydro-ecological modelling of a hypersaline lagoon ecosystem (Pulicat) in Southeast coast of India. *Estuarine Coastal and Shelf Science, 304*, Article 108833. https://doi.org/10.1016/j.ecss.2024.108833

Karl, D. M., & Björkman, K. M. (2014). Dynamics of dissolved organic phosphorus. In *Elsevier eBooks* (pp. 233–334). https://doi.org/10.1016/b978-0-12-405940-5.00005-4

Khan, M. N., & Mohammad, F. (2014). Eutrophication: Challenges and solutions. In *Springer eBooks*, 1–15. https://doi.org/10.1007/978-94-007-7814-6_1

Krasakopoulou, E., & Pagou, K. (2011). Seasonal steady-state budgets of nutrients and stoichiometric calculations in an Eastern Mediterranean lagoon (Papas Lagoon-Greece). *Mediterranean Marine Science, 12*(1): 21–41. https://doi.org/10.12681/mms.51

Krishnan, P., & Sampath, V. (1973). A survey of Pulicat Lake and its fisheries, Dept. of Fisheries. Govt. of Tamil Nadu.

Kumar, A. Y., & Reddy, M. V. (2009). Assessment of seasonal effects of municipal sewage pollution on the water quality of an urban canal - a case study of the Buckingham canal at Kalpakkam (India): NO_3, PO_4, SO_4, BOD, COD and DO. *Environmental Monitoring and Assessment, 157*, 223–234.

Kumar, N. V., Nagarjuna, A., Reddy, D. C., Rajasekhar, M., Reddy, K. M., & Rao, A. N. (2008). Satellite remote sensing and field studies on a sea mouth in the northern part of Pulicat Lake. *Current Science, 95*(10), 1405–1406.

Liu, M., He, Y., Cao, L., Zhi, Y., He, X., Li, T., Wei, Y., Yuan, X., Liu, B., He, Q., Li, H., & Miao, X. (2022). Fate of dissolved inorganic nitrogen in turbulent rivers: The critical role of dissolved oxygen levels. *Environmental Pollution, 312*, Article 120074. https://doi.org/10.1016/j.envpol.2022.120074

Liu, S. M., Hong, G., Zhang, J., Ye, X. W., & Jiang, X. L. (2009). Nutrient budgets for large Chinese estuaries. *Biogeosciences, 6*(10), 2245–2263. https://doi.org/10.5194/bg-6-2245-2009

Lønborg, C., Müller, M., Butler, E. C., Jiang, S., Ooi, S. K., Trinh, D. H., Wong, P. Y., Ali, S. M., Cui, C., Siong, W. B., Yando, E. S., Friess, D. A., Rosentreter, J. A., Eyre, B. D., & Martin, P. (2021). Nutrient cycling in tropical and temperate coastal waters: Is latitude making a difference? *Estuarine Coastal and Shelf Science, 262*, Article 107571. https://doi.org/10.1016/j.ecss.2021.107571

Mardani, N., Suara, K., Fairweather, H., Brown, R., McCallum, A., & Sidle, R. C. (2020). Improving the accuracy of hydrodynamic model predictions using Lagrangian calibration. *Water, 12*(2), 575.

Menció, A., Madaula, E., Meredith, W., Casamitjana, X., & Quintana, X. (2023). Nitrogen in surface aquifer—Coastal lagoons systems: Analyzing the origin of eutrophication processes. *The Science of the Total Environment, 871*, Article 161947. https://doi.org/10.1016/j.scitotenv.2023.161947

Mondal, N., Rajkumar, M., Sun, J., Kundu, S., Lyla, P. S., Khan, S. A., & Trilles, J. P. (2010). Biodiversity of brackish water amphipods (crustacean) in two estuaries, southeast coast of India. *Environmental Monitoring and Assessment, 171*(1–4), 471–486. https://doi.org/10.1007/s10661-009-1292-z

Muduli, P. R., Kanuri, V. V., Robin, R., Kumar, B. C., Patra, S., Raman, A., Rao, G. N., & Subramanian, B. (2013). Distribution of dissolved inorganic carbon and net ecosystem production in a tropical brackish water lagoon, India. *Continental Shelf Research, 64*, 75–87. https://doi.org/10.1016/j.csr.2013.05.014

Nagarjuna, A., Kumar, N. N., Kalarani, V., & Reddy, D. C. (2010). Aquatic and avian biodiversity of Pulicat brackish water lake and ecological degradation. *World J. Fish Mar. Sci, 2*(2), 118–123.

Newton, A., Icely, J., Cristina, S., Brito, A., Cardoso, A. C., Colijn, F., Riva, S. D., Gertz, F., Hansen, J. W., Holmer, M., Ivanova, K., Leppäkoski, E., Canu, D. M., Mocenni, C., Mudge, S., Murray, N., Pejrup, M., Razinkovas, A., Reizopoulou, S., & Zaldívar, J. (2013). An overview of ecological status, vulnerability and future perspectives of European large shallow, semi-enclosed coastal

systems, lagoons and transitional waters. *Estuarine Coastal and Shelf Science, 140*, 95–122. https://doi.org/10.1016/j.ecss.2013.05.023
Nguyen, P. Q. A., Gourbesville, P., Vo, N. D., Audra, P., Abily, M., & Ma, Q. (2020). Modelling NH 4+ dispersion from wastewater of urban drainage to the coastal area of Danang City, Vietnam. In *Advances in hydroinformatics: SimHydro 2019-models for extreme situations and crisis management* (pp. 767–778). Springer.
Nixon, S. W. (1988). Physical energy inputs and the comparative ecology of lake and marine ecosystems. *Limnology and Oceanography, 33*(2), 1005–1025.
Orellana, J., Waller, U., & Wecker, B. (2014). Culture of yellowtail kingfish (Seriola lalandi) in a marine recirculating aquaculture system (RAS) with artificial seawater. *Aquacultural Engineering, 58*, 20–28.
Ortega-Arbulú, A. S., Pichler, M., Vuillemin, A., & Orsi, W. D. (2019). Effects of organic matter and low oxygen on the mycobenthos in a coastal lagoon. *Environmental Microbiology, 21*(1), 374–388.
Padedda, B. M., Lugliè, A., Ceccherelli, G., Trebini, F., & Sechi, N. (2010). Nutrient-flux evaluation by the LOICZ Biogeochemical Model in Mediterranean lagoons: The case of Cabras Lagoon (Central-Western Sardinia). *Chemistry and Ecology, 26*(2), 147–162.
Panda, U. S., & Mohanty, P. K. (2008). Monitoring and modeling of Chilika environment using remote sensing data. In *Proceedings of Taal 2007: The 12th world lake conference* (pp. 617–638).
Patten, B. C. (2010). Natural ecosystem design and control imperatives for sustainable ecosystem services. *Ecological Complexity, 7*(3), 282–291.
Prabhakar, K. N., & Zutshi, P. L. (1993). Evolution of southern part of Indian East Coast basins. *Journal of the Geological Society of India, 41*(3): 215–230. https://doi.org/10.17491/jgsi/1993/410303
Ramesh, R., Chen, Z., Cummins, V., Day, J., D'elia, C., Dennison, B., Forbes, D. L., Glaeser, B., Glaser, M., Glavovic, B., Kremer, H., & Wolanski, E. (2015). Land–ocean interactions in the coastal zone: Past, present & future. *Anthropocene, 12*, 85–98.
Ramteke, D., Paramasivam, K., Viswanathan, C., Abhilash, K., Jeevamani, J. J. J., Samuel, V. D., Sankar, R., Muruganandam, R., Purvaja, R., & Ramesh, R. (2023). Assessment of benthic habitats of highly threatened oyster reefs of Pulicat Lake, India. *Regional Studies in Marine Science, 66*, Article 103122. https://doi.org/10.1016/j.rsma.2023.103122
Reddy, B. S. R., Jayaraju, N., & Reddy, K. R. (2012). Anthropogenic impact on the Pulicat lagoon monitoring with foraminifera, East Coast of India. *Marine Science, 2*(5), 66–76.
Redzuan, N. S., & Underwood, G. J. (2021). The importance of weather and tides on the resuspension and deposition of microphytobenthos (MPB) on intertidal mudflats. *Estuarine Coastal and Shelf Science, 251*, Article 107190. https://doi.org/10.1016/j.ecss.2021.107190
Rengarajan, R., & Sarma, V. (2015). Submarine groundwater discharge and nutrient addition to the coastal zone of the Godavari estuary. *Marine Chemistry, 172*, 57–69. https://doi.org/10.1016/j.marchem.2015.03.008
Rysgaard, S., Risgaard-Petersen, N., & Sloth, N. P. (1996). Nitrification, denitrification, and nitrate ammonification in sediments of two coastal lagoons in Southern France. In *Coastal lagoon eutrophication and ANaerobic processes (CLE AN.) nitrogen and sulfur cycles and population dynamics in coastal lagoons a research programme of the environment programme of the EC (DG XII)* (pp. 133–141).
Segura-Noguera, M., Cruzado, A., & Blasco, D. (2016). The biogeochemistry of nutrients, dissolved oxygen and chlorophyll—An in the Catalan Sea (NW Mediterranean Sea). *Scientia Marina, 80*(S1), 39–56. https://doi.org/10.3989/scimar.04309.20a
Smith, J. K., & Anderson, R. S. (2018). Tidal and freshwater discharge effects on flood/ebb tide dominance: A case study in coastal dynamics. *Estuarine, Coastal and Shelf Science, 215*, 102–115.
Spivakovskaya, D., Heemink, A. W., Milstein, G. N., & Schoenmakers, J. G. M. (2005). Simulation of the transport of particles in coastal waters using forward and reverse time diffusion. *Advances in Water Resources, 28*(9), 927–938.

Srichandan, S., Kim, J. Y., Kumar, A., Mishra, D. R., Bhadury, P., Muduli, P. R., Pattnaik, A. K., & Rastogi, G. (2015). Interannual and cyclone-driven variability in phytoplankton communities of a tropical coastal lagoon. *Marine Pollution Bulletin, 101*(1), 39–52.

Srinivasan, K., & Natesan, U. (2013). Spatio-temporal variations in water quality of Muttukadu backwaters, Tamil Nadu, India. *Water Environment Research, 85*(7), 587–595.

Swaney, D. (2011). Biogeochemical budgeting in estuaries. In *Elsevier eBooks* (pp. 343–362). https://doi.org/10.1016/b978-0-12-374711-2.00513-1

Torregroza-Espinosa, A. C., Restrepo, J. C., Escobar, J., Brenner, M., & Newton, A. (2020). Nutrient inputs and net ecosystem productivity in the mouth of the Magdalena River, Colombia. *Estuarine, Coastal and Shelf Science, 243*, Article 106899.

Tosi, L., Da Lio, C., Cosma, M., & Donnici, S. (2024). Vulnerability of tidal morphologies to relative sea-level rise in the Venice Lagoon. *Science of the Total Environment, 931*, Article 173006.

Tri, D. Q., Linh, N. T. M., Thai, T. H., & Kandasamy, J. (2018). Application of 1D–2D coupled modeling in water quality assessment: A case study in Ca Mau Peninsula, Vietnam. *Physics and Chemistry of the Earth Parts a/B/C, 113*, 83–99. https://doi.org/10.1016/j.pce.2018.10.004

Upton, S., Reichstein, M., Gans, F., Peters, W., Kraft, B., & Bastos, A. (2024). Constraining biospheric carbon dioxide fluxes by combined top-down and bottom-up approaches. *Atmospheric Chemistry and Physics, 24*(4), 2555–2582.

Vardy, A. E., & Tijsseling, A. S. (2022). On Boussinesq and Coriolis coefficients and their derivatives in transient pipe flows. *Applied Mathematical Modelling, 111*, 454–470.

Vybernaite-Lubiene, I., Zilius, M., Bartoli, M., Petkuviene, J., Zemlys, P., Magri, M., & Giordani, G. (2022). Biogeochemical budgets of nutrients and metabolism in the Curonian Lagoon (southeast Baltic Sea): Spatial and temporal variations. *Water, 14*(2), 164.

Wang, J., de Swart, H. E., & Dijkstra, Y. M. (2021). Dependence of tides and river water transport in an estuarine network on river discharge, tidal forcing, geometry and sea level rise. *Continental Shelf Research, 225*, Article 104476.

Weyhenmeyer, G. A. (2024). Toward a fundamental understanding of ecosystem metabolism responses to global warming. *One Earth, 7*(10), 1886–1898. https://doi.org/10.1016/j.oneear.2024.07.019

Williams, J. J., & Esteves, L. S. (2017). Guidance on setup, calibration, and validation of hydrodynamic, wave, and sediment models for shelf seas and estuaries. *Advances in Civil Engineering*, 1–25. https://doi.org/10.1155/2017/5251902

Wong, J., Münnich, M., & Gruber, N. (2024). Column-compound extremes in the global ocean. *AGU Advances, 5*(3) https://doi.org/10.1029/2023AV001059

Geochemical and Microfaunal Studies of Sediments from Off-Shore Krishna-Godavari Basin: Implications for Palaeo-Environment and Hydrocarbon Prospecting

Ramesh Chandra Phani Pothuri

1 Introduction

The Krishna-Godavari (KG) basin is a vast deltaic plain formed by two major rivers of southern India namely Krishna and Godavari in Andhra Pradesh. The basin extends to Bay of Bengal on the eastern side into which these rivers discharge their waters. The KG basin extends both onshore and offshore, and the basin trends in NE-SW alignment extend from Kakinada to Bapatla along the coast of Andhra Pradesh. The KG offshore basin has drawn the attention of petroleum geoscientists owing to its proven hydrocarbon resources over the past three and a half decades, ever since the Oil and Natural Gas Corporation Limited, India (ONGC) discovered hydrocarbon deposits in the basin. Microfossils have been instrumental in palaeoenvironmental, palaeoceanographic, and palaeoclimatic studies and petroleum explorations (Jones, 1996; Singh, 2008). Generally, the study of foraminifera plays a vital role in reconstructing the palaeo-environment and palaeo-climate (Gupta & Satapathy, 2000; Gupta & Srinivasan, 1992; Gupta et al., 2001). Micropalaeontological and palynological studies of sedimentary rocks of the KG offshore basin were carried out for correlation to determine the age and palaeo-environment of the strata in the KG Basin by many researchers. Owing to extensive work carried out by the ONGC, the basin has reported a number of hydrocarbon prospects offshore by several workers (Govindan, 2004; Gupta, 2006; Murthy et al., 2015; Prasad, 1999; Raju & Dave, 1996; Reddy et al., 2000; Singh, 2008;). Age and palaeoenvironment of Mesozoic sedimentary strata of the KG Basin were reported by several workers have attempted geochemical characterization of sediments of the KG basin (Mazumdar et al., 2015;

R. C. P. Pothuri (✉)
Geoscience Division, Cyient Limited, Hyderabad, India
e-mail: Ramesh.Pothuri@cyient.com

P. Thambidurai and A. K. Dikshit (eds.), *Coastal Environments of India*,
https://doi.org/10.1007/978-3-031-97341-3_4

Peketi, 2012). Despite enormous research, the KG Basin still attracts palaeontologists, stratigraphers, sedimentologists, palaeo-environmentalists, palaeoclimatologists, and alike. The present study aims at deciphering deep-water oxygenation and palaeoenvironmental conditions during the Miocene based on benthic foraminiferal, pteropod faunal studies, and further geochemical analyses for components such as organic carbon (C_{org}), total organic carbon (TOC), total organic matter (TOM) and phosphorous (P) of a sediment core from a well of ONGC located in the offshore KG basin within the Bay of Bengal.

2 Location and General Stratigraphy

The studied well is situated ~ 5 km off the coast of Amalapuram, Andhra Pradesh. The general geological map and location of the well sampled in this investigation are shown in Fig. 1. A 760 m thick column of Miocene-age sedimentary rock has been encountered in this well (NDR, 2015). The general stratigraphy and gross lithology of the KG basin are shown in Table 1. The KG basin occurs on the Dharwar Craton of southern India made up of highly metamorphosed Precambrian rocks and was formed in the Jurassic age. The KG basin occurs on the margin of the Indian Craton and thus classified as a peri-cratonic basin (Gupta, 2006; Kumar, 1983; Rao, 2001). The KG basin extends beyond the mid-continental slope, extending between Visakhapatnam and Ongole from north to south occupying an area of 280000 km^2 and 145000 km^2 onshore and offshore respectively. At the northwest of Rajahmundry, this coastal basin abuts against NW–SE trending Pranahita- Godavari Gondwana graben, and towards the SE, the basin extends into the Bay of Bengal. At the basin margin on land, the outcrops include the Chintalpudi Formation of Permo-Triassic, Gollapalli, Raghavapuram, and Tirupati Formations of early Cretaceous, lava flows of late Cretaceous to early Eocene and Rajahmundry sandstone of Mio-Pliocene. The recent alluvium covers the entire coastal basin (Kumar, 1983). The onshore part of the KG basin comprises a number of NE-SW trending horsts and graben structures, some of which extend towards off-shore (Sahu, 2005). The graben portions in the onshore basin are filled with 3–6.5 km thick volcanic lava flows embedded with intertrappeans such as clay, limestone and sand beds (Rao, 2001; Rao & Mani, 1993; Sahu, 2005). The offshore basin covering the shelf and mid-continental slope region is filled with Carboniferous to Holocene sediments (Rao, 2001). The deeper portions of the basin inclusing the lower slope and further offshore is filled with sediments of Cretaceous to Recent (Curray et al., 1982; Pateria et al., 1992). The dominant litho units in the offshore are claystone with sandstone and siltstone beds. The basement is overlain by the lower Cretaceous and Palaeocene shales with intertrappean volcanic flows (Gupta, 2006; Rao, 2001; Rao & Mani, 1993; Sahu, 2005). All the sediments form the early Cretaceous basement (Pateria et al., 1992). Upon these formations of Palaeocene, litho units such as Vadaparru shale (Eocene), thin clay stone (Oligocene), Ravva sandstone (Miocene early Pliocene), and Godavari clay (Pleistocene to Recent) are present (Bastia et al., 2006; Rao, 2001). Enrichment of foraminifer populations is observed in the shallow geological successions composed of clay/silt dominant sediments (Collet et al., 2008). Basement configuration revealed several hydrocarbon prospects in the KG basin (Shandilya & Kumar, 2013).

In total, twenty-seven core samples of 10 cm length (~ 300 gms) of 4″ diameter size were collected from the core repository of ONGC Rajamahendravaram, Andhra Pradesh. The samples were divided into two halves to study the geochemical aspects and faunal assemblages, namely foraminiferal and pteropod species. Geochemical

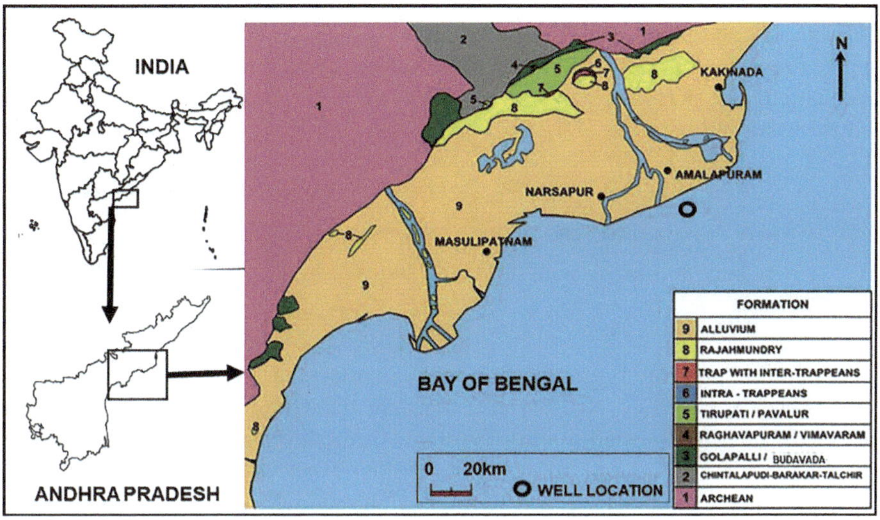

Fig. 1 Geological map of Krishna-Godavari basin and studied well location (Modified after Rao, 2001)

Table 1 Simplified stratigraphy of the studied well (Rao, 2001)

Depth (m)	Litho column	Formation	Lithology	Age
1890		Godavari clay	Clay	Pliocene and Younger
2650		Ravva Formation	Claystone, shale with sandstone beds	Miocene
2700		Matsyapuri Formation	Sandstone	Oligocene
		Vadaparru Formation	Shale	Eocene

analyses were carried out to determine organic carbon (C_{org}), TOC, TOM, and Phosphorous (P). The C_{org} and TOM were estimated by the Walkley-Black method of 1934, adopting standard procedure (Glosolan, 2019). The core sample was powdered to < 2 mm particle size, and one gram of the sample was taken in a 500 ml conical flask and further added 10 ml of potassium dichromate ($K_2Cr_2O_7$) solution and Conc. H_2SO_4. The solution was placed in the darkroom for about 30 min. In addition, 10 ml of Ortho-phosphoric acid (H_3PO_4) and 200 ml of distilled water with 10–15 drops of diphenylamine (indicator) were added. Adding the indicator imparts a violet colour to the solution. A blank solution was also prepared to use as a standard. Both standard and sample solutions were treated against 0.5 N ferrous ammonium sulfate (FAS) to give the endpoint (violet-blue-green) colouration, and finally, the burette readings were noted. The percentage of C_{org} and, in turn, the organic matter is given by the following formula.

$$C_{org}\% = \{10 \times (B - A)/B\} \times (0.003 \times 1.33/\text{wt. of sample}\} \times 100$$

Where A = burette reading with sample solution

B = burette reading with the blank solution

$$TOM = C_{org}\% \times 1.724$$

The other part of 1 gm (< 2 mm) was digested in a three-acid mixture (HNO_3:H_2SO_4:$HClO_4$ = 10:1:4) at a constant temperature of 100 °C for 5–6 h in a sand bath. The digested solution is filtered through Whatman (No: 1) filter paper and made up the volume to 50 ml, with distilled water. A quantity of 5 ml of this filtrate was taken in a test tube, and 10 ml of Ammonium Molybdo Vanadate solution was added, which was made up to 50 ml using distilled water. The solution was fed to a Calorimeter (Model SPECTRONIC 20) at 420 nm to obtain the transmittance (%). By comparing the values with the standard plot between Phosphorous (μg) versus transmittance, the content of phosphorous in the sediment was determined in μg, which is converted to transmittance percentage.

For conducting faunal studies, the sample was soaked in hydrogen peroxide (30% H_2O_2) for 12–24 h. After disaggregation, the sample was washed with water through a 230 ASTM (63 μm) sieve and oven-dried at 70 °C. A few soft shale samples were disaggregated instantly with the evolution of heavy fumes due to reaction with H_2O_2. A 2 gms of each dried sample of < 149 μm fraction was examined for both foraminifera and pteropod fauna. The faunal populations were picked from each sample under the stereo-microscope and mounted on the microfaunal slides. The faunal population was identified and counted as absolute abundances. These counts were used to plot the depth versus absolute abundance curves.

3 Results

The present study demonstrates interesting trends in geochemical and faunal population data of sediment core in the offshore KG Basin during the late Miocene. In general, most of the sediment core samples were rich in organic matter and pyrite, which are visually recognizable in hand specimens. A few shale samples were laminated and fissile. Lignite patches have been observed in samples from KG7 to KG21 (2197–2206 m). Pyrite framboides were also common components in most samples observed through microscopic examination. However, in the samples KG22 and KG24 (2519 and 2521 m), pyrite was less common or absent, with less organic matter/organic carbon. Heavy minerals like magnetite, ilmenite, garnet, zircon, rutile, spinel, and tourmaline were also observed. Fibrous sponge spicules are yet another common feature observed.

3.1 Geochemical Results

The values of C_{org}, TOM and TP determined in the present study are shown in Table 2. The processes of sedimentation, lithification, and diagenesis play a significant role in the formation of petroleum (Tissot et al., 1974). The downhole variations of these three parameters are shown in Fig. 3. The observations made from the analysis suggest that the C_{org} and, in turn, TOM values are controlled by lithology. The data is plotted depth-wise, and it is observed that the C_{org}, TOM, and TP trends are strikingly similar (Fig. 2a–c).

The geochemical data has been plotted, and it is observed that the Corg, TOM and phosphorous display a strikingly similar trend (Fig. 2). The C_{org} values range from 0.46 to 2.94%, and those of TOM from 0.79% to as high as 5.07%. The lowest values are recorded in sandstone samples (KG5), and the highest (KG7) are in grey to black shales.

The percentage of TP in the studied sediments ranges from zero (KG2 and KG20) to 0.36 (KG7). From the plot (Fig. 2) between TP and depth it can be inferred that TP follows the trend of C_{org} or TOM in most samples. A major peak of C_{org} or TOM occurs at 2197 m.

3.2 Foraminifera and Pteropoda

In total, thirty-three foraminiferal and five pteropod species were identified from the core samples. Among foraminifera, twenty-nine were benthic, and four were planktic. The specimens of the fauna are either wholly or partly pyritised with a shiny golden appearance. The reason for pyritization could be all the foraminifera seem to be quickly buried and preserved thereafter. Foraminiferal and pteropod species found

Table 2 Geochemical parameters of the core samples

Sample No.	Depth (m)	C_{org} (%)	TOM (%)	TP (%)
KG1	1891	1.13	1.95	0.03
KG2	1892	1.13	1.95	–
KG3	1893	1.59	2.74	0.04
KG4	1894	1.33	0.29	0.01
KG5	1895	0.46	0.80	0.01
KG6	1896	2.34	4.03	0.04
KG7	2197	2.94	5.07	0.36
KG8	2197.5	1.13	1.95	0.00
KG9	2198	2.60	4.48	0.01
KG10	2198.5	2.27	3.93	0.05
KG11	2199	1.95	3.37	0.10
KG12	2199.5	2.32	3.99	0.05
KG13	2200	1.43	2.47	0.22
KG14	2200.5	1.61	2.78	0.02
KG15	2201	1.15	1.98	0.00
KG16	2201.5	1.17	2.01	0.00
KG17	2202	0.99	1.70	0.06
KG18	2203	1.25	2.15	0.07
KG19	2204	1.49	2.57	0.06
KG20	2205	2.18	3.75	–
KG21	2206	1.51	2.61	0.05
KG22	2519	1.03	1.77	0.02
KG23	2520	0.97	1.67	0.01
KG24	2521	1.53	2.64	0.11
KG25	2523	1.21	2.08	0.05
KG26	2524	1.73	2.99	0.07
KG27	2526	1.13	1.95	0.09

in the sediments are listed in Table 3. Hand-drawn illustrations of some important foraminifera and pteropoda identified using a stereo microscope are shown in Figs. 3 and 4, respectively. The trends of various faunal representatives are shown in Fig. 5. In the samples KG3 to KG6 (1894–1899 m), the faunal representatives are absent, which could be due to high organic content or due to intervening sandstone layers. Numerous pyritised foram bodies are observed in these samples. However, an increase in the faunal count was observed in sample no. KG7 (2197 m). The samples KG7 to KG21 (2197–2206 m) have recorded various benthic foraminiferal species. In sample KG11(2199 m), the benthic faunal count goes up to 39, showing a trend in which maximum planktics were present, whereas organic carbon decreased.

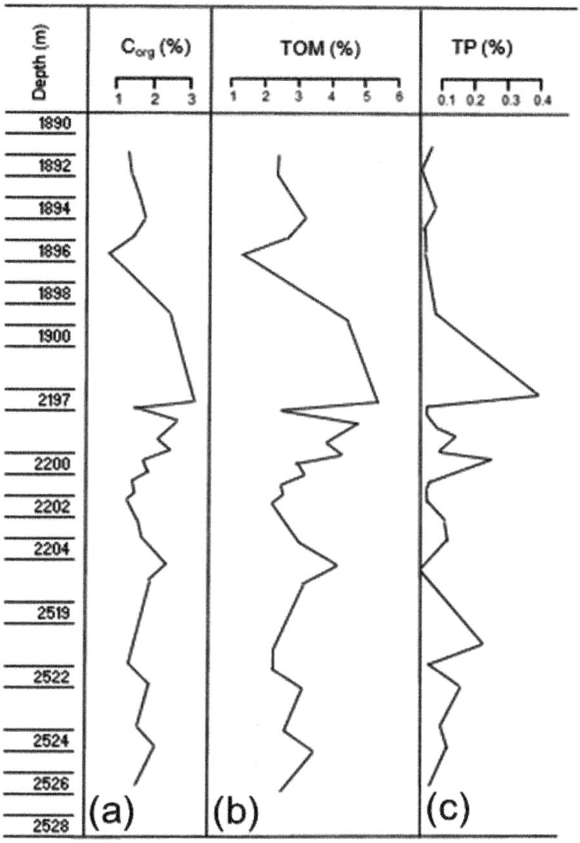

Fig. 2 Down the core geochemical parameters of sediment samples

Benthic foraminifera attain a maximum count in sample KG24 (2521 m) wherein the specimen count has strongly been influenced by the occurrence of abundant *Cibicides pseudoungerianus*. A good number of planktic foraminifera also characterizes this depth interval. The absolute abundances of all faunal assemblages show that foraminiferal species are more in their population than pteropod fauna (Fig. 5a–e). The trends of some selected foraminifera and pteropods (*Limacina inflata*) show their respective populations down the hole (Fig. 6a–j). With the help of microfaunal studies, it is observed that varieties of benthic species were more than planktic. The occurrence of a wide variety of benthic species compared to planktic is attributed to the diversity of benthic populations dominated during the Miocene in the northeastern part of the Indian Ocean, especially in deep-water regimes (Gupta et al., 2001).

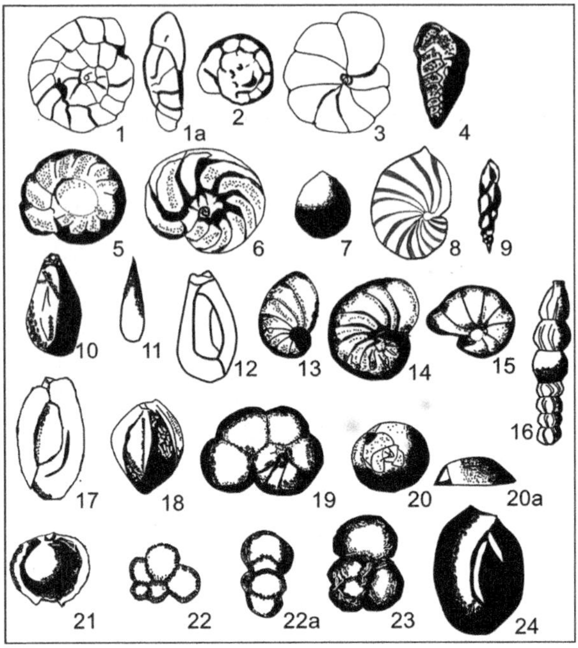

Fig. 3 Sketches of foraminifer fauna in sediments of present study. 1. *Ammonia beccarii*; a. apertural view. 2. *A. batavus*, 3. *Anomalinoides* sp. 4. *Bolivina robusta*. 5. *Cibicides pseudoungerianus*. 6. *Cibicides wuellerstorfi*. 7. *Fissurina* sp., 8. *Florilus* sp., 9. *Fursenkoina davisi*, 10. *Guttulina pacifica*, 11. *Lagena* sp. 12. *Milliammina lata*. 13. *Nonion scaphum*, 14. *N. asterizans* 15. *N. depressulum* 16. *Orthomorphina challengeriana* 17. *Quinqueloculina* sp. 18. *Triloculina tricarinata* 19. *Vulvulinaria rugosa* 20. *Rosalina cocina* 20a. *R. cocina* side view 21. *Pyrgo murrhyna* 22. *Globigerinella aequilateralis*; a. *G. aequilateralis* apertural view 23. *Globigerina bulloides* 24. *Quinqueloculina seminulum*

4 Discussion

Based on geochemical and faunal studies, palaeo-environmental aspects such as deep-water oxygenation, organic carbon flux, and provenance are discussed.

4.1 Deepwater Oxygenation

The less dominance/abundance of fauna, enrichment of organic carbon, pyrite, and presence of pyritised fauna support that the sediments in the studied well site were deposited in anoxic deep-water environmental conditions (e.g., Chandra et al., 1993). The abundance of organic matter in sediments produces high concentrations of CO_2 level and a correspondingly low pH level at or below the sediment–water interface.

Table 3 List of microfaunal (foraminifera and pteropoda) species in the present study

Benthic foraminifera	
Alveolophragmium aff.wiesneri (Parr)	*Florilus de Montfort*
Ammonia batavus Hofker?	*Fursekoina davisi* (Chapman and Parr)
Ammonia beccarii (Brady)	*Guttulina pacifica* (Cushman and Ozawa)
Ammonia compacta	*Miliammina lata*
Anamolinoides, (Brotzen)	*Nonion asterizans* (Fichtel and Moll)
Astrononion tum*idium*	*Nonion depressulum* (Walker and Jacob)
Bolivina robusta (Brady)	*Nonion scaphum* (Fichtel and Moll)
Buliminella sp.	*Orthomorphina challengeriana* (Thalman)
Calcarina venusta (Brady)	*Pyrgo murrhyna, l.c.supra*
Cancris aff. indicus (Cushman)	*Quinqueloculina seminulum* (Linnaeua)
Cibicides pseudoungerianus (Cushman)	*Quinqueloculina sp.*
Cibicides wuellerstorfi Hofke	*Rosalina concinna (Brady),* Cushman
Discorbis sp. (Cushman)	*Triloculina tricarinata (*d'Orbigny)
Fissurina	*Vulvulinaria rugosa* (Cushman)
Planktic foraminifera	Pteropoda
Globigerinella aequilateralis (Brady and Cushman)	*Clio cuspidata (Bosc)*
	Clio pyramidata (L. f. lanceolata Lesueur)
Globigerinoides, (Cushman)	*Limcina inflata* (d'Orbigny)
Globorotalia scitula (Brady)	*Limacina trochiformis* (d'Orbigny)
Globigerina bulloides (d'Orbigny)	*Styliola subula* (Quoy and Gaimard)

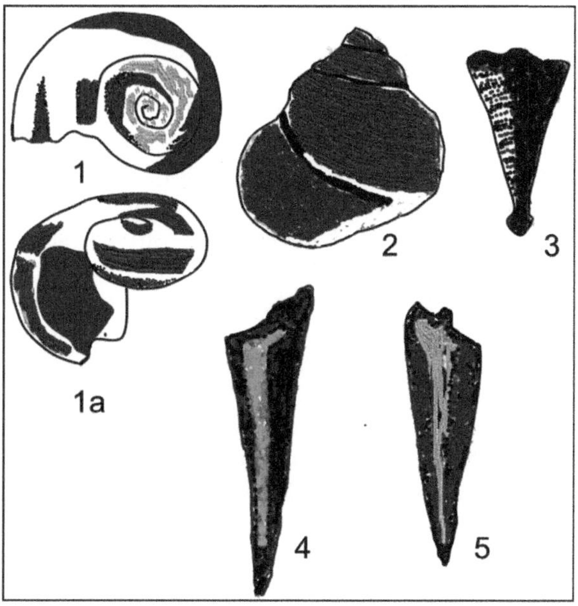

Fig. 4 Sketches of pteropod fauna in sediments of present study. 1. *Limacina inflata; a. apertural view.* 2. *Limacina trochiformis* 3. *Clio cuspidate (protoconch)* 4. *Clio pyramidata* 5. *Styliola subula (embryonic part of shell)*

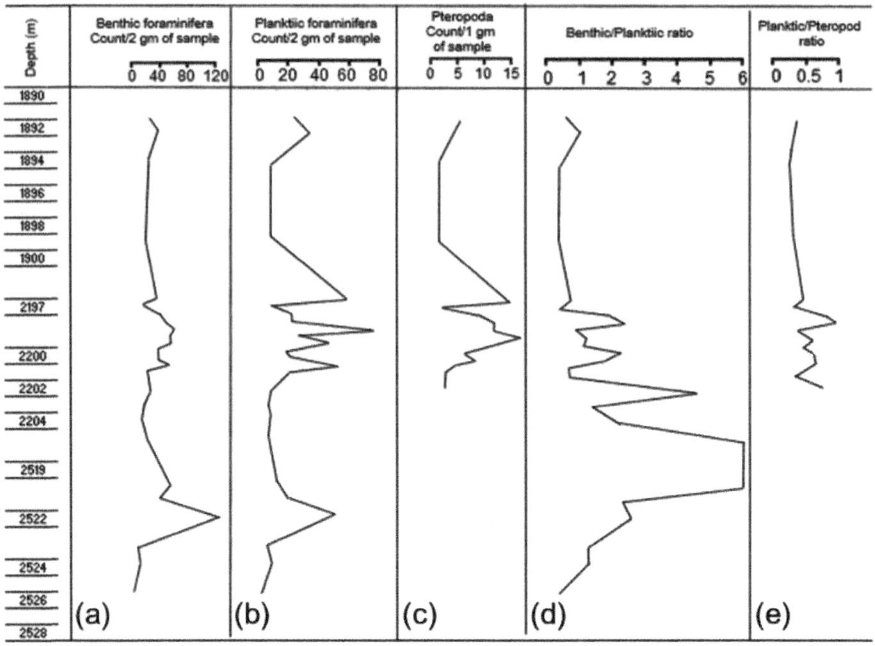

Fig. 5 Benthic and planktic foraminiferal and pteropod faunal abundances in core samples

These kinds of conditions influence the shell composition and may lead to the pyritization of faunal shells. According to ONGC, the studied sediments broadly belong to the Miocene age (Venkatarengan & Ray, 1993). However, the ubiquitous presence of a planktic foraminifer, *Globigerinella aequilateralis*, a tropical to temperate form that ranges in age from 12.5 Ma to Recent, strongly suggests that these sediments belong to the late middle to late Miocene age. The pteropod specimens recorded also suggest that these sediments belong to the late Miocene age (*Janssen, pers. commn.*). The presence of *Cibicides wuellerstorfi* and *C. pseudoungerionus* in the bottom portion of the studied section indicates a deeper depositional regime; hence, the sedimentation might have occurred in the deep-water marine environment. The sediments of the studied well are characterized by the presence of *Bolivina indicating* regions of high productivity and low oxygen environments with sustained flux of organic matter (Gooday, 1993; Phleger & Soutar, 1973). The *Ammonia* species is dominant in estuaries, bays, and inner neritic zones of open marine shelf. The largest population of *Ammonia beccari* (Fig. 5) was observed at a salinity range of 24–28.5% and a temperature range of 23.3–28.5 °C and dissolved oxygen content of 5.0–6.7 ml/l (Rao & Rao, 1974). The specimens of *A. beccarii* are associated with a sandy substrate with low organic carbon content. The disappearance of *Ammonia beccarii* at a 2200 m depth level might be due to the gradual increase in the organic carbon content (Fig. 6g). The *Triloculina tricarinata* (Fig. 6h) and *Miliammina lata* are well-adopted to brackish water conditions. The samples KG24 and KG26 (2521 and 2524 m) have recorded

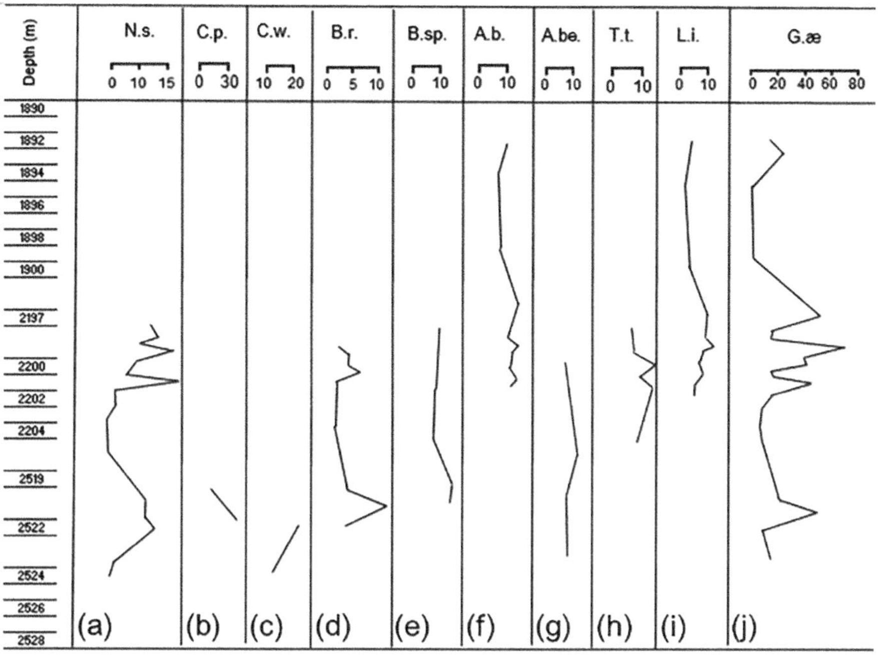

Fig. 6 Absolute abundances of selected foraminiferal and pteropod species. **a** Nonion scaphum (N.s.), **b** *Cibicides pseudoungerianus* (C. p.), **c** *Cibicides wuellerstorfi* (C. w.), **d** *Bolivina robusta* (B. r.), e) *Buliminella sp.* (B. sp.), **f** *Ammonia batavus* (A. b.), **g** *Ammonia beccarii* (A. be.), **h** *Triloculina tricarinata* (T.t.), **i** *Limacina inflate* (L. i.), **j** *Globigerinella aequilateralis* (G. æ.)

Cibicides wuellerstorfi (Fig. 6c), whereas samples 22 and 24 (2519 and 2521 m) are characteristically dominant in *Cibicides psendoungerianus* (Fig. 6b). Owing to these aspects, the sediments at these depth levels might reflect the prevalence of oxic-environmental conditions. This is also suggested by the absence of pyrite at these depth levels. *Cibicides wuellerstorfi* is an epibenthic species which does not require high organic carbon flux and can withstand bottom currents (Linke & Lutze, 1993). The pteropods *Limacina inflata* and *Styliola subula* are sub-tropical, cold-tolerant forms, whereas *Limacina trochiformis* is a sub-tropical, warm-tolerant form (Chen & Be, 1964). The *Clio pyramidata* is characteristic of temperate waters (Chen, 1968). The species *L. inflata* and *S. subula* are infrapelagic and mesopelagic forms. The *Clio pyramidata* is an eurythermic, polymorphic, and mesopelagic species with a temperature range of 7–27.8 °C and a salinity range of 36.1 – 36.5% (Bucheri, 1984). All the pteropod specimens recorded from the present study are juvenile and wholly pyritized. The high pteropod/planktic foraminiferal ratio indicates less stability of aragonite shells of pteropods, which are solution-susceptible (Fig. 5e). A tropical to temperate, warm-water climatic condition may be elucidated for the sediments under the present study based on the pteropods and planktic foraminifera.

4.2 Organic Carbon Flux

The significance of organic carbon (C_{org}) flux and its quality in controlling the distribution of benthic foraminifera in the marine realms has been realized in several investigations (Gooday, 1988; Berger & Diester-Haas, 1988). The nature and amount of organic matter and C_{org} incorporated in the sediments depends on various biological and geological factors such as organic matter supply, palaeowater depth, palaeobathymetry, sedimentation rates, and syn-post depositional tectonics. The net effect of organic matter accumulation and preservation is an interplay of all the above factors. The deposition of organic carbon-rich sediments is controlled by factors such as (a) an increased production of marine organic carbon (ii) an enhanced preservation of organic carbon and (c) an increased supply of terrigenous organic carbon (Canfield, 1994). The organic geochemical proxies act as important tools to study the influx of organic matter from the near-shore terrestrial sources and its deposition in a strongly reducing and low oxygen regime (Mani et al., 2016). In general, the potential for C_{org} deposition and its preservation and thus formation of hydrocarbon source rock is more favourable in near-shore areas than the central open environments which have more oxidizing conditions. The amount of organic matter and, in turn, C_{org} is governed by several factors (Stein, 1991). Benthic foraminifera were used as indicators of organic carbon content in sediments to decipher depositional environments (Cappelli et al., 2019). The correlation between foraminiferal populations and organic carbon content in sediments was established by many workers (Setty & Nigam, 1982; Chaturvedi et al., 1996; Martins et al., 2015). In addition to C_{org}, oxygen penetration depth, which is a function of oxygen concentration in the bottom water, has been suggested as another vital factor influencing benthic assemblages (Setty & Nigam, 1982). The C_{org} values in the present study suggest that they are strongly dependent on the lithology. The C_{org} was found to be very low (0.4634) in the sandstones and high in shales (0.9874–2.9421). The low C_{org} values are attributed to low rate of sedimentation, low involvement of organic matter, rapid decay, arid condition, and lack of river flux. The increase in the C_{org} at a depth of 2197 m (sample KG7) reflects an increase in the supply rate of organic carbon from coastal regions, rivers, and shelves exposed by regression, as well as a change in circulation patterns and ocean fertility.

4.3 Provenance and Sedimentary Pyrite Formation

The macroscopic and microscopic examination revealed that the sediments contain several terrigenous mineral constituents such as zircon, garnet, tourmaline, ilmenite, monazite, spinel, rutile, and magnetite. This indicates that these sediments have an Archaean igneous-metamorphic provenance in the Eastern Ghat Mobile Belt of the Indian peninsula. In the case of siliciclastic marine sediments, sulfur content is mainly sourced from pyrite. This leads to the dissolution of the oxygen present in the sediment. The dissolved oxygen in the pore waters of sediments is completely dried up, resulting in aerobic bacterial decomposition causing anaerobic conditions (Berner, 1984); this process applies to the sediments of the present study. The result of an anaerobic environment, when abundant dissolved sulfate is present, enables the growth of sulfate-reducing bacteria, which utilize organic matter to convert sulfate to hydrogen sulfide. Under these conditions, the formation of pyrite occurs and becomes the most stable form of iron. The pyritized tests of foraminifera observed in the present study are consistent with this process and imply an oxygen-depleted and/or anaerobic environment. However, the samples KG22 and KG24 (2519 and 2521 m siltstones) are devoid of pyrite, indicating the prevalence of aerobic conditions at those depths.

4.4 *Phosphorous and Organic Carbon*

Yet another geochemical parameter, phosphorous (P), seems to have control over faunal abundance. The supply of phosphorous to the oceans from continental through weathering plays a key role in regulating oceanic productivity and sediment burial, especially organic carbon (Delaney & Filippeli, 1994). In this study, the total phosphorous content follows the trend of C_{org} in almost all the samples (see Fig. 2). Once the sediments are buried below the sediment–water interface, phosphorous (P)I s released to interstitial waters during the process of degradation of organic matter. The P thus released could be absorbed to sediment grain surfaces and/or captured by iron oxyhydroxides. In this scenario, the in situ P is ultimately attains mineralised form such as authigenic mineral carbonate fluorapatite (CFA) (Kim et al., 1999). The degradation of organic matter may not be the only source of P that drives authigenic CFA formation. In the case of continental margin sediments, the depth of reduction and dissolution of iron oxyhydroxides for the release of bound phosphorous is much shallower (e.g., Paytan & McLaughlin, 2007). The primary sources of phosphorous in marine shelf sediments could be from the continental rocks. The increase in the phosphorous content reflects the increased river fluxes due to continental weathering. Phosphorous reaches the ocean realm mainly in the form of P associated with organic matter in the form of sinking particulate matter associated with various inorganic and organic substances (Benitez-Nelson, 2000). The parallel trends between total phosphorous and C_{org} suggest that increased organic carbon in the KG basin resulted from terrigenous flux due to increased continental erosion during a low stand of sea level. Local surface productivity has had, if at all, very little impact.

5 Implications for Hydrocarbon Prospecting

Understanding the chemistry of organic carbon has evidenced great development in the past decades mainly as a consequence of interest in shale as a source rock of petroleum (Tissot & Welte, 1978; Hunt, 1995; Hemmati-Sarapardeh et al., 2020). The strength of a shale to generate hydrocarbons is governed by the amount of organic matter present in it (Fu et al., 2021). The minimum amount of organic matter for the generation of a significant proportion of hydrocarbons is about 0.5%, with known source beds averaging 2.2% (Tissot & Welte, 1978). The source rock potential is expressed in terms of poor, good, and very good with respect to C_{org} content in the rock. There exist several factors such as sediment volume, degree of thermal maturation, organic content, type of organic constitutents and the degree of degradation of the organic constituents preserved, to govern the petroleum generation potential of sediment. When copared to terrestrial and deltaic sediments, the marine sediments are more oil-prone for a given volume (Tissot & Welte, 1978). The primary marine organic constituents such as phytoplankton are more oil-prone than terrestrial organic constituents such as plant fragments. In addition to hydrocarbon-prone organic matter, there is also a lower percentage of gas or non-productive decayed or recycled organic matter preserved in the marine sediments. As a result, marine sediments are more likely to be oil-bearing while flusial or deltaic sediments are more gas-prone (Ibach, 1982). Some marine sediments may produce less oil than anticipated owing to a high content of gas-prone or non-generating organic matter. These marine sediments are characteristically found close to deltaic environments, where there is a large clastic input, higher proportion of terrestrial and recycled organic matter. In such environments, rapid degradation of organic matter occurs from intermittent exposure to subaerial environment and intensified bioturbation activity (Ibach, 1982). The less occurrence of oil and dominance of gas in the KG basin can be attributed to this.

Turbidity currents may also transport terrestrial and recycled organic matter in deep-water environments. In addition to terrigenous components, oil-prone marine organic constituents may also be present which may be volumetrically significant. The larger plant fragments may be deposited along with coarser sediment portions owing to hydraulic equivalency, thus reducing the gas-prone terrigenous component of fin-grained sediments. If the primary productivity of phytoplankton and zooplankton in the surface water overlying turbidite and marine deltaic depositional regimes, the quantity iof oil-prone organic constituents would also be high. The upwelling of nutrient rich water along the continental slope or from the nutrient supply by rivers and seasonal nearshore upwelling causes high primary productivity (Sverdrup et al., 1942). Although oil-producing organic constituents are preserved in sediment, the TOC proportion may be inferior to generate commercial quantities of oil under congenial conditions of thermal maturation. Owing to this, the petroleum source rocks are classified based on the TOC content in the sediments (Ibach, 1982). The preservation of organic matter is principally controlled by the microbial degradation additionally contributed by bioturbation which eliminate functional groups on the organic molecules (Coleman et al., 1979; Tissot & Welte, 1978). Typically, long hydrocarbon chains are a prerequisite for the formation of liquid hydrocarbon, the reduction of such long chains by microbial activity not only diminishes TOC content within the sediment but also reduced the ability of organic matter to generate oil as thermal maturation progresses Tissot & Welte, 1978). Thus, the preservation of organic-rich sediments capable of producing oil depends on conditions that reduce the impact of microbial degradation. Microbial and benthic activity decrease with increasing depth so that the most intense degradation of organic matter occurs in the near-surface oxidation zone and in the sulfate reduction zone. The vertical extent of these zones is linked by the diffusible limit of O_2 and SO_4 through the sediment column that is controlled by the composition of sediment (Tissot & Welte, 1978). Therefore, surplus organic matter with higher oil-generating quality is preserved in fine-grained sediments and in anoxic sediments where the aerobic oxidation zone is eliminated. This typical scenario is envisaged for the well-site of the present study. However, recent studies reveal that a high organic carbon content in sediments is not necessarily a prerequisite for the character of an oil source. Additional aspects of evidence, such as hydrogen enrichment (i.e. Hydrogen Index) of humic substance, kerogen type, and overall soluble lipid content of the sediments, are all required to assess the hydrocarbon source rock potential. The enhanced preservation of hydrogen-rich and lipid-rich organic matter in sediments deposited under anoxic water is important for general oil source beds. A higher than normal organic matter concentration observed under anoxic waters might be a consequence of higher rates of sedimentation and slower level of aerobic degradation (Demaison & Moore, 1980, Mishra & Pande, 1996).

The overall study shows that the oil source potential of the shales at sampled depth appears to be 'good to very good' with more than 0.5% $C_{org.}$ Moreover, the C_{org} and, in turn, TOM are controlled by the lithology in this case. In addition, the total organic carbon (TOC) by weight per cent increases from calcareous sediments to calcareous siliceous sediments to siliceous sediments to black shales. In the Miocene section under the present investigation, sandstone has the lowest C_{org} value (0.4634%), and the shale (grey to black in colour) has higher C_{org} values ranging from 0.9874 to 2.9421%, indicating good source rock potential. Hence, the higher C_{org} values than the minimum cut-off (0.5%) and the anoxic nature of the sediments indicate that these sediments may form good source facies for hydrocarbon. However, the type of kerogen, type, quality, and maturity of organic matter, hydrogen richness, and several other factors are all required to confirm this statement. Moreover, based on the general agreement that the Miocene onshore and offshore sedimentary horizons in the KG Basin have been proven to be potential hydrocarbon sources, the presently studied section may also be expected for plausible hydrocarbon source facies. The observations made in this study are consistent with the exploration results announced much later by ONGC. The studied well-started production of oil and gas in 2011 with 1.52 mmscmd with a projection of 0.029 mmscmd in 2023. The well is expected to produce 9400 barrels per day of crude oil in 2023 (DGH, 2019). The well is the first deep-water oil field of ONGC and holds a promising future for further expansion (Upstream, 2021).

6 Conclusions

Based on geochemical and faunal studies of the sediments of the oil well site in the KG offshore basin, the following conclusions have been drawn. The organic carbon content is dependent on the lithology, and the higher (i.e., > 0.5%) organic carbon values suggest that the studied sediments may form source rock facies for hydrocarbons, especially oil, which was also proven during exploration in the year 2011 by ONGC (India). The total phosphorous content of the sediments follows the trend of organic carbon suggesting that both are associated with each other. Increased phosphorous content reflects the increased terrigenous flux to the ocean and increased continental weathering. The lower abundance of benthic faunal suits and enrichment of organic carbon and pyrite suggest that the sediments were formed under an anoxic or oxygen-depleted environment. The pyritization of faunal species clearly indicates an anaerobic environment of sedimentation. The presence of Bolivia also supports the low-oxygen environment. The presence of lignite in a few samples indicates the calorification of organic carbon. The planktic foraminifer viz., *Globigerinella aequilateralis* suggests a late middle to late Miocene age to these sediments, indicating a marine environment in tropical to temperate climates. The majority of the benthic fauna recorded, which are typical of shallow water regimes, indicate shallow marine deposition for the sediments. However, the occurrence of *Cibicides wuellerstorfi* and *C. pseudoungerionus* in the bottom portion of the section suggests a deeper depositional regime. Pteropod and planktic foraminiferal assemblages and characteristics of warm waters strongly suggest a tropical to temperate climate. The presence of heavy minerals like zircon, spinel, rutile, garnet, tourmaline, ilmenite, and magnetite indicates an igneous-metamorphic provenance of the Indian sub-continent through terrigenous river fluxes.

Acknowledgements The author dedicates this article to his teacher and guide, Prof. Anil. K. Gupta, who is superannuating in the year 2025 from the Department of Geology & Geophysics, Indian Institute of Technology, Kharagpur after spending four decades in academics and continued high quality research in micropalaeontology, palaeoclimate and palaeoceanography. The author gratefully acknowledges Prof. A.K. Gupta for introducing him to micropaleontology and guidance during his M. Tech. dissertation work, the source for this article. Thanks are due to Sh. I. B. Raina, Sh. S. S. Yalamarthy and Dr. D.C. Rao of Oil and Natural Gas Corporation Ltd., Rajahmundry, for issuing core samples and arranging transport to their Core House during the years 1997–98. According to ONGC officials, this was the first time, in 1997, that core samples were issued by ONGC Ltd., Rajahmundry to a student for carrying out dissertation work. The finding of this study in the year 1998 that the studied well-site can emerge as a prospective oil field was later proven through detailed exploration by ONGC Ltd in 2011. Faunal illustrations were hand-drawn by the author. Thanks to Dept. of Geology & Geophysics and Dept. of Agricultural & Food Engineering, IIT, Kharagpur for extending laboratory facilities. Thanks to Cyient Limited, Hyderabad for their support. The opinions expressed in this article are of the author and not necessarily of his employer organization. Thanks are due to anonymous reviewers for their valuable comments. This work received no funding from any Institute or organisation.

Declarations The author declares there is no conflict of interest with any individual or organization.

References

Bastia, R., Singh, P., & Nayak., P. (2006). Linking shelf delta to deep water; Krishna-Godavari Basin. *Journal Geological Society of India, 47*(2), 619–628.

Berger, W. H., & Diester-Haas., L. (1988). Paleoproductivity: The benthic/planktonic ratio in foraminifera as a productivity index. *Marine Geology, 81*, 1–a.

Berner, R. A. (1984). Sedimentary pyrite formation: An update. *Geochimica Et Cosmochimica Acta, 48*, 605–615.

Bucheri, G. (1984). Pteropods as climatic indicators in quaternary sequences: A lower-middle pleistocene sequence outcropping in Cava Puleo (Ficarazzi, Palermo, Italy). *Palaeogeography, Palaeoclimatology, Palaeoecology, 45*, 75–86.

Benitez-Nelson, C. R. (2000). The biogeochemical cycle of phosphorous in marine systems. *Earth-Science Reviews, 51*(1–4), 109–135.

Canfield, D. E. (1994). Factors influencing organic carbon preservation in marine sediments. *Chemical Geology, 114*(3–4), 315–329.

Cappelli, E. L. G., Clarke, J. L., Smeaton, C., Davidson, K., & Austin, W. E. N. (2019). Organic-carbon-rich sediments: Benthic foraminifera as bio-indicators of depositional environments. *Biogeosciences, 16*, 4183–4199. https://doi.org/10.5194/bg-16-4183-2019

Chandra, K., Raju, D. S. N., & Mishra, P. K. (1993). Anoxic events, organic content and source rock potential of the Cretaceous sequences of Cauvery Basin. In B. J. Katz, & L. M. Pratt (Eds.), *Source rock in sequence stratigraphic framework* (Vol. 37, pp. 131–146). AAPG studies in Geology.

Chaturvedi, S. K, Nigam, R., & Khare, N. (1996). Ecological response of foraminiferal component in the sediments of Kharo Creek, Kachchh (Gujarat), West Coast of India. *ONGC Bulletin, 372*, 1–11.

Chen, C. (1968). Pleistocene pteropods in pelagic sediments. *Nature, 219*, 1145–1149.

Chen, C., & Be, A. (1964). Seasonal distributions of Euthecosomates Pteropods in the surface waters of 5 stations in the NW Atlantic. *Bulletin of Marine Science Gulf Carib, 14*, 185–220.

Coleman, M. L., Curtis, D. C., & Irwin, H. (1979). Burial rate, a key to source and reservoir potential. *World Oil, 4*, 83–92.

Collett, T. S., Riedel, M., Cochran, J. R., Boswal, R., Kumar, P., & Sathe, A. V. (2008). Indian continental margin gas hydrate prospects: Results of the Indian National Gas Hydrate Program (NGHP) Expedition 01. In *Proceedings of the 6th international conference on gas hydrates* (ICGH 2008) (Vol. 11).

Curray, J. R., Emmel, F. J., Moore, D. G., & Raitt, R. W. (1982). Structure, tectonics, and geological history of the Northeastern Indian Ocean. *The Ocean Basins and Margins, 9*, 399–450.

Delaney, M. L., & Filippeli, G. M. (1994). An apparent contradiction in the role of Phosphorous in Cenozoic chemical mass balances for the world ocean. *Palaeoceanography and Plaeoclimatology, 9*(4), 513–527.

Demaison, G. J., & Moore, G. T. (1980). Anoxic environments and oil source bed genesis. *AAPG Bulletin, 64*, 1179–1209.

DGH. (2019). *India's hydrocarbon outlook, a report on exploration & production activities* Directorate General of Hydrocarbons, Ministry of Petroleum & Gas, Government of India, 196.

Fu, J., Li, X., Sun, Y., Huo, Q., Gao, T., Liu, Y., Dong, S., & Fan, H. (2021). A new evaluation method of total organic carbon for shale source rock based on the effective medium conductivity theory. *Geofluids, 2021*, Article ID 9030311, 12. https://doi.org/10.1155/2021/9030311

Gooday, A. J. (1988). A response by benthic foraminifera to the distribution of phytodetritus in the deep sea. *Nature, 332*, 70–73.

Gooday, A. J. (1993). Deep-sea benthic foraminiferal species which exploit phytodetritus: Characteristic features and controls on distribution. *Marine Micropaleontology, 2*, 187–205. https://doi.org/10.1016/0377-8398(93)90043-W

Govindan, A. (2004). Miocene deep water agglutinated foraminifera from offshore Krishna-Godavari Basin, India. *Micropalaeontology, 50*(3), 213–252. https://doi.org/10.2113/50.3.213

Gupta, A. K., Joseph, S., & Thomas, E. (2001). Species diversity of Miocene deep-see benthic foraminifera watermass stratification in the northeastern Indian Ocean. *Micropalaeontology, 47*(2), 111–124.

Gupta, A. K., & Satapathy, S. K. (2000). Latest Miocene–Pleistocene abyssal benthic foraminifera from west-central Indian Ocean DSDP Site 236: Paleoceanographic and paleoclimatic inferences. 46:32–48.

Gupta, A. K., & Srinivasan, M. S. (1992). Deep-sea benthic foraminiferal changes and terminal Miocene event at tropical Indian Ocean DSDP Site 214. *Journal of Geological Society of India, 78*, 262–278.

Gupta, S. K. (2006). Basin architecture and petroleum system of Krishna Godavari Basin, east coast of India. *The Leading Edge, 25*(7). https://doi.org/10.1190/1.2221360

Hemmati-Sarapardeh, A., Larestani, A., Amar, M. N., & Hajirezaie, S. (2020). Application of intelligent models in exploration engineering, Chapter 6. In *Applications of artificial intelligence techniques in the petroleum industry* (pp. 279–294).

Hunt, J.M. (1995). Petroleum Geochemistry and Geology. WH Freeman and Company (Formerly Macmillan), 1–743.

Ibach, L. E. (1982). Relationship between sedimentation rate and total organic carbon content in Ancient Marine sediments. *AAPG Bulletin, 66*(2), 170–188.

Jones, R. W. (1996). *Micropaleontology in petroleum exploration* (p. 416). Oxford University Press.

Kim, D., Schuffert, J. D., & Kastne, M. (1999). Francolite authigenesis in California continental slope sediments and its implications for the marine P cycle. *Geochimica Et Cosmochimica Acta, 63*(19–20), 3477–3485.

Kumar, S. P. (1983). Geology and hydrocarbon prospects of Krishna-Godavari and Cauvery Basins. *Petroleum Asia Journal, 6*(4), 57–65.

Linke, P., & Lutze, G. F. (1993). Microhabitat preferences of benthic foraminifera a static concept or a dynamic adaptation to optimize food acquisition? *Marine Micropaleontology, 20*, 215–234.

Mani, D., Ratnam, B., Kalpana, M. S., Patil, D. J., & Dayal, A. M. (2016). Elemental and organic geochemistry of Gondwana sediments from the Krishna-Godavari Basin, India. *Geochemistry, 76*(1), 117–131.

Martins, M. V. A., Silva, F., Laut, L. L., Frontalini, F., Clemente, I. M., Miranda, P., Figueira, R., Sousa, S. H., & Dias, J. M. A. (2015). Response of benthic foraminifera to organic matter quantity and quality and bioavailable concentrations of metals in Aveiro Lagoon (Portugal). *PLoS One, 10*(2), e0118077.

Mazumdar, A., Kocherla, M., Carvalho, M. A., Peketi, A., Joshi, R. K., Mahalaxmi, P., Joao, H. M., & Jisha, R. (2015). Geochemical characterization of the Krishna-Godavari and Mahanadi offshore basin (Bay of Bengal) sediments: A comparative study of provenance. *Marine and Petroleum Geology, 60*, 18–33.

Mishra, P. K., & Pande, M. N. (1996). Palaeoenvironmental significance of source rock deposition in Krishna-Godavari Basin—A case study. *IJPG, 5*(2), 59–66.

Murthy, D. N., Nawaz, S., & Sarma, B. P. (2015). Strati-structural and stratigraphic prospect identification in East Godavari Sub Basin in Krishna Godavari-Basin: A case study. *Geohorizons, 14*(2), 1–7.

NDR. (2015). National data repository, KG-Basin. https://www.ndrdgh.gov.in/NDR/?page_id=647

Pateria, M. L., Rangaraju, M. K., & Raiverman, V. (1992). A note on the structure and stratigraphy of Bay of Bengal sediments. In *Recent geoscientific studies in the Bay of Bengal and the Andaman Sea* (Vol. 29, pp. 21–23). Special Publication Series of the Geological Survey of India.

Paytan, A., & McLaughlin, K. (2007). The oceanic phosphorus cycle. *Chemical Reviews, 107*, 563–576.

Peketi, A. (2012). *Marine sedimentary Sulfur-Carbon-Iron systematics: A study from Krishna Godavari Basin, Bay of Bengal*. Unpublished PhD Thesis.

Phleger, E. B., & Soutar, A. (1973). Production of benthic formainifera in three east pacific oxygen minima. *Micropal, 19*, 110–115.
Prasad, B. (1999). Palynological characterisation, biostratigraphy and depositional environment of PreTertiary litho-units of Krishna-Godavari basin, India. *Geoscience Journal, 20*, 12–33.
Raju, D. S. N., & Dave, A. (1996). Oligocene to Pleistocene Uvigerinidae and their value in reconstruction of paleodepth in Krishna—Godavari and Cauvery basins. In J. Pandey et al. (Eds.), *Proceedings on vol. XI Indian Colloq. Micropal. Strat. Dehradun* (pp. 185–203).
Rao, G. N. (2001). Sedimentation, stratigraphy, and petroleum potentla1 of Krishna-Godavari basin, East Coast of India. *AAPG Bulletin, 85*(9), 1623–1643.
Rao, G. N., & Mani, K. S. (1993). A study on generation of abnormal pressures in Krishna Godavari Basin, India. *Indian Journal of Petroleum Geology, 2*(1), 20–30.
Rao, T. V., & Rao, M. S. (1974). Recent foraminifera of Suddagedda estuary, east coast of India. *Micropaleontology, 20*(4), 398–419.
Reddy, A. N., Jaiprakash, B. C., & Bhaktavatsal, K. V. (2000). Sequence stratigraphy: A case study from the Godavari Shelf, India. *Geological Society of India, 56*, 183–198.
Sahu, J. N. (2005). Deepwater Krishna-Godavari basins and its potential. *PetroMin, 2005*, 26–34.
Setty, A. P., & Nigam, R. (1982). Foraminiferal assemblages and organic carbon relationship in benthic marine ecosystem of Western Indian continental shelf. *Indian Journal of Geo-Marine Sciences (IJMS), 11*, 225–232.
Shandilya. A., & Kumar, T. (2013). Basement exploration in KG Basin—The untouched frontier. In *10th Biennial international conference & exposition* (Vol. 351, pp. 1–6).
Singh, A. (2008). Micropaleontology in petroleum exploration. In *P-288, 7th International conference and exposition of petroleum geophysics* (pp. 1–7).
Standard operating procedure for soil organic carbon Walkley-Black method (1934). Titration and colorimetric method (pp. 1–27).
Stein, R. (1991). Organic carbon accumulation in Baffin Bay and paleoenvironment in High-Northern Latitudes during the past 20 m.y. *Geology, 19*, 356–359.
Sverdrup, H. O., Johnson, N. W., & Fleming, R. H. (1942). *The oceans: Englewood cliffs* (p. 1059). Prentice Hall Inc.
Tissot, B. P., & Welte, D. H. (1978). *Petroleum formation and occurrence, new approach to oil and gas exploration* (p. 539). Springer-Verlag.
Upstream, (2021). *ONGC moves to expand potential of its first deep-water field*. https://www.upstreamonline.com/exclusive/ongc-moves-to-expand-potential-of-its-first-deep-water-field/2-1-1046310.
Venkatarengan, R., & Ray, D. (1993). Geology and petroleum systems, Krishna-Godavari Basin. In S. K. Biswas, A. Dave, & P. Garg, (Eds.), *Proceedings of the second seminar on petroliferous basins of India*. Indian Petroleum Publishers.

High-Energy Deposits in the Shadow Zone of Rameswaram Island, Southeast Coast of India

P. Udayaganesan, A. R. Gujar, V. J. Loveson, Sridhar D. Iyer, Xavier Ignaci Muthu, and P. Thambidurai

1 Introduction

The coastlines of India often encounter several kinds of extreme tropical monsoon events such as storms, cyclones, flooding, over-wash, and seldom tsunamis, which are non-monsoonal. Barring tsunamis, the frequency of such events is more along certain coastal segments, especially along the east coast of India. Several field-based and predictive modelling types of research have been carried out along the Indian coasts on the causative factors of extreme events, such as their intensity, periodicity, and the resultant effects on the coastal geomorphology and oceanographic parameters, amongst others (Carayannis, 2006; Dawson & Shi, 2000; Rastogi, 2007). These works dealt with different aspects of meteorology, oceanography, disaster mitigation and management, and so forth. However, reports on sediment sequences due to extreme events and textural and compositional changes in sediments were sparse, especially along the Indian coast, till 26th December 2004, the day an unprecedented tsunami that was generated in the Indian Ocean hit several places along the east coast. The tsunami of December 26, 2004 (2004 Indian Ocean tsunami or 2004 IOT) was one of the worst ever catastrophic events that the earth had experienced so far. The tsunami triggered the interest of many scientists in India to describe and catalogue the deposits of extreme events in general and of tsunamis and paleo-tsunamis in particular. A portion of the southern Tamil Nadu coast from Rameswaram Island

P. Udayaganesan (✉) · X. I. Muthu
Alagappa Govt. Arts College (Alagappa University), Karaikudi, Tamil Nadu, India
e-mail: udayaganesan1@gmail.com

A. R. Gujar · V. J. Loveson · S. D. Iyer
Formerly with CSIR-National Institute of Oceanography, Dona Paula, Goa, India

P. Thambidurai
Department of Coastal Disaster Management, School of Physical, Chemical and Applied Sciences, Pondicherry University - Port Blair Campus, Sri Vijaya Puram, Andaman and Nicobar Islands, India

P. Thambidurai and A. K. Dikshit (eds.), *Coastal Environments of India*,
https://doi.org/10.1007/978-3-031-97341-3_5

in the north to Tuticorin in the south along the east coast of India for a stretch of about 140 km forms a 'shadow zone' with respect to the Java-Sumatra tsunamigenic source region. This is because of the presence of the island nation of Sri Lanka which diffracted the waves and reduced the energy and velocity of the tsunami-generated waves that entered the shadow zone. Shadow regions are known to experience lesser inundation by tsunami waves. Explorations for Tsunami signatures from proxies, hitherto, have not been carried out in shadow zones as they are believed to have no characteristic sedimentation. That has created a lacuna in the understanding of the dynamics of tsunami waves in shadow zones. Tsunami waves refracted by physical features differ in hydrodynamic conditions and leave significant imprints on the shadow beaches. These imprints are distinctly different from those caused by the normal waves. The study describes the presence of palaeo-high energy deposits in the north and north-eastern part of the shadow zone of Rameswaram Island.

2 Material and Methods

2.1 Overview of Study Area

The Rameswaram Island is situated in the extreme southeast of Tamil Nadu, India. The island forms the northern entity of the Gulf of Mannar Biosphere Reserve and is a historical town and one of the important pilgrimage centres of India. The island is surrounded by Palk Strait on the north, northeast, and north-western sides and the Gulf of Mannar on the south, southeast, and southwestern sides (Fig. 1). The coast of Rameswaram island is about 140 km long and located 1800 km WNW of the tsunamigenic source region, the Andaman–Sumatra-Sunda (ASS) trench at the sheltered northern fringe of the Gulf of Mannar. Tsunami waves that were generated from the ASS subduction zone travelled WNW towards the study area before getting diffracted at the shores of Sri Lanka. Subsequently, the waves further propagated in a slightly varied direction but still maintained an overall trend of WNW and hit the northern sector of the Rameswaram coast with reduced energy.

The tide gauge at Tuticorin Port by 9.57 am, the time at which the first arrival of the tsunami wave was recorded on December 26, 2004, measured a wave height of 1.8 m, which was only marginally higher than the high-tide levels (Sheth et al., 2006).

Geology

At Rameswaram Island, marine terraces and platforms of calcareous composition occur as hardpans with admixtures of quartz and heavy mineral concentrations of garnet and ilmenite. Huge exposures of marine sandstone on either side of the Palk Bay and Gulf of Mannar occur up to an altitude of 2 m above sea level (ASL). They

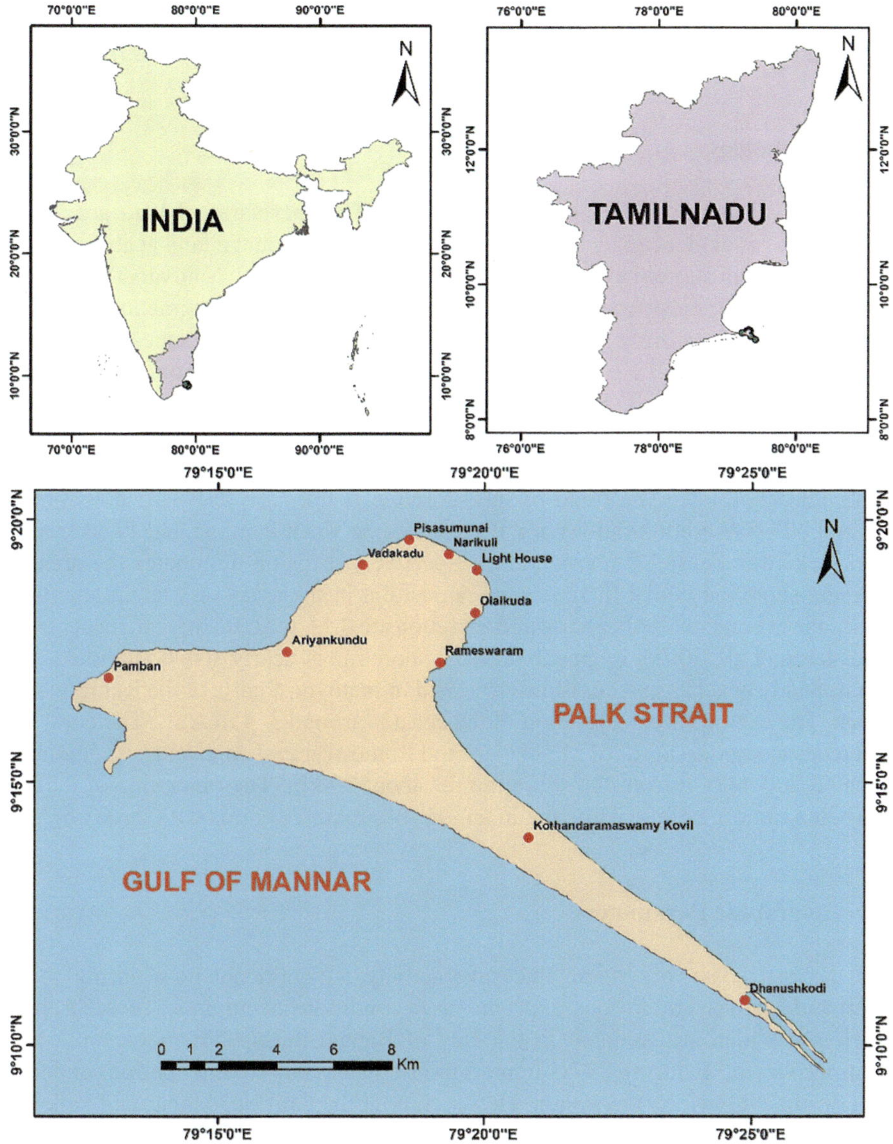

Fig. 1 Study area map

were formed during middle Holocene high sea-level positions around Mandapam and Rameswaram regions and are called 'Mandapam sandstone' (Sahayam et al., 2010; Vaz et al., 2008). The sandstone is well-exposed around the tip of the Mandapam foreland for about a stretch of 3 km towards the west along the coastline of Palk Bay and along the coast of the Gulf of Mannar and further south. No other rock exposures of either crystalline or sedimentary nature occur on the coastlines of Mandapam and

Rameswaram Island. Tertiary-era sandstones intercalated with clay formation occur beneath the dead coral reefs (Sivasankar & Ramachandramoorthy, 2009).

Geomorphology

The relatively straight and smooth shorelines of Rameswaram Island are 'emergence type' as evidenced by the presence of beach ridges on the land at about 0.5 km from the present shoreline (Stoddart & Gopinadha Pillai, 1972; Vaidyanathan, 1993). The coasts of Rameswaram Island can be classified into two segments viz. North and South. The northern coast extends from Pamban through Rameswaram town to Dhanushkodi and the southern coast extends from Pamban through Ramakrishnapuram to Danushkodi (Fig. 1). Cuspate shoreline is found on the northern coast between Peikkarumbu and Rameswaram town and on the southern coast between Pamban and Ayyanthopu. The areas between the cuspate features generally have a straight shoreline. The trend of the shoreline is normally NW–SE however, E-W and NE–SW configurations are also present between Pamban and Peikkarumbu and Peikkarumbu and Rameswaram, respectively. Erosion dominates the areas of cuspate shoreline whilst there is sand deposition in the areas of straight shoreline. Since the average rates of erosion and accretion are 0.11 and 0.06 m/year, respectively (Baskaran et al., 2003), it appears that the shoreline is nearly stable and in equilibrium. Sandy beaches and coralline terraces dominate most parts of the Rameswaram coast. The average mean elevation of the coast is around 3–4 m ASL. The coral reef stretches along the shores of Mandapam and Rameswaram Island and is sporadically located 200–600 m from the shoreline for about 30 km. The outer side of the reef harbours ramose corals, while the inner side has massive corals with large polyps.

Oceanographic Parameters

On the east coast in general, the wave activity is significant both during southwest and northeast monsoons. Extreme wave conditions occur under severe tropical cyclones, which are frequent in the Bay of Bengal during NE monsoon. Waves approach from SE during SW (summer) monsoon and fair-weather periods and from NE during NE (winter) monsoon. However, both the Gulf of Mannar and Palk Bay have varied oceanographic setups with regard to current direction, sediment transport, tidal conditions, etc. In the Gulf of Mannar, the Indian Ocean waters that enter through the head of the Gulf are piled up like a cushion (Silas, 1962). In Palk Bay, the waters that move in through the Palk Strait have a major influence on the hydrographic conditions, while the Gulf of Mannar waters influence Palk Bay to a minor extent (Murty & Varma, 1964). In the Gulf of Mannar, the wave heights are more during SW monsoon and the maximum wave height is 5.4 m. Both wind seas and swells dominate the Gulf, and the high energy waves are from SE and SW. The sea state is 'very rough' according to the Douglas Sea scale. The wave spectra are multi-peaked due to the prevalence of two or more wave systems and swells (Gowthaman et al., 2013). In Palk Bay, high energy waves are noticed during NE monsoon, and the wind seas prevail in all directions (NNE during November-March and SE-SW during the remaining period) while swells are absent. The maximum wave height is 3.6 m with a 'rough' sea state. The wave spectra are mainly single-peaked, as it is during storms and extreme events, because of the presence of only the wind seas. The tides of the region are semi-diurnal, with the spring tide being 0.60 m and the neap tide 0.16 m (Gowthaman et al., 2013). The currents in the Gulf of Mannar and Palk Bay are swift. It is interesting to note that during the NE monsoon season (October–December), there is less water throughout the space to the south of the Pamban Pass. During the SW monsoon (June–September), the reverse phenomenon occurs, and the water levels are higher or lower in direct proportion to the wind force. The currents passing through the Pamban Pass frequently record a velocity of about 6 knots/hr. The water depths differ from 1 to 3 m in the Gulf of Mannar, while in the Palk Bay, the depth varies from 2 to 10 m (Subba Rao et al., 2008). Both Palk Bay and the Gulf of Mannar are the major sinks for sediments (Chandramohan et al., 2001). The Palk Bay covers an area of 12,285 km^2 and is fed by sediments brought in by Vaishali and Vaigai rivers; littoral drifts from northern Tamil Nadu and the Gulf of Mannar through Pamban Pass and Adam's Bridge, rivers of Sri Lanka that debouch into Palk Bay and currents from the Bay of Bengal. Low wave action and protection from southerly waves encourage the deposition of transported material in the Bay. This is evident from sediment accretion at Mandapam and Rameswaram (Natesan & Subramanian, 1993). The sediments move towards the south during the NE monsoon and north during the SW monsoon. The annual longshore sediment transport during the NE monsoon (October–February) is around $0.273 \times 106\ m^3$ towards the south and $0.175 \times 106\ m^3$ towards the north during the rest of the year. The gross sediment transport is around $0.448 \times 10^6\ m^3$ in Palk Bay. The net long shore sediment transport is between $0.098 \times 10^6\ m^3$ towards South in the Bay (Gowthaman et al., 2015).

The Gulf of Mannar, which sprawls for 263.2 km^2 receives sediments from the rivers Tambraparani, Vembar, Vaippar and Guntar. Sediments are also contributed from the west coast of India all the way, encircling Kanyakumari, the southern tip of the Indian peninsula, and from the Palk Bay. Chandramohan et al. (2001) have estimated $2.6 \times 10^6\ m^3$ as the average annual rate of sedimentation in this region. The longshore current transports the sediments from the Gulf of Mannar to Palk Bay and back through Adam's Bridge during monsoon and non-monsoon periods. The net annual sediment transport ($0.024 \times 10^6\ m^3$) is from the Gulf of Mannar to Palk Bay (Subba Rao et al., 2008). It is estimated that Palk Bay receives an average of 98,000 m^3 of sediments through longshore sediment transport. Sediment movement from Palk Bay to the Gulf of Mannar and vice-versa is also affected by tidal currents. The net annual transport of sediment from the Gulf of Mannar to Palk Bay through Adam's Bridge and Pampan Pass by tidal waves is estimated to be $0.0195 \times 10^6\ m^3$ (Subba Rao et al., 2008).

Meteorology

The Palk Bay and Gulf of Mannar regions experience a tropical climate. The SW monsoon contributes very little towards the annual rainfall while rainfall is moderate to heavy during NE monsoon (October–December). The mean annual rainfall varies from 762 to 1270 mm. The sea is rough between April and August, calm during September and stormy during June–August. The tidal amplitude is about half a meter. The study region is influenced by both SW and NE monsoons, and hence, the physical, chemical and biological characteristics are different from other areas. Light penetration also varies from season to season. Temperature oscillation is unique in the study region, with a steady increase from January to May that reaches up to 32 °C but declines to 25 °C during December. The wind velocity is generally high and is north/north-easterly from June to December and westerly during the rest of the period. The wind speed is lowest (6.9 km/h) in November and highest (17.6 km/h) in August (Subba Rao et al., 2008). The study area is prone to be affected by high-intensity storms and cyclones during NE monsoons, and the last cyclone was in 1964 AD. The devastating tidal waves that were 7 m high attacked this region and wiped out Dhanushkodi, a vibrant coastal town. No other event of such magnitude has been reported from this region so far, except for seasonal cyclones.

Neotectonism and Sea-Level Oscillations

Neotectonic movements and glacio-eustatic sea-level changes have caused fluctuations all along the east coast of India in general and in the study area in particular (Banerjee, 2000; Bruckner, 1988, 1989; Loveson & Rajamanickam, 1999; Stoddart & Gopinadha Pillai, 1972). Vaz et al. (2006) have identified five thermal springs under artesian conditions along E–W trending faults around Manamelkudi in central Palk Bay. Due to the post-formational neotectonic activity, palaeo-high shoreline positions are found to differ from one coastal sector to another. Evidence of neotectonic activity in Palk Bay, the Gulf of Mannar coastal regions and around Dhanushkodi in Rameswaram Island have been documented through marine geophysical surveys (Vaz et al., 2007). Their study revealed activation of WNW–ESE trending Holocene fault within unconsolidated sediments parallel to the coastline of the Gulf of Mannar at the southeastern part of the island (Dhanushkodi foreland) with a vertical displacement of ~ 5 m during 1948–49 led to the subsidence of the southern part of erstwhile Dhanushkodi town. This has been the most recent neotectonic activity ever recorded along the east coast of India, especially in the study region. Uplifted unconsolidated seabed sediments also occur along these regions (Vaz et al., 2008). The study region is one of the important locations along the east coast of India and boasts spectacular evidence for the palaeo-sea-level stands of the Late Pleistocene and Holocene ages. Various geochronological studies carried out on the samples collected from the carbonate marine terraces and shells present in the beach rocks of this region have confirmed that the oldest Holocene high stand was about 3 m above from the present-day shoreline during 7300 yr B.P. (Stoddart & Gopinatha Pillai 1972; Bruckner, 1988,

1989; Rajamanickam & Loveson, 1990; Banerjee, 2000) and during mid-Holocene, the sea-level was at 2 m above the present datum between 5650 and 6110 yr B.P. (Vaz et al., 2008).

2.2 Field Work

Field works in three phases (December 2004; May 2008; September 2008) were conducted in the study area to locate the prospective zones where the 2004 IOT and palaeo-high energy waves could have caused sedimentation. During an initial phase, evidence for the landward limit of the 2004 tsunami run-up was observed, and the landward inundation distance was demarcated after consulting a few local fisher folks who had witnessed the tsunami event. The preliminary report published immediately after the tsunami was also referred (Singarasubramanian et al., 2006). Satellite images taken before and after the tsunami were used to elicit information regarding the extent of inundation. The elevations at different places of inundation were measured using a SOKKIA Auto level-C310. Much attention was given to areas where conspicuous sediment deposition brought about by the tsunami was observed. The geographic positions of such areas were compared with the epicentre of the M 9.3 Indian Ocean earthquake (3.316° N, 95.854° E) and the direction of propagation of the resultant tsunami waves. A model was created in the Geographical Information System (GIS) and animated using power point 2007. Based on the results, an intensive survey was carried out during the second and third phases, and probable locations of high-energy deposits were delineated. The tsunami deposits were identified on the surface of the beach, while sub-surface deposits of palaeo-high energy events were unearthed by trenching. Lithological variations in the trenches (1–1.5 m deep; 0.5–1 m wide) were measured, described, photographed and sampled. Wherever necessary to trace the extension of a particular sediment layer, additional short trenches were also dug. Different techniques were used for trenching wet and dry sediments. For trenches with wet sediments, a vertical exposure was created using a shovel and trowel. Transects with dry sediments were carefully wetted by percolating water through a cloth placed on the surface prior to trenching to avoid slumping. The sedimentological observations in the trench include the colour and grain size of sediments, layering pattern and their thickness, presence or absence of sedimentary structures, and datable materials and nature of contacts. A total of 15 trenches were made, and 75 sediment samples were collected for laboratory analysis. However, significant sedimentological characteristics from only selective trenches are presented and discussed here. Topographic beach profiles were measured using a SOKKIA Auto level-C310, staff and measuring tape. Elevations were referenced to the mean swash elevation at the time of the survey. Later these elevations were corrected for tidal variations. For textural analysis, ASTM sieves were used in a Ro-tap sieve shaker at ¼ Φ interval. GPR profiles were collected perpendicular to the coastline and up to the low tide limit. Ground Penetration Radar (Model SIR-20,

Geophysical Survey Systems Inc., GSSI) system with a 200 MHz antenna was used, and sub-surface profiles of 4 and 10 m depth penetration were generated.

3 Results and Discussions

3.1 Sedimentation Due to the 2004 IOT

The 2004 IOT inundated to a maximum distance of about 20–25 m in the Palk Strait region (Kumaraguru et al., 2005). According to the fishermen of Rameswaram Island, the maximum inundation was noticed in the coastal segment between Olaikuda and Pisasumunai. Contrarily, in the Gulf of Mannar, the sea retreated for about 500–600 m and resumed after a while. Inundation mapping carried out along the study region also attested to the same (Singarasubramanian et al., 2006). The effect of tsunami waves in the Palk Strait region was more along the NE part of the island and flat beaches. The shallow-shelf beaches were not affected by the tsunami waves. Beach erosion was also prominent on the flat beaches. Thick and extensive sedimentation was observed at places where the coast is slightly projected towards the sea in the NE part of the island, i.e., the area between NE Olaikuda and Pisasumunai (Fig. 1). In general, the tsunami sedimentation in the study area indicates that they were deposited at different time intervals and at different inundation distances and by successive waves. This fact is corroborated by prominent multiple grading of sediment deposition. Only along the coast between NE Olaikuda and Pisasumunai in the Palk Strait region the study observed characteristic geomorphological and geological changes due to tsunami waves. Because of the steep slope of this region and the presence of a coralline terrace for about 1.60 m above ground level and about 3 m ASL (Fig. 2a and b), no evidence of inundation was observed beyond the coastal dune complexes which occur parallel to the coast. The maximum sedimentation was restricted to a few metres away from the shoreline.

However, there is a drastic change in the thickness of sedimentation. In NE Olaikuda, Narikuli and Pisasumunai the sedimentation was maximum. The deposits extent for about 2.5 km. with 10–12 m width and 4–20 cm. thickness (Fig. 3a). In other regions like Olaikuda south and Vadakadu no visible accumulation of sediments was noticed (Fig. 3b). From field observations it is inferred that the sedimentation due to the 2004 IOT is a function of coastline configuration, direction of propagation of tsunami waves and the nature and compositions of sediments available nearshore for transportation by the tsunami waves. Deposition by the 2004 IOT was characteristic and visible only along the coast between NE Olaikuda and Pisasumunai facing the Palk Strait. Deposition is less and occurs in some places as carpets along the major portions of the coastlines facing the Gulf of Mannar.

However, wherever the energy level of the tsunami waves was relatively higher owing to a reduction in the local slope of the shore, an appreciable amount of sediment deposition was noticed. The sediments have a thick concentration of heavy minerals

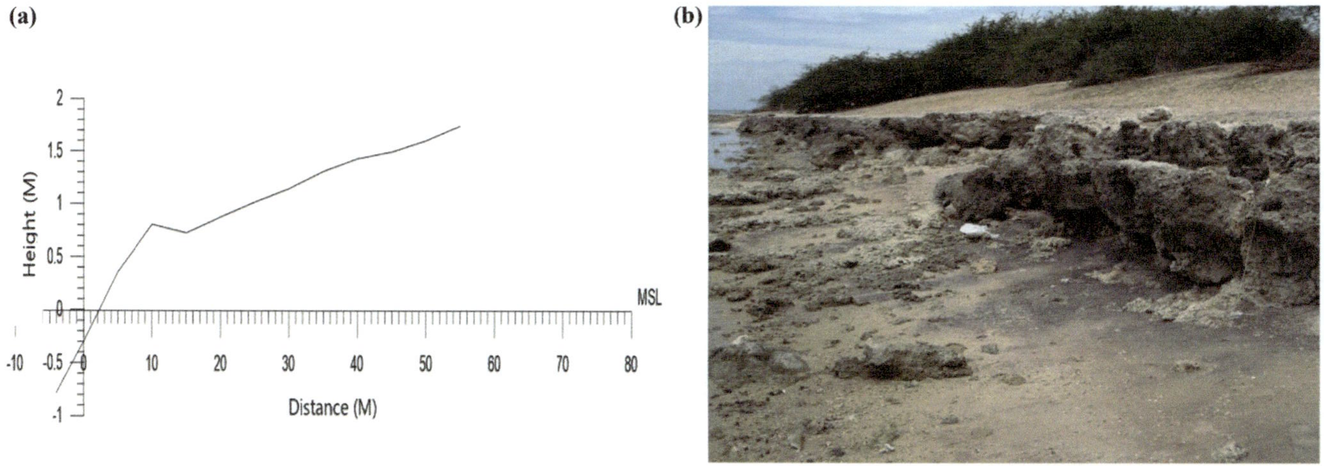

Fig. 2 **a** Profile at Narikuli-Pisasumunai region, **b** Coral terraces and sand dunes in Narikuli – Pisasumunai region

Fig. 3 **a** Heavy mineral carpet in Narikuli, **b** No visible accumulation of Tsunami sediments Vadakadu region

such as garnet, magnetite and ilmenite. Similar observations have also been made along the east coasts at Nagore, Nagapattinam, Karaikal and Poompuhar in Tamil Nadu (Rajamanickam & Prithviraj, 2006) and along the west coast in Thotapalli and Valiazhikal in Kerala (Babu et al., 2007). The unusual distribution of heavy mineral-rich sands along the inundated area is a unique phenomenon that hitherto has been reported globally, neither after a storm surge nor after a tsunami (Rajamanickam & Prithviraj, 2006). These heavy mineral seams form a belt parallel to the coast above the high tide limit (Fig. 3a) between NE light-house and Pisasumunai. Interestingly, in all the heavy mineral enriched areas, the configuration of the coast is NW–SE. In some other places of this region, the heavy mineral deposition is restricted to the intertidal zone where the coralline terrace is exposed. The presence of coralline terraces as a natural sea wall in this region has blocked the advancement of tsunami waves and resulted in the deposition of sediments in front of the physical barrier (Fig. 4).

Along the other coastlines that run NE-SW between Pisaumunai and Ariyankundu, and Olaikuda and Remeswaram, no visible accumulations of sediments, particularly heavy minerals, observed (Fig. 5). Along the steep slope of the beach, since inundation by wave uprush is restricted, even the minor quantity of sediments carried by the tsunami waves formed a thick sediment sequence when they were transported back by backwash along the down slope. A unique coexistence of steeply sloping beaches and coral terraces is a primary geomorphological factor for the observed maximum sedimentation for more than 1 m depth with an aerial extension of greater than 1 km in the regions between the lighthouse and Pisasumunai. In Kothandaramasamy kovil area of the Palk Strait region, according to reports of eyewitnesses, the 2004 IOT tsunami waves inundated laterally less than 5 m from the high tide line. But along the line of zero velocity, a small ridge-like shore parallel structure built by fine white sands was seen. Beyond this zero-velocity line further landward, the spilling waves had caused sedimentation for less than a

Fig. 4 Garnet-enriched heavy mineral sand deposited in front of the coral terrace

cm on marshy land. No heavy mineral accumulation either as surface deposits or as laminae/layers within the subsurface sediment sequences could be identified (Fig. 5). In Dhanushkodi beach facing the Gulf of Mannar, tsunami wave height might have been more than 2 m, as evidenced by the deposition of sands out of suspension atop a palm-tree stump ~ 1.5 m above the ground (Fig. 6) and, the role of storm waves in it is ruled out as the range of spring tide and neap tide in this region is 0.6 and 0.16 m, respectively (Gowthaman et al., 2013).

The site-specific or coastline configuration-specific presence or absence of tsunamigenic sediments, and if present, their aerial extent, textural and compositional variations, including the presence or absence of heavy minerals, the thickness of sediment layers and a kind of physical sedimentary structures present in them have revealed very complex patterns of coastal run-up and sediment deposition in the study region. Local variations in the configuration and bathymetry of individual wave inlets, sediment erosion and reworking during periods of backwash along the steep slope, presence of physical structures like coral terraces that obstruct the advancement of waves as well as nearshore sediment supply have selectively played a combined role in the deposition of tsunami sediments in the study region. A corroboration of observations made from the trenches excavated at Olaikuda, Narikuli, Pisasumunai, Vadakadu and Kothandaramasamy kovil area augments the fact that the sedimentological characteristics of the tsunamigenic sequences on land were dependent on the erosion, nature and composition of sediments available offshore and transported by the tsunami waves (Goff et al., 2004, Jaffe & Gelfenbaum, 2007, Phantuwongraj & Choowong, 2012). Besides, the influence of local variation in the configuration and reworking of sediments by backwash, which is hitherto less known from the tsunami deposits described worldwide (Dawson, 1994), in the deposition of tsunamigenic sediments in the study area has been clearly brought out from this study.

Fig. 5 Depth profile in Kothandaramasamy Kovil

Fig. 6 Sediment deposition atop a palm tree stump of $\approx$1.5 m height

3.2 Palaeo High-Energy Deposits

The only conspicuous deposition by the 2004 IOT in the region between Olaikuda and Pisasumunai further gave ample scope to explore palaeo-high-energy deposits by similar extreme events along this particular coastal segment. Interestingly, the coastal stretch between Olaikuda and Pisasumunai is the only site to have sedimentological signatures for the occurrence of high-energy events, while the other parts of the 70 km stretch of the Rameswaram coast revealed no distinct evidence. Considering the characteristics of sediment sequences, textural and compositional variations and micro/macro-faunal assemblages within the trenches made in the targeted region, the possibility of occurrence of palaeo-high-energy event in this unique region is discussed below. The trench made at Narikuli parallel to the coastline (NW–SE) revealed five different sedimentary layers, A–E (Fig. 7). Layer A is a dark pasty peat mud (PM) that unconformably overlies the hard-underlying fringing coral reef with a thickness of about 40 cm. Abundant foraminiferal and ostracod tests, along with unbroken gastropod shells, are well preserved, indicating the *in-situ* origin of the PM. Intra-clasts of coral gravels from the substrate are incorporated within the PM. During the regressive course of sea level after the Holocene transgression around 6000 years BP, the fringing reefs had undergone a slow death, and the palaeo-shallow marine area was converted slowly first into an intertidal flat and then to a mudflat. Layer B is a buff or light-yellowish white sand (YWS) of 8 cm thickness. Layer C was partially developed and has an erosional contact with layer B and contains sands enriched with red grains of garnet sands and is 1 cm thick. Lying above C is layer D, which is 2 cm thick and contains white sands (WS) with no appreciable concentration of heavy minerals. Layer E, with a thickness of 1 cm, lies at the top of the sedimentary sequence and is highly enriched with grains of red garnet. The dark coloured silt is enriched with abundant shells of molluscs and admixture with less amount of sand. Intra clasts of coral gravels from the substrate are incorporated within the overlying dark grey sandy silt. No Alcyonarians in silty clay were observed. A buff/light yellow-coloured sandy layer lies over the dark silt layer. The contact between the dark-coloured silt and yellow-coloured sand is clearly erosional. The sandy layers C, D and E that lie above the yellowish sand (layer B) and at the top of the sedimentary sequence were fresh and might have been deposited by the recent 2004 Indian Ocean Tsunami, as evidenced by the abundant Alcyonarians.

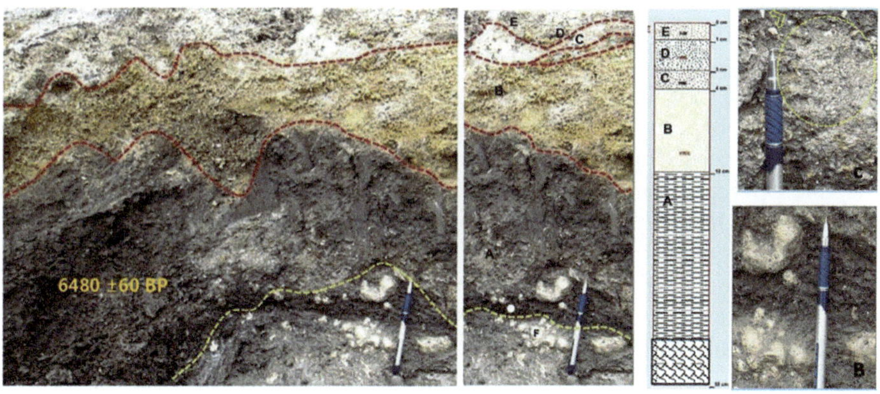

Fig. 7 Sedimentary characteristics and evidence of tsunami deposition in Narikuli

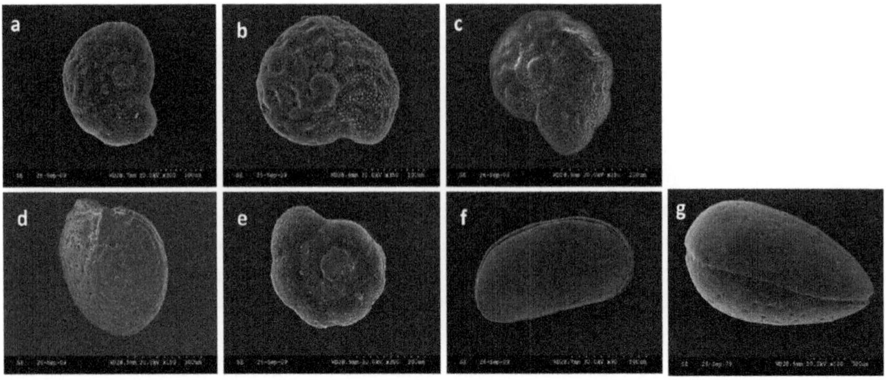

Fig. 8 Side view of Protoelphidium (**a** and **b**), Elphidium excavatum (**c**), Elphidium crispum (**d**) and dorsal view of Ammonia beccarii var. tepida (**e**) Ostracod (**f** and **g**)

The presence of abundant unbroken marine gastropod shells and well-preserved foraminiferal tests of *Ammonia beccarii var. tepida* (Fig. 8a), *Elphidium excavatum* (Fig. 8b), *Elphidium crispum* (Fig. 8c), *proto elphidium* (Fig. 8d and e) *and ostracods* (Fig. 8f and g) within the PM mixed with about 5% of fine sand supplements this inference.

The abrupt invasion of high-energy waves on the palaeo-mud flat (PM) in this region deposited coarse sands ($M_Z = 0.95\ \Phi$) that have abundant broken marine shells (YWS) from the adjacent inner shore region. The resultant contact between the layers PM and YWS is evidently erosional. The granulometric characteristics like polymodal grain size distribution, poor sorting, and leptokurtic nature of YWS (Table 1, Fig. 9a and b) corroborate their deposition by high-energy waves. In Fig. 9, layers C, D, and E have a significant fining upward trend from 10, 1.25Φ & 1.5 Φ—deposition by the recent Tsunami. Layer B: Gravelly sand: sub-samples show fining upward within the coarse class (– 4 Φ, – 1.50, – 0.250, 0.750), poorly sorted, very coarse skewed, very leptokurtic.

The grain size of the sub-samples collected within the YWS shows a fining upward trend (– 4 Φ, – 1.5 Φ, – 0.25 Φ, 0.75 Φ), confirming once again the role of high-energy in the deposition of YWS. The presence of abraded and broken tests of *Ammonia beccarii* (Fig. 10a), *Quinqueloculina seminulum* (Fig. 10b), *Elphidium crispum, Operculina* and *Amphistegina* in YWS further supports the transportation of sands by high energy waves. Due to oxidation after deposition, the sands acquired a buff or light-yellow colour, which is supported by the presence of some tests of foraminifera and gastropods that have dissolved or pitted surfaces.

The YWS that got deposited over PM mud might have been the product of the 1964 cyclone that devastated the entire Dhanushkodi. In Fig. 11, the sandy layers C, D and E that lie above B (YWS) and at the top of the sediment sequence are fresh and deposited by the 2004 IOT. Layers C, D and E, Alcyonarian, an extinct soft coral characteristic of the tropical sheltered environment (Fig. 10c and d), is dominant, whereas in YWS, it is rarely present, and in PM, it is absent. The abundant presence

Table 1 Textural parameters of sediments in the trenches

Sediment layers in Trench	Folk and Ward (1957)				Median	Qualitative scale			
	Mean	Sorting	Skewness	Kurtosis		Mean	Sorting	Skewness	Kurtosis
Figure 7									
E	1.59	0.90	– 0.08	0.80	1.65	Medium sand	Moderately sorted	Near symmetrical	Platy kurtic
D	1.14	1.14	– 0.03	1.20	1.01	Medium sand	Poorly sorted	Near symmetrical	Lepto kurtic
C	1.25	0.92	0.10	0.89	1.15	Medium sand	Moderately sorted	Near symmetrical	Platy kurtic
B	0.95	0.80	0.01	1.13	0.97	Coarse sand	Moderately sorted	Near symmetrical	Lepto kurtic
A	1.33	1.25	0.05	1.39	1.16	Medium sand	Poorly sorted	Near symmetrical	Lepto kurtic
Figure 11									
E	1.54	0.35	0.48	0.96	1.41	Medium sand	Well sorted	Very fine skewed	Meso kurtic
D	1.50	0.44	0.36	0.95	1.42	Medium sand	Well sorted	Very fine skewed	Meso kurtic
C	0.98	0.26	0.06	1.39	0.96	Coarse sand	Very well sorted	Near symmetrical	Lepto kurtic
B 2	1.63	0.50	0.53	1.01	1.38	Medium sand	Well sorted	Very fine skewed	Meso kurtic
B 1	0.72	1.17	– 0.43	2.29	1.00	Coarse sand	Moderately sorted	Very coarse skewed	Very lepto kurtic
A	0.91	0.91	– 0.20	1.67	0.96	Coarse sand	Moderately sorted	coarse skewed	Very lepto kurtic

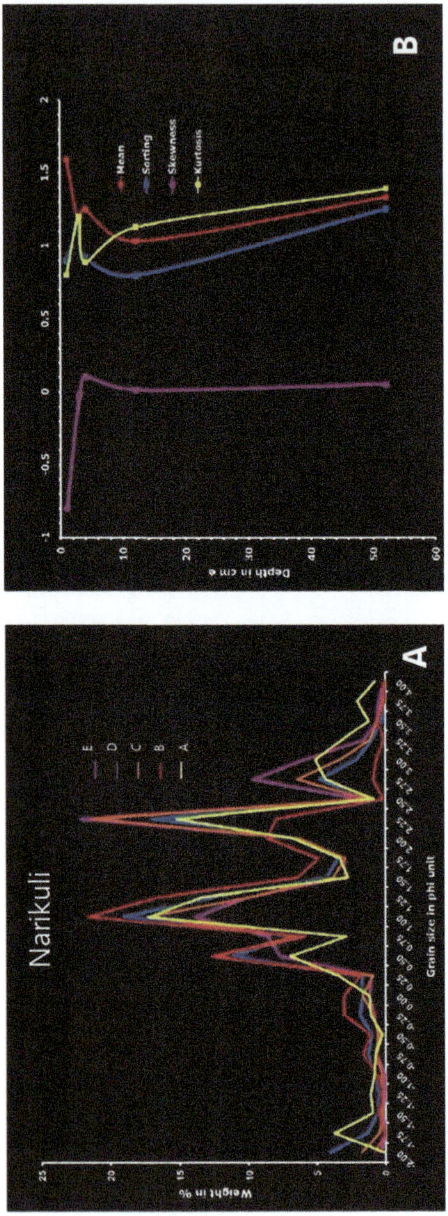

Fig. 9 Polymodal, **a** poor sorting and leptokurtic, **b** nature of sediments

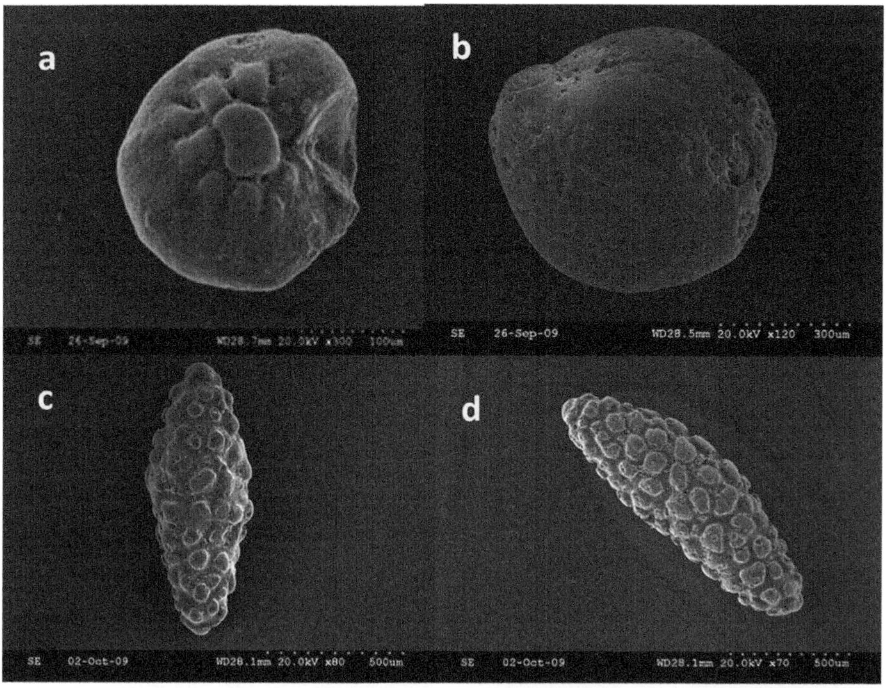

Fig. 10 **a** Ammonia beccarii var.tepida (ventral view), **b** Quinqueloculina seminulum (side view), **c** Aclyonarian sp.1, **d** Aclyonarian sp.2

of endangered/extinct soft coral in the overlying sandy layers (C, D and E) reveals that they have been transported to the shore by the 2004 IOT from the palaeo-sediments that carpet the shallow sheltered regions. This is evidenced by the abundance of foraminifera *Amphistegina, Elphidium* and *Calcarina calcar* (Table 2), which are characteristic of lagoonal, coral reef inner shelf environment (Hussain et al., 2006; Razzhigaeva et al., 2006).

Figure 11 shows the subsurface sediment sequences within another trench made along the coast about 20 m north of the trench shown in Fig. 7. The trench has six sedimentary layers, A-E, with two divisible layers, B1 and B2. Layer 'A' is a PM that extends from the previous trench (Fig. 7), showing the same characteristics, and rests on the dead coral basement. B1 and B2 are sandy layers lying above A, have distinct textural differences, and are conspicuously visible along the boundary of B1 and B2. Sand in B1 is coarse in size (Mz = 0.72) while in B2 it is medium-sized (Mz = 1.63), showing an upward fining in grain size (Fig. 12a and b). *Quinqueloculina seminulum* and *Alcyonarian* are dominantly present in B1 and B2, along with the occasional presence of *Cellanthus craticulatum, Ammonia gibbosa, Calcarina calcar* and *Operculina ammonoides* (Table 3).

Young (Holocene-age, deposited within 10,000 years BP) cohesionless saturated or partially saturated loose low-density granular coastal sands and silts of similar

Fig. 11 High energy deposits with soft sediment structures (liquefaction) at Narikuli

Table 2 Faunal assemblages in sediment layers of trench are shown in Fig. 7

Sediment layer	Faunal assemblages
E	*Alcyonarians* dominant. Foraminifera poor and represented by *Q.seminulum, C.craticulatum* and *A. lessoni* Rare coral and gastropod shells
D	*Alcyonarians* abundant. Foraminifera rare and represented by *Quinqueloculina* and *Elphidium*. Few coral, gastropod and mollusk shells
C	*Alcyonarians* abundant. Forams rare, *Quinqueloculina, Amphistegina*and *C,craticulatum*. Few coral and gastropod shells
B	Rare *alcyonarians*. Foraminiferal tests mostly with dissolved surfaces. *Elphidium crispum, A.beccarii, Operculina, Quinqueloculina* (broken) and *Amphistegina* Rare corals and mollusk shells with dissolved and often abraded surfaces
A	No *alcyonarians*. Foraminifera abundant and tests are well preserved, rarely broken. Both *A.beccarii, A.beccarii var. tepida* are dominant. *E.excavatum, Q.seminulum* and *C.craticulatum* and abundant smooth ostracodes (dominantly closed carapace with rare single ones) Unbroken gastropod shells with coral fragments

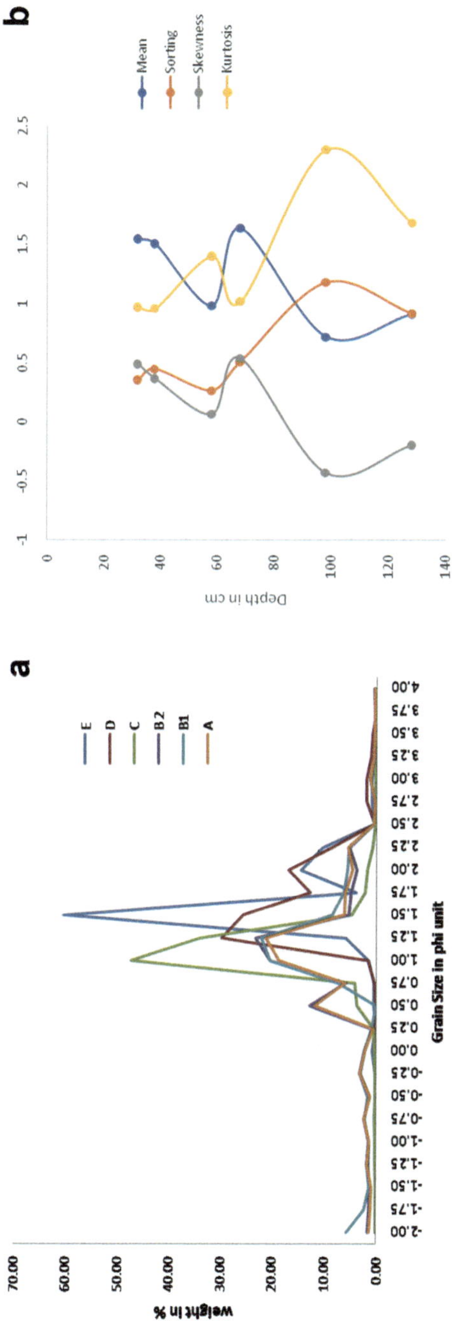

Fig. 12 Textural characteristics of sandy layers B1 and B2

Table 3 Faunal assemblages in sediment layers of trench (Ref: Fig. 11)

Sediment layer	Faunal assemblages
E	*Eponides schreibersi, Quinqueloculina seminulum, Cellanthus craticulatum* along with very rare coral and gastropod fragments
D	*Q.seminulum* is dominant followed by *C.craticulatum, Amphistegina gibbosa* and *Operculina* (broken) Very rare corals and few gastropods and alcyonarians Foraminifera dominant over alcyonarians
C	*Q.seminulum* (dominant), *Q.parker*i and *C.craticulatum* Few mollusk and gastropod shells No coral fragments. Alcyonarians abundant
B2	*Q.seminulum* (broken tests), rare *C.craticulatum* Few Coral and gastropods. Alcyonarians abundant
B1	Dominantly *Q.seminulum* followed by occasional *C.craticulatum, A.gibbosa, Operculina ammonoides (Gronovius), Calcarina calcar* with abraded tests of *Ammonia beccarii* Few coral and gastropod shells. Alcyonarians dominant
A	Solitary *Q. Q.seminulum* (broken, surface eroded), *Ammonia beccarii* (surface dissolved), few eroded gastropod, alcyonarian and coral fragments

grain size with better sorting markedly lose strength and stiffness in response to applied stress and behave like a liquid. Such sediments are most susceptible to liquefaction-induced soft-sediment deformation (SSD) structures. Many SSD structures are related to the presence of a gravitationally unstable profile of bulk density within a sedimentation unit or sequence of units. Heavy mineral laminae and bands in quartz sands, when sufficiently close vertically, become deformed in response to density instabilities when all are liquidized (Lowe, 1975; Selley, 1964; Stewart, 1963). Such depositional condition has caused SSD structure in layer (Fig. 11, B2) wherein, during liquefaction, relatively high-density red garnet sands (2406 kg/m^3) flowed down through the low-density beach sands (1500–1600 kg/m^3) forming garnet sags and drops. The rapid loading and unloading caused by wave runup and backwash may generate high enough hydraulic gradients to fluidize the nearly saturated sandy deposit near the bed surface leading to liquefaction failure of the coastal slopes. A similar observation has been made from the southern tip of the Dive Agar beach on the west coast of India, where the presence of SSD structure in the coastal sedimentary sequence has been attributed to the 2004 IOT (Meshram et al., 2011). Several types of SSD structures have been observed in the meta greywacke-argillite strata (Gor Group) along the northern beaches of Goa. These represent the Bouma sequence and were formed in a deltaic environment.

Quartz and lateralized calcareous pebbles were seen impregnated into the PM and occurred at the bottom of layer B1. Quartz pebbles vary in size from $1.5 \times 1 \times 0.5$ cm to $1 \times 0.5 \times 0.5$ cm, while calcareous pebbles range in size from $4.5 \times 3 \times 1$ cm to $5.5 \times 3.5 \times 2$ cm. The presence of calcareous pebbles could be attributed to the preponderance of coral reefs in the nearby marine environment. However, the co-existence of quartz pebbles in the same layer 'A' is intriguing because not only in the study region but for the entire Rameswaram Island, there are no reports of occurrence either of crystalline rocks with such large quartz grains or any major and long streams that could have transport quartz pebbles from elsewhere to the site of deposition (Vaz et al., 2008). Those authors studied the cross-bedded shelly calcareous Mandapam sandstone formation exposed along the coast of Palk Bay about 15 km east of the present study region and reported the presence of quartz pebbles with abundant heavy mineral concentrations. The Mandapam sandstone formation extends eastward towards the study region at sub-surface levels and forms a basement for the growth of younger coral reefs. Hence, the Mandapam sandstone might have been the source of quartz pebbles and plenty of heavy minerals, like garnet and ilmenite, in the study region. The Palk Bay experiences heavy sediment influx from the adjoining landmass through several rivers, and this could transform the bay into a lagoon in the future. The above observations amply demonstrate the fluvial influence in the study region and substantiate the presence of quartz pebbles in the nearby coral debris-dominant marine environment. Hence, quartz pebbles might be transported products of a palaeo-course of the Vaigai River. This observation is supported by Prabakaran and Anbarasu (2010), who found that the Holocene transgressive and regressive events associated with dystrophic movements have caused the shifting of Vaigai River from down south near Tuticorin to Mimisal in the north through Mandapam near the study region. They have also reported that the ancient channel of the Vaigai River might have extended further east into Rameswaram island through the present study region. Therefore, the quartz pebbles at the bottom of the sedimentary sequence might have pre-existed, over which the high-energy waves deposited coralline pebbles and sands. The coarse-grained, poorly sorted polymodal sands that lie unconformably over the nearshore-derived pebbles with fining upward sequence characteristics show their deposition by high-energy waves under turbulent conditions. This sediment layer with evidence of a high degree of oxidation might be the deposition by a cyclonic storm that hit this region in December 1964. Layers A, B and C, with irregular bedding planes, are abundant with garnet sands, which are the deposits of the 2004 IOT. High energy deposits on mud surface with anomalous enrichment of red garnet, red garnet with black sands and black sands as lag deposits with varying thickness were observed in northern and north eastern regions (Olaikuda, Narikkuli, Pisasumunai and Vadakadu) of the study area facing the Palk Bay (Figs. 2a, 3a, 4). But on the southern side, adjoining the Gulf of Mannar (Danuskodi check post and southern fringe of Danuskodi spit), there is no specific evidence of higher energy deposits in the form of heavy mineral lag deposits (Figs. 3b and 5) The entire pile of sediments shows no characteristic change either in texture or in mineralogy.

3.3 Ground Penetrating Radar (GPR) Profiling

Several GPR profiles were acquired to complement the trench made at Narikuli coast to investigate the subsurface structure of various layers and their inter-relationships. The main focus of the GPR study is to identify the geometry of the erosional surface over which the deposits of extreme events were observed. The GPR study involved analysis of the geological record for an erosional signature within the sedimentary sequence presumed to be a result of an extreme event.

About 30 m long subsurface profiles using 200 MHz were generated with 4 and 10 m depth of penetrations. GPR data were normalized and interpreted. The surface normalized profile is presented in Fig. 13a. Due to the saline water table up to around 2.0 m depth, the signal loss was appreciably noticed below the water table and represented as (1) Reworked storm sediments, (2) Tsunami deposit and (3) Paleo-beach sediment in profile legend. There were three distinct zones observed within 2 m depth with significant signal characteristics due to the physical aspects of the materials in the layers (Fig. 13a). A portion of dune sediments was noticed on the upper right side in Fig. 13a, and the remaining parts of the dune could not be observed due to the high water-saturated condition below onwards. High energy deposits and part of reworked materials are seen up to 2 m in thickness above the buried dune sediments. The data were visualised through an elevation correction algorithm and presented in Fig. 13b. It is observed that the earlier dune was disinformed by high energy waves, possibly by the 2004 IOT and new sediments were deposited subsequently. Further, the sediments occurring at the end of the dune were reworked, as revealed by the disturbed pattern of occurrence towards HTL. The sequence of sediment layers, as observed in the trench (Fig. 7), was well corroborated with the GPR observations. Another GPR profile along the same survey line with 4 m depth penetration further illustrates the depositional environment described more vividly than the earlier profile of 10 m depth.

Figure 14 denotes the basement over which the sediments of high energy waves were deposited to a thickness of nearly 1 m depth. Huge waves had destroyed the exterior portion of the sand dune and washed away the top layer of materials. The sediments deposited by the high energy waves at the top of the present-day dune were significantly clear from other deposits and had no typical pattern of layering due to relatively quick deposition. But, while observing the middle part of the above figure, the sediments had comparatively enough time to settle. This clearly illustrates that the yellowish-white sand (YWS) sediment sequence in the trench made at Narikuli (Fig. 7). Near the high tide level (HTL), the deposits slid, reworked further by the last spell of waves and a balanced slope was attained. A thin veneer of sand occurs all over the study site which could be the resultant effect of the present-day prevailing wind action. The GPR profiles significantly depict the disturbed area over a relict sand dune. This could have resulted due to the sudden invasion of waves with abnormal energy like a storm/ tsunami. The disturbance was noticed up to 1 m depth. The near-shore region is calm and characterized by the presence of fringing reefs, and the wave action is normally obstructed by these reefs. The significant disturbance in the dune

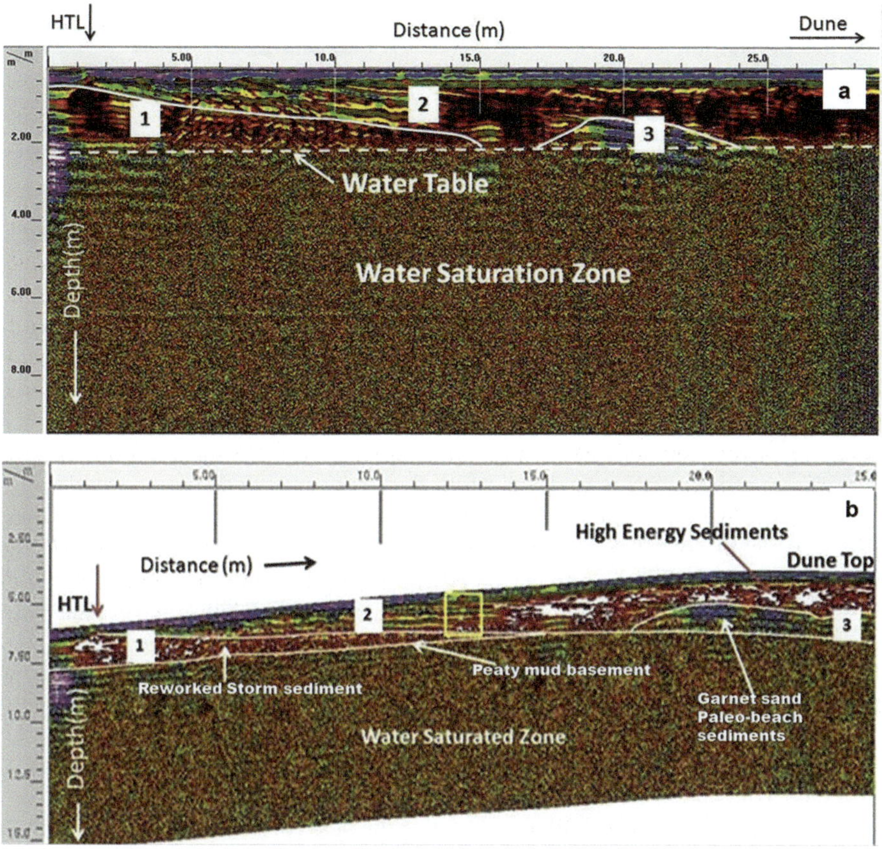

Fig. 13 GPR profile in Narikuli coast, **a** surface normalized, **b** elevation normalized

sediments up to a depth of 1 m could be possible only due to high energy waves that perhaps crossed the near shore obstructions and left behind noticeable changes in the dune. The GPR profiles reveal the presence of dead corals in a basement, over which pasty organic matter-rich mud (PM) is deposited. The thickness of the PM thins landward and thickens seaward. This depositional set-up once again supports the fact that the environment might have been a shallow marine when the shoreline was above 2 m from the present (Vaz et al., 2008), and the fringing reefs would have actively grown. This is further confirmed by the abundant presence of *Alcyonarians,* a soft coral in the sediment sequence.

The lowering of sea level due to neo-tectonic movements and glacio-eustatic sea level changes (Vaz et al., 2008) has gradually changed the shallow marine environment into a tidal flat region with the deposition of organic matter-rich fine sediments over the fringing reefs. The deposition of sediments, together with the lowering of sea level, are plausible causes for the death of the corals in this region. The textural characteristics of sandy sediments that lie above PM, erosional contact between PM

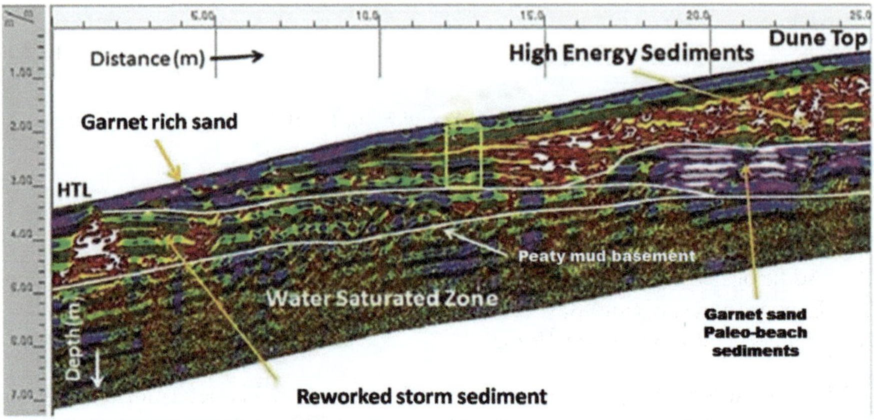

Fig. 14 GPR profile of 4 m depth penetration at Narikuli coast (Elevation normalized)

and YWS and the presence of coarse pebbles at the base of the sequence indicate that during the receding phase of the sea, sea waves with an abnormal energy condition would have brought the reworked coarse shelly sands and pebbles from near shore and deposited them over the peaty mud that capped the coralline basements. The fresh sands with abundant enrichment of garnet in the same area as deposits of the 2004 IOT is a unique feature. To corroborate the above view, we plotted the standard deviation (sorting) against the mean grain size of the analysed sediments (Table 1). This plot (Fig. 15) helps to distinguish sediments derived from the swash zone, shoreface, tsunami, and berm/dune deposits (Srisutam & Wagner, 2010). It is seen that the sediments from Narikuli (Fig. 7) fall in the fields of swash and tsunami, while the sediments that show liquefaction and SSD (Fig. 11) lie nearer to the dune, except for one which is in the swash zone. Therefore, it appears that the swash sediments were reworked and overlain by the 2004 IOT sediments, while the dune sediments were liquified and deformed to produce various SSD structures.

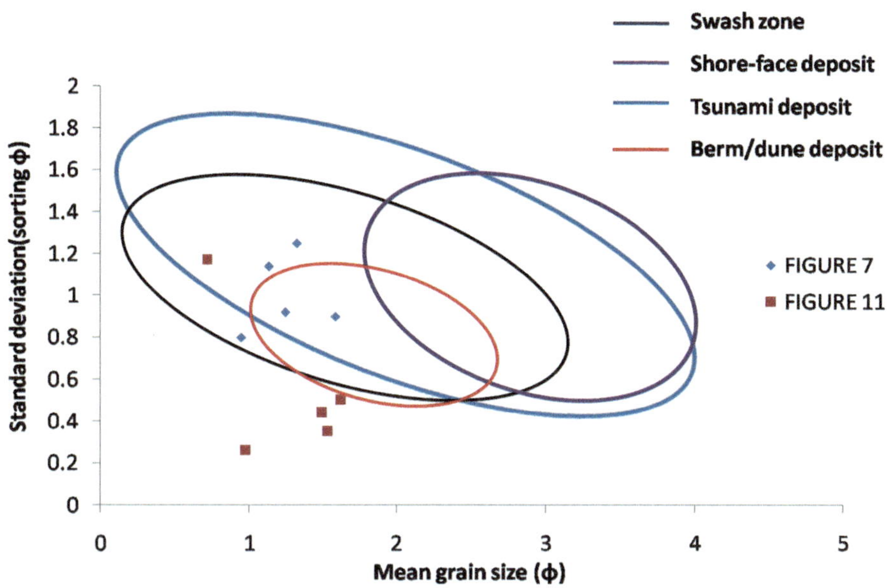

Fig. 15 Plot of Standard deviation versus mean grain size to differentiate sediment origins in Narikuli

4 Conclusion

Geological imprints in the sedimentological records of trenches studied at the NE part of Rameswaram Island have clearly indicated that the sedimentary deposits have been produced by sea waves of a high-energy environment. A combination of multiple pieces of evidence unearthed through this study, like fining upward sediment sequences, erosional contacts between litho-layers, presence of broken molluscan shells and benthic/nearshore microfossils with broken tests in sandy layers, and anomalous enrichment of heavy minerals correlate well with signatures reported world over for both storm and paleo-tsunami deposits making it challenging to distinguish between the sedimentary deposits left by tsunamis and those left by storms though the underlying processes are different. One of the important criteria of tsunami deposits that distinguishes it from storms is the fining upward and thinning landward sediment sequence due to relatively more inundation of tsunami waves than that of storms. Since the study area is a sheltered coast with respect to tsunamigenic sources, the diffracted tsunami waves with relatively reduced energy can alone reach and inundate the land to a lesser distance. Considering the sediment textural characteristics, the occurrence of SSDS and corroborating GPR data, we opine that the nearshore/swash sediments were transported and covered by the sediments of the 2004 IOT. This also resulted in the liquefaction of the dune sediments, but these are not overlain by the tsunami deposits. The presence of longitudinal dunes very near the coastline in the study region further arrests the landward inundation of tsunami waves, prohibiting the formation of landward thinning sedimentary layers, a piece of typical field evidence for tsunamigenic sediments.

Acknowledgements The authors thank Dr. Shailesh Nayak and Dr. Satheesh Shenoi, former Directors and associated authorities of the Indian National Centre for Ocean Information Services (INCOIS), Ministry of Earth Sciences, Govt. of India, Hyderabad for financial assistance in the form of a major research project (INCOIS/283/2008/3413/08, 29.02.2008) under the 'Palaeotsunami Reconstruction Network Programme'. We also thank the Directors, staff of Geological Oceanography Division, CSIR-National Institute of Oceanography, Goa and the Principals, Alagappa Govt. Arts College (Alagappa University), Karaikudi, for having provided the necessary facilities to carry out the research. Authors express their sincere thanks to Dr. Ashok Kumar Singhvi, Former Dean of Physical Research Laboratory, Ahmedabad and the National co-ordinator of the project, for his technical advice, assistance in 14C dating from ANU, Canberra and moral support. Gratitude is due to Dr. N. Rajeshwara Rao, Former Professor of Geology at the University of Madras, Chennai, for his help in identifying the microfossils, Mrs. M. Suvedha and Mrs. M. Suganya for their assistance in the field and the laboratory, and in the preparation of the figures and tables. We thank the reviewers for their comments that helped to improve the presentation. We acknowledge the editors for having given us an opportunity to contribute to this chapter.

References

Babu, N., Babu, D. S., & Das, P. M. (2007). Impact of tsunami on texture and mineralogy of a major placer deposit in southwest coast of India. *Environmental Geology, 52*(1), 71–80.

Banerjee, P. K. (2000). Holocene and Late Pleistocene relative sea level fluctuations along the east coast of India. *Marine Geology, 167*(3–4), 243–260.

Baskaran, R., Anbarasu, K., Ferozekhan, M., Selvaraj, S., & Rajamanickam, G. V. (2003). Geomorphology as a guide to mangrove afforestation in Rameshwaram Island, Tamilnadu. *Indian Jour. Geomorphology, 8*(1&2), 117–122.

Bruckner, H. (1988). Indicator for formerly high sea levels along the east coast of India on the Andhaman island. *Hamburger Geographischestudie Heft, 44*, 47–72.

Bruckner, H. (1989). Late quaternary shorelines in India. In *Late quaternary sea-level correlation and applications: Walter S. Newman Memorial* (pp. 169–194). Springer.

Carayannis, G. P. (2006). The potential of tsunami generation along the Makran Subduction Zone in the northern Arabian Sea: Case study: The earthquake and tsunami of November 28, 1945. *Science of Tsunami Hazards, 24*(5), 358–384.

Chandramohan, P., Jena, B. K., & Kumar, V. S. (2001). Littoral drift sources and sinks along the Indian coast. *Current Science*, 292–297.

Dawson, A. G. (1994). Geomorphological effects of tsunami run-up and backwash. In *Geomorphology and natural hazards, Geomorphology* (Vol. 10, pp. 83–94).

Dawson, A. G., & Shi, S. (2000). Tsunami deposits. *Pure and Applied Geophysics, 157*, 875–897.

Goff. J., Mc Fadgen., B. G., Chague Goff. C. (2004). Sedimentary differences between the 2002 Easter Storm and the 15th century Okoropungo tsunami, southeastern North Island, New Zealand. *Journal of Marine Geology, 204*(1–2), 235–250.

Gowthaman, R., Kumar, V. S., Dwarakish, G. S., Mohan, S. S., Singh, J., & Kumar, K. A. (2013). Waves in Gulf of Mannar and Palk Bay around Dhanushkodi, Tamil Nadu, India. *Current Science*, 1431–1435.

Gowthaman, R., Kumar, V. S., Dwarakish, G. S., Shanas, P. R., Jena, B. K., & Singh, J. (2015). Nearshore waves and longshore sediment transport along Rameshwaram Island off the east coast of India. *International Journal of Naval Architecture and Ocean Engineering, 7*(6), 939–950.

Hussain S. M, Krishnamurthy R., Suresh Gandhi, M., Illayaraja, K., Ganesan, P., & Mohan, S. P. (2006). Micropalaeontological investigation on tsunamigenic sediments of Andaman Islands. *Current Science, 91*(12), 1655–1667.

Jaffe, E. B., & Gelfenbaum, G. (2007). A simple model for calculating tsunami flow speed from tsunami deposits. *Sedimentary Geology*, 347–361.

Kumaraguru, A. K., Jayakumar, K., Wilson, J. J., & Ramakritinan, C. M. (2005). Impact of the tsunami of 26 December 2004 on the coral reef environment of Gulf of Mannar and Palk Bay in the southeast coast of India. *Current Science*, 1729–1741.

Loveson, V. J., & Rajamanickam, G. V. (1999). Geomorphic evolution of Rameshwaram Island, South India, Indian. *Journal of Geomorphology, 3*, 59–74.

Lowe, D. R. (1975). Water escape structures in coarse grained sediments. *Sedimentology, 22*, 157–204.

Meshram, D. C., Sangode, S. J., Gujar, A. R., Ambre, N. V., Dhongle, D., & Porate, S. (2011). Occurrence of soft sediment deformation at Dive Agar beach, west coast of India: Possible record of the Indian Ocean tsunami (2004). *Natural Hazards, 57*, 385–393.

Murty, A. V. S., & Varma, P. U. (1964). The hydrographical features of the waters of Palk Bay during March, 1963. *Journal of the Marine Biological Association of India, 6*(2), 207–216.

Natesan, U., & Subramanian, S. (1993). Seasonal shoreline oscillation of Tamil Nadu coast. *Current Science, 65*(9), 667–668.

Phantuwongraj, S., & Choowong, M. (2012). Tsunamis versus storm deposits from Thailand. *Natural Hazards, 63*, 31–50.

Prabakaran, K., & Anbarasu, K. (2010). Coastal geomorphology and evolution of Rameswaram Island, Tamil Nadu, India. *Research Journal of Geoscience, 2*(2), 30–35.

Rajamanickam, G. V., & Loveson, V. J. (1990). Results of Radiocarbon dating from some beach terraces around Rameshwaram Island. In G. Victor (Ed.), *Sea level variation and its impacts of coastal environment* (pp. 389–396). Tamil University Publication.

Rajamanickam, G. V., & Prithviraj, M. (2006). Great Indian Ocean Tsunami: Indian prospective. In G. V. Rajamanickam, B. R. Subramaniyan, M. Baba, R. Ramesh, L. Elango, & M. Prithviraj (Eds.), *26th December 2004 tsunami causes, effects, remedial measures, pre and post tsunami disaster management: A geoscientific perspective*. New Academic Publishers.

Rastogi, B. K. (2007). A historical account of the earthquakes and tsunamis in the Indian Ocean. *The Indian Ocean Tsunami*, 3–18.

Razzhigaeva, N. G., Ganzei, L. A., Grebennikova, T. A., Ivanova, E. D., & Kaistrenko, V. M. (2006). Sedimentation particularities during the tsunami of December 26, 2004, in northern Indonesia: Simelue Island and the Medan coast of Sumatra Island. *Oceanology, 46*, 875–890.

Sahayam, J. D., Chandrasekar, N., Kumar, S. K., & Rajamanickam, G. V. (2010). Distribution of arsenic and mercury in subtropical coastal beachrock, Gulf of Mannar, India. *Journal of Earth System Science, 119*, 129–135.

Selley, R. C. (1964). The penecontemporaneous deformation of heavy mineral bands in the Torridonian sandstone of northwest Scotland. *In Developments in Sedimentology, 1*, 362–367.

Sheth, A., Sanyal, S., Jaiswal, A., & Gandhi, P. (2006). Effects of the December 2004 Indian Ocean tsunami on the Indian mainland. *Earthquake Spectra, 22*, 435–473.

Silas, E. G. (1962). Tuna fishery of the Tinnevelly coast. *Gulf of Mannar, Symposium on Scombroid Fishes, Part, III*, 1083–1121.

Singarasubramanian, S. R., Mukesh, M. V., Manoharan, K., Murugan, S., Bakkiaraj, D., John Peter, A., & Seralathan, P. (2006). Sediment characteristics of the M-9 tsunami event between Rameswaram and Thoothukudi, Gulf of Mannar, southeast coast of India. *Science of Tsunami Hazards, 25*(3), 160–172.

Sivasankar, V., & Ramachandramoorthy, T. (2009). An investigation on the pollution status of holy aquifers of Rameswaram, Tamil Nadu, India. *Environmental Monitoring and Assessment, 156*, 307–315.

Srisutam, C., & Wagner, J. F. (2010). Reconstructing tsunami run-up from the characteristics of tsunami deposits on the Thai Andaman Coast. *Coastal Engineering, 57*(5), 493–499.

Stewart, A. D. (1963). On certain slump structures in the Torridonian sandstones of Applecross. *Geological Magazine, 100*(3), 205–218.
Stoddart, D. R., & Gopinadha Pillai, C. G. (1972). Raised reefs of Ramanathapuram, south India. *Transactions of the Institute of British Geographers—The Royal Geographical Socity*, 111–125.
Subba Rao, D., Rao, K. S., Iyer, C. S. P., & Chittibabu, P. (2008). Possible ecological consequences from the Sethu Samudram canal project. *India. Marine Pollution Bulletin, 56*(2), 170–186.
Vaidyanathan, R. (1993). Quaternary of the east coast of India. *Current Science, 64*, 11–12.
Vaz, G. G., Hariprasad, M., & Rao, B. R. (2008). Evidences and radiocarbon (^{14}C) ages of palaeo-high sea-level position around Mandapam, southeast coast of India. *Current Science*, 786–790.
Vaz, G. G., Hariprasad, M., Rao, B. R., & Rao, V. S. (2007). Subsidence of southern part of erstwhile Dhanushkodi township, Tamil Nadu–evidences from bathymetry, side scan and underwater videography. *Current Science*, 671–675.
Vaz, G. G., Rao, V. S., & Ravikumar, V. (2006). Thermal springs in Indian coastal areas of the Palk Bay: Their implications in relation to lineaments, coastal morphology and seismicity. *Geological Society of India, 68*(4), 593–596.

Recent Trend Analysis of Groundwater Water Level Fluctuations in Coastal Aquifers of Lower Palar Basin, Tamil Nadu, India

V. Thirukumaran and N. Manjunathan

1 Introduction

Groundwater is the most important component of any ecosystem. It is not only an essential component of human life but also a precious natural gift for other living organisms (IWMI, 2007a, 2007b). The groundwater resources are over-exploited in many countries, which in turn leads to a sudden decline in the groundwater level (UNDP GWF, 2014). Due to the increase in economic development and population every year, groundwater usage has increased. Groundwater usage accounts for more than 67% of Irrigation 22% is used for domestic (Drinking and Sanitation), while 11% is used for Industrial purposes (UNEP, 2006, 2008). The United Nations Development Programme reported that groundwater resources are over-exploited in many countries, which indicates that in the twenty-fifth century, two-thirds of the human population will face an acute shortage of fresh water across all parts of the world (UNDP Annual Report, 2023). In order to meet the ever-increasing water needs in recent years, the scientific investigation on the advanced development and management of freshwater resources has gained much importance. Because of dire necessity, an attempt was made to make a systematic basin-wise hydrogeological study, i.e., ARIMA modelling to deduce controlling Geo-parameters for water level fluctuations in coastal aquifers of the Lower Palar Basin, which was taken for research studies. Numerous studies on groundwater assessment and identification have been conducted at the minor and major levels (Manjunathan et al., 2021; Singh et al., 2015). The rainfall distribution map was created using geostatistical and geospatial techniques, such as the Thiessen Polygon method (Preeja et al., 2011; Singh et al., 2015). Time series autoregressive integrated moving average (ARIMA) models in their systematic search for identification, estimation, and diagnostic checking (Scott

V. Thirukumaran (✉) · N. Manjunathan
Department of Geology, Government Arts College (Autonomous), Salem, India
e-mail: mailkumarann75@gmail.com

P. Thambidurai and A. K. Dikshit (eds.), *Coastal Environments of India*,
https://doi.org/10.1007/978-3-031-97341-3_6

Armstrong, 2001; Thirumurugan et al., 2016a, 2016b; Manjunathan & Thirukumaran, 2019), least square technique was applied in earlier studies (Pankratz, 1983; Chambers et al., 1983; Cimbala, 2013). ArcGIS methods were extensively used for groundwater evidence (Krishnamurthy et al., 1996; Srivastava & Bhattacharya, 2006). Kanchipuram district is one of the most important districts in Tamil Nadu, India. It contributes almost 80% of the rainfall in the northeast monsoon seasons, with the remaining around 20% falling in the southwest pre-monsoon. The Autoregressive Integrated Moving Average (ARIMA) model for forecasting future trends has been widely applied in anticipating social, engineering, economic, and hydrological problems.

Despite the wide application of groundwater fluctuation trends forecasting models from different parts of the world. However, there is no reference study of groundwater trend analysis in the Lower Palar Basin of Kanchipuram district in Tamil Nadu. Therefore, the present study aimed to 1. Analyse the historical groundwater level data to identify trends and patterns 2. Evaluate the performance of different forecasting models in predicting groundwater levels, 3. Identify the most suitable model for groundwater trend analysis in the study area, 4. Investigate the impact of climatic and anthropogenic factors on groundwater fluctuations.

2 Study Area

Kanchipuram district, also known as Temple City, is situated on the northern East Coast of Tamil Nadu and about 70 km south of Chennai. Lower Palar River Basin is one of the most important basins it is also known as Kshiranadhi (Milk River). It originates from Nandi Durga, Chikkaballapuram district, at an elevation of 1428 m above MSL in the eastern part of Karnataka state and reaches into the Lower Palar River Basin via Kanchipuram, Walajabad, Chengalpattu, and finally mixes with the Bay of Bengal near Sadurangapattinam (Kalpakkam town). The intermediate tributaries of Cheyyar, Killiyar, and Vegayar join with main the Palar River. The Lower Palar River Basin covers the geographical area of 443,210 Ha. It lies between 78°48′00″ to 80°50′00″ and 12°00′00″ to 13°53′00″ (Fig. 1). The length of the river is about 45 km in Tamil Nadu. In the study area, there are two types of monsoons: NEM and SWM. The SWM basin receives about 30 to 40% of total annual rainfall, while the NEM basin receives about 45% (CGWB, 2014 and DPAP technical report, 2008). The lower Palar River basin is bordered north by Tiruvallur district, east by the Bay of Bengal, west by Vellore district and south by Viluppuram district. The mapology of the district further divides Kanchipuram district into Kanchipuram and Chengalpattu. The average rainfall in the district varies from 1805 to 2614 mm (TNSCCC report, 2015). Because of their proximity to the Bay of Bengal, the coastal belts of Mamallapuram, Pankatucheri, Madurangam, and Kelambakkam receive more rainfall than the other stations.

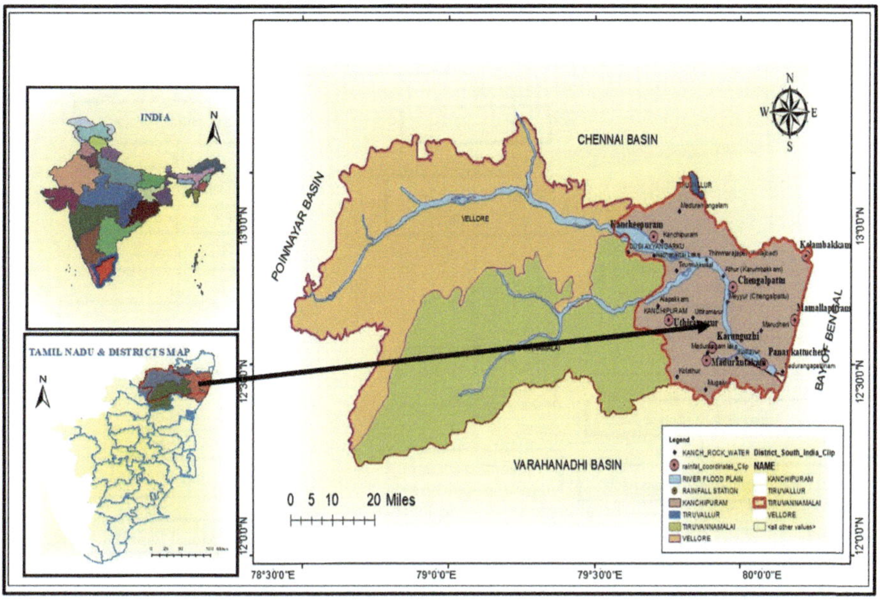

Fig. 1 Key map of the Lower Palar River Basin

3 Materials and Methodology

The goal of this study is to show that the groundwater level changes in the coastal aquifer of the Lower Palar River Basin (LPRB), and that there is a strong link between rainfall (which is dependent) and water level (which is independent). To do this, we will use Thiessen polygons, ARIMA models, and ArcGIS techniques. We did thorough field investigations to find out how accurate visual interpretations were and to make a map of all the wells (three dug wells, 31 irrigation wells, 17 monitoring wells, and two public wells) in the lower PRB, Kanchipuram district. The wells included monitoring wells, irrigation wells, dug wells, and public wells. The Public Works Department of the Government of Tamil Nadu, India, supplied secondary data on rainfall from 11 locations and water levels from 53 stations annually for a 14-year study period (2005–2018). Time series Auto-Regressive Integrated Moving Average (ARIMA) models were made with SPSS software. The results were then imported into the ArcGIS platform and turned into different types of maps, such as a geological evidence map, a lineament evidence map, a geomorphological evidence map, a drainage evidence map, and a slope evidence map (Fig. 2). We used the Thiessen polygon technique to create the geographical distribution of the research regions. It is one of the most frequently employed methods, as it yields more accurate and satisfactory results (Thiessen, 1991). It is also a well-liked deterministic method that produces excellent accuracy and results. Rainfall (observed) and water level

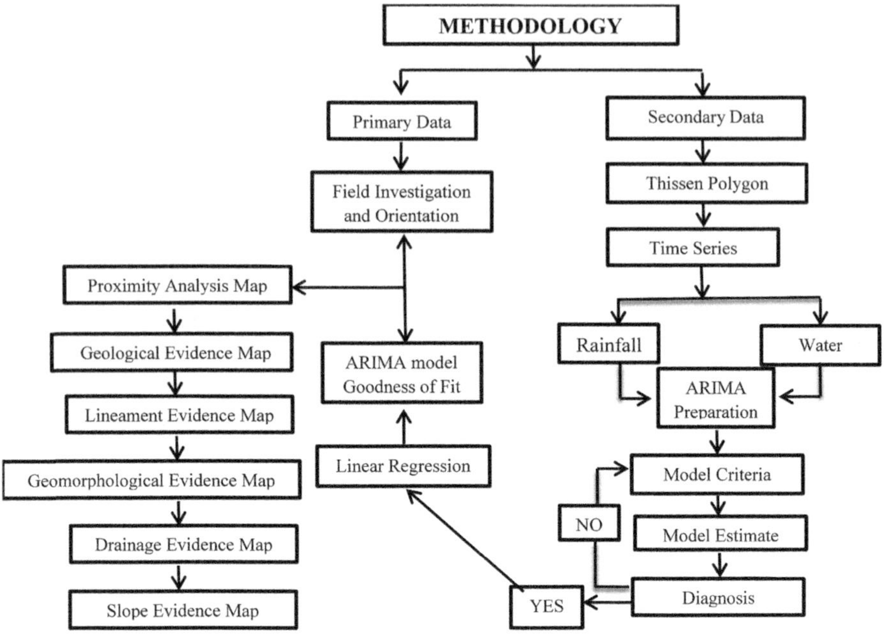

Fig. 2 Flowchart of the methodology adopted in the study

(dependent) data are investigated using the Thiessen Polygon (TP) method of Inverse Distance Weightage (IDW) (Ragunath, 1987).

The polygons consist of segments that link to proximity stations (rainfall stations) and other pertinent stations (such as water level stations), as illustrated in Fig. 3. We need to update the polygon each time we add or remove a station from the network. We used the Thiessen polygon technique to create the geographical distribution of the research regions. This method is among the most commonly utilized, providing enhanced accuracy and satisfactory outcomes (Thiessen, 1991). It is also a well-liked deterministic method that produces excellent accuracy and results. Observed rainfall and dependent water level data are investigated using the Thiessen Polygon method of Inverse Distance Weightage (IDW) (Ragunath, 1987). These polygons consist of segments that link to adjacent stations (e.g., rainfall stations) and other pertinent stations (e.g., water level stations), as illustrated in Fig. 3. Whenever a station is included in or removed from the network, the polygon must be revised accordingly. Time series analysis is extensively utilized across numerous scientific fields, including hydrology, engineering, and environmental management (Haan, 1977; Pankratz, 1983; Box et al., 1976). The time series ARIMA model serves as a valuable instrument for extracting comprehensive information from a data structure. The SPSS software was used to create time series Auto Regressive Integrated Moving Average (ARIMA) models; the most acceptable way to constrain the effective recharge as a function of rainfall and water level is to use the Box-Jenkins ARIMA

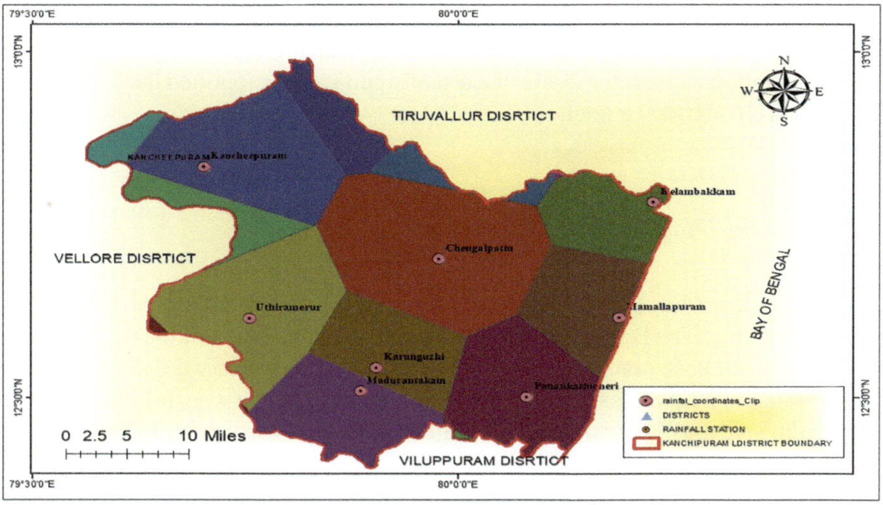

Fig. 3 Kanchipuram district, Lower Palar River basin—Thissen polygon method (Rainfall vs. Water Level)

model (Box et al., 2015). The systematic computation of identification, estimation, and diagnostic checking is a major benefit of the ARIMA model, explained below.

3.1 Model Identification

The first step used to demonstrate stationarity and the significance of seasonality is the identification of the model stage. The ARIMA model (P, D, Q) is computed using the parameters. We note that there are a few outliers and spikes in the autocorrelation function (ACF) and partial autocorrelation function (PACF). This information will help the higher-order autoregressive plots (Chitra et al., 2015; Aflatooni & Mehdi, 2011).

3.2 Model Estimation

The second stage, defined as the "best fit" estimation of the model stage, was calculated by keeping the model statistics at the lowest level for the Bayesian Information Criterion (BIC).

$$\mathrm{BIC(p, q)} = \ln\sigma^2 \tag{1}$$

where (σ^2 is a sample covariance function, and maximum for the R2 parameter (Aflatooni & Mehdi, 2011). Additionally, the "t" statistic was used to test the model for computational correctness, and the test of significance supported the development of the best-fit ARIMA model.

3.3 Model Diagnostics Checking

The third stage, diagnostics checking, was carried out to determine the ARIMA model's validity. The Box-Ljung Q statistic and autocorrelation of the residuals were computed with 95% confidence levels. All of the models that were developed kept their minimum BIC values, maximum R2 values, outliers that were minimum or zero, maximum statistics, and minimum significations for stationary residual error values (Brockwell & Davis, 1987). The models' and the ACF and PACF plots' outliers are well-maintained to be below or close to zero, confirming the model's valid Box-Jenkins' diagnostic character. The ARIMA model diagnosis revealed the best-fitting model across the entire study area.

3.4 Linear Regression Analysis

After completion of the diagnostic tests, the ARIMA model found the mean water level (MWL) of each water level well, which was compared through geostatistical analysis (goodness of fit). The linear regression least squares method was used in geostatistical analysis to find out the groundwater levels in each monitoring well character depending on how much rain fell in the Kanchipuram region (Chambers et al., 1983; Cimbala, 2013). ARIMA model parameter PDQ with (3,1,3), (2,1,2), (1,0,1), or (1,1,1) was used to compute the ARIMA model of all 53 well locations. The upper confidence level (UCL m bgl), lower confidence level (LCL m bgl), and mean water level (MWL m bgl) are explained in Table 1. Scatter plots were created in order to analyze trends, compute the linear regression equation, and identify the best fit.

4 Results and Discussion

Geological characteristics have been used to constrain the effective recharge in the Lower Palar River Basin. The study area's rainfall and water level records are statistically computed to understand the behavior, and time series ARIMA model analysis is also performed to understand the fluctuation. Time series linear regression and correlation are effectively considered in time series ARIMA models. The Upper Confidence Level (UCL m bgl), Lower Confidence Level (LCL m bgl), and Mean

Table 1 Proximity analysis of the Lower Palar River Basin—Kanchipuram district

S. No.	Rainfall station name	Water level station name/ village	Latitude	Longitude	Geology	Geomorphology	Well use	Sample collection	Stratigraphy
1	Chengalpattu	Chengalpet-police Qtrs	12.6967	79.9761	GGN				Archaean crystalline formation
2		Ichankaranai	12.7142	80.0172				WS	
3		Karumbakkam	12.7039	80.0889		BP		WLM	
4		Nenmeli	12.6658	80.0175		PE			
5		Pudupakkam	12.6311	79.9736				RS	
6		Reddikuppam	12.6972	80.0550		BP			
7		Senneri	12.6992	80.0467				WS	
8		Vadakkupattu	12.8106	79.9378		PE			
9		Venbakkam	12.7689	79.9572					
10		Arunkundram	12.7369	79.8678			IW		Archean
11		Kilakadi	12.6533	79.8922				WLM	
12		Konnamangalam	12.5156	79.3714					
13		Malaiyur	12.4939	79.4717	CGN			WLM	
14		Nedungunam	12.4686	79.3869	CNK	PE		WLM	
15		Nerkunam	12.3636	79.4944	CNK	PL			
16	Kanchieepuram	Ayyamicheri	12.8667	79.7197	CNK		IW		
17		Chettiyarpettai	12.9083	79.7219	SST	BP		WLM	Tertiary
18		Kancheepu Collector Off	12.8264	79.6944	ALV	AP			Sedimentary
19		Kancheepuram	12.8425	79.6894	AVM		DW		Recent
20		Kancheepuramwarehouse	12.8247	79.7117	SND			PS	
21		Kattavakkam	12.8336	79.8519	CNK		IW		Archean

(continued)

Table 1 (continued)

S. No.	Rainfall station name	Water level station name/ village	Latitude	Longitude	Geology	Geomorphology	Well use	Sample collection	Stratigraphy
22		Muthialpet	12.8097	79.7536	ALV				Sedimentary
23		Puliyambakkam	12.7833	79.8447			MW		Recent
24		Thangi	12.7986	79.7742			IW	PS	Sedimentary
25		Thenampakkam	12.8086	79.7447	SST	BP			tertiary
26		Uthukadu	12.8092	79.8297	CNK	PE			Archean
27	Karunkuzhi	Kinar	12.5264	79.9367	CNK			PS	
28	Kelambakkam	Ellalur	12.7333	80.1667		BP	MW	PS	ACF
29		Mambakkam	12.8375	80.1667	SDM	PL	Irrigation		Recent
30	Madurantakam	Kannimangalam	12.3969	79.8908	SND	PP		PS	
31		Pekkaranai	12.3847	79.8608					
32		Thenpakkam	12.3839	79.8117				WS	
33		Acharapakkam Mosque	12.4036	79.8164	CNK				
34		Chitravadi-3	12.4631	79.9078		PE			
35		Velamur	12.4769	79.7656					
36	Mamallapuram	Amur	12.6611	80.1408	GGN		MW		Sedimentary
37		Kottamedu	12.7217	80.1106		BP			ACF
38	Padappai	Pondur	12.7919	80.0864	GGN	PE	MW	WLM	
39		Kannivakkam	12.8017	80.0872	CNK	PP	PW		
40	Pankatucheri	Naduvakkarai	12.5519	80.0750		PE	MW		Sedimentary
41		Nathankariyachery	12.5714	80.1086		BP			
42		Nedumaram	12.4347	80.0881	SND	PP	IW		Recent

(continued)

Table 1 (continued)

S. No.	Rainfall station name	Water level station name/ village	Latitude	Longitude	Geology	Geomorphology	Well use	Sample collection	Stratigraphy
43		Palur	12.4492	80.0364	CNK	SP			Archean
44	Uthiramerur	Laddakaranai	12.5200	79.7769	CNK		IW	PS	
45		Tittalam	12.5522	79.7869				PS	
46		Malayankulam	12.7067	79.8089	TSS	BP			Tertairy
47		Marudam	12.6650	79.7797	GGN	PE			Archean
48		Pattancheri	12.6044	79.7597	CNK			PS	
49		Perunagar-2	12.6508	79.6497	GGN	PD			
51		Thirupulivanam-2	12.6508	79.7631		PE			
52		Koolamandal	12.6500	79.6667			MW	WLM	
53	Sriperumdur	Navettukulam	12.8625	79.8756	SND	SP	DW	PS	Recent

Water Level (MWL m bgl) were calculated using these three phases. The findings of the time series ARIMA model were used to compute the linear regression least squares methods. Both results are added to the ArcGIS platform so that the geological details of different theme maps can be added. These include the geological evidence map, the geomorphological evidence map, the lineament evidence map, the drainage evidence map, the slope evidence map, and the soil evidence map. The proximity analysis of the wells in the study area was carried out in order to determine the effective recharge from spatially dispersed rainfall stations. It is imperative to understand the underlying geological and geomorphological conditions to augment efficient recharge, though rainfall is the prime controlling factor. So, a spatial correlation study was intended based on the proximity of the well to the rain gauge station. For this, 5 and 10 km buffers were drawn, and the goodness of fit of the well is visually interpreted for underlying geology, geomorphology, lineament, slope, proximity to drainage, and soil. Table 1 exhibits geological parameters that control the recharge for excellent fit, good fit, poor fit, and no fit categories.

4.1 Model Derivation and Classification

The ARIMA model derived MWL (m) bgl, LCL (m) bgl and UCL (m) bgl with the goodness of fit were statistically computed from the ARIMA model obtained from Tables 1. Further, it is explained that the study area receives very little rainfall, and the water levels are recorded as time series on a monthly interval. In the Kanchipuram district, almost all rainfall stations have an average MCL water level of 3.90 m (bgl), an average LCL water level of 1.69 m (bgl), average UCL water level of 6.41 m (bgl). For example, Chengalpattu has an average MCL of 3.93 m (bgl), a LCL of 1.45 m (bgl), and a UCL of 7.34 m (bgl). Madurantakam has an average MCL of 3.85 m (bgl), LCL of 1.48 m (bgl), and UCL of 6.23 m (bgl), Kanchipuram has an average MCL of 3.63 m (bgl), a LCL of 2.06 m (bgl), and a UCL of 5.21 m (bgl). Padappai has an average MCL of 4.53 m (bgl), an LCL of 2.27 m (bgl) and a UCL of 6.8 m (bgl). Kelambakam has an average of 4.28 m (bgl), an LCL of 1.74 m (bgl), and a UCL of 6.81 m (bgl), Pankatucheri has an average MCL of 4.12 m (bgl), LCL of 1.73 m (bgl) UCL of 6.51 m (bgl) and Uthiramerur has an average of 3.8 m (bgl), LCL of 1.50 m (bgl) UCL of 6.10 m (bgl), According to the aforesaid statistics, Kanchipuram district demonstrates a good water table. For 18 well locations have no goodness of fit (no fit) and hence do not have any dependency on the rainfall of the region, 8 locations have poor fit on the linear regression scale and hence may have a relationship to the rainfall, 25 well locations have a good fit in this location hence may having the relationship between the rainfall and water level recorded and also geological formation of the region, only 2 locations have excellent fit hence in this area reveal a relationship between rainfall and water level. Thematic maps for MWL, LCL, and UCL were generated using the Geographical Information Software platform (ArcGIS) and are presented in the Fig. 4a, b and 3c, for the spatial and visual understanding of the Kanchipuram district. Groundwater goodness of fit results.

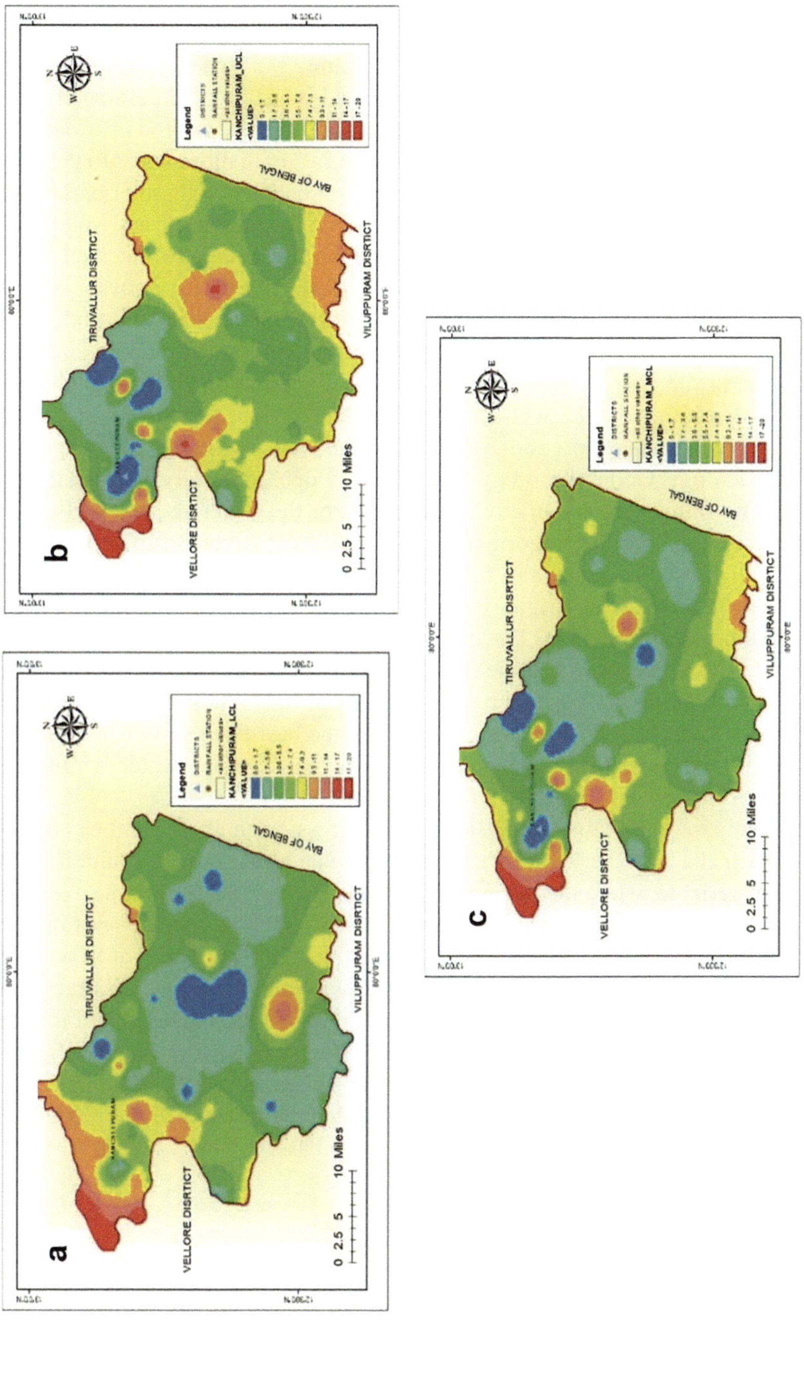

Fig. 4 **a** Mean confidence level, **b** upper confidence level, **c** lower confidence level

4.2 Proximity Analysis

The proximity analysis of wells to rain gauge stations is an arbitrary assumption that each well is equally influenced by the amount of rainfall in a particular area marked by the Thissen polygon, and is definitely controlled by geological parameters. The proximal wells get more water than the distant ones. That means the upstream one gets less water than the downstream one. The fractures drain more water to the wells and fall into the fractures. If these conditions are not encountered, other parameters like slope, lithology, lineament, geomorphology, drainage pattern and soil would have their influence in single or collective. The proximity analysis of the wells in the study area was carried out in order to determine the effective recharge from spatially dispersed rainfall stations (Fig. 5). It is imperative to understand the underlying geological and geomorphological conditions to augment efficient recharge, though rainfall is the prime controlling factor. So, a spatial correlation study was intended based on the proximity of the well to the rain gauge station. For this, 5 and 10 km buffers were drawn, and the goodness of fit of the well is visually interpreted for underlying Geology, Geomorphology, Lineament, Slope, proximity to Drainage and Soil. Table 1 exhibits geo parameters which control the recharge for excellent fit, good fit, poor fit, and no fit categories. In the Lower Palar River Basin geostatistical modelling, 11 rainfall stations and 53 water level stations were used, i.e., 3 wells are dug wells, 31 wells are irrigation wells, 17 wells are monitoring wells, and 2 wells are public wells. Of these, 23 wells are within a 5 km buffer zone, 18 wells are within a 10 km buffer zone, and the remaining 12 are outside of the 10 km buffer zone. The MWL (m) bgl, LCL (m) bgl, and UCL (m) bgl values produced from the ARIMA model were statistically computed using the best-fit method's goodness of fit. Rainfall is the primary source of groundwater recharge, where water is infiltrated by fractures into the subsurface.

Scatter plots were used to compute the linear regression analysis and the goodness of fit between the time series rainfall and the water level station. Rainfall is the independent variable, while the mean water level is the dependent variable, resulting in scatter plots between the two variables. The least squares method was used to produce the linear regression plot as well as to estimate the goodness of fit between rainfall and water levels. The standard error R2 was kept as low as feasible when interpreting the plots with the linear regression trend and the linear equation in the quadratic sense for successful validation of the interdependency between rainfall and water level, as goodness of fit. Most well locations demonstrate an excellent match to no fit, as seen by the results. It further explains that the research area is dependent on rainfall and water levels measured on a monthly basis.

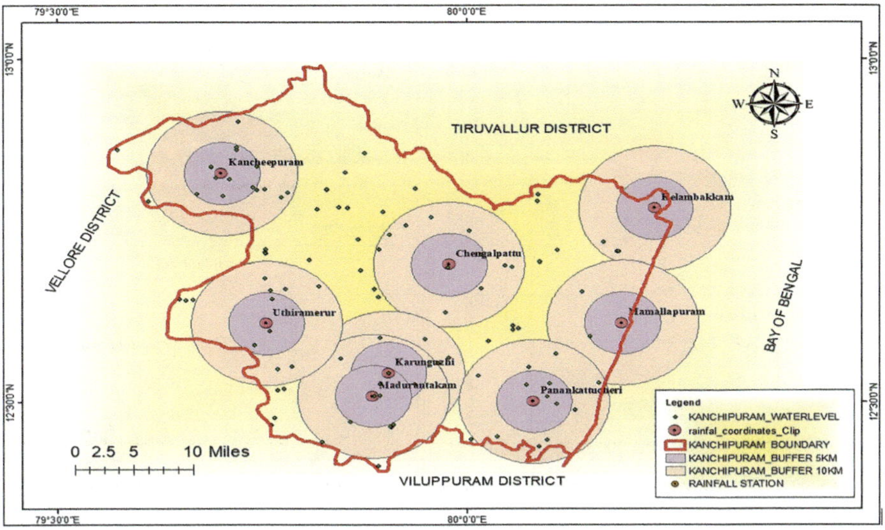

Fig. 5 Map of proximity analysis

Geological Evidence

Geologically, groundwater recharge in the study area depends on the rock occurrences in the region (Fig. 6a). The groundwater occurs in sedimentary formations, i.e., sandstone, silt, clay, fossiliferous shale, conglomerate, alluvium, and gravel that have less porosity and high permeability. The soft rocks, which are either loose or compact in nature, are more prominent in the Chengalpattu, Madurantakam, Pankatucheri, Uthiramerur, and Kanchipuram areas. However, soft rocks are considered an excellent groundwater recharge zone. The study area is mostly made up of sedimentary formations like the Gondwana formation. The results also show that in the Kanchipuram district, water level stations such as irrigation wells, dug wells, and monitoring wells are affected by rainfall stations that are spread out in different areas. Soft rock formations sink these wells. Generally, sedimentary rocks have excellent primary porosity and permeability. Excessive rainfall, coupled with the area's geological structure, results in excellent groundwater recharge zones. Conversely, locations with less rainfall and the area's geological structure result in poor or no groundwater recharge zones.

Geomorphological Evidence

Geomorphology is an important factor in the evolution of groundwater recharge. In the eastern part of the study area (Fig. 6b), you can find alluvial plains and valley-fill landforms. The porosity and permeability of these types of land are moderately good in the Maduranthakam, Uthiramerur, Vayalur, Karunkuzhi, and Kanchipuram areas.

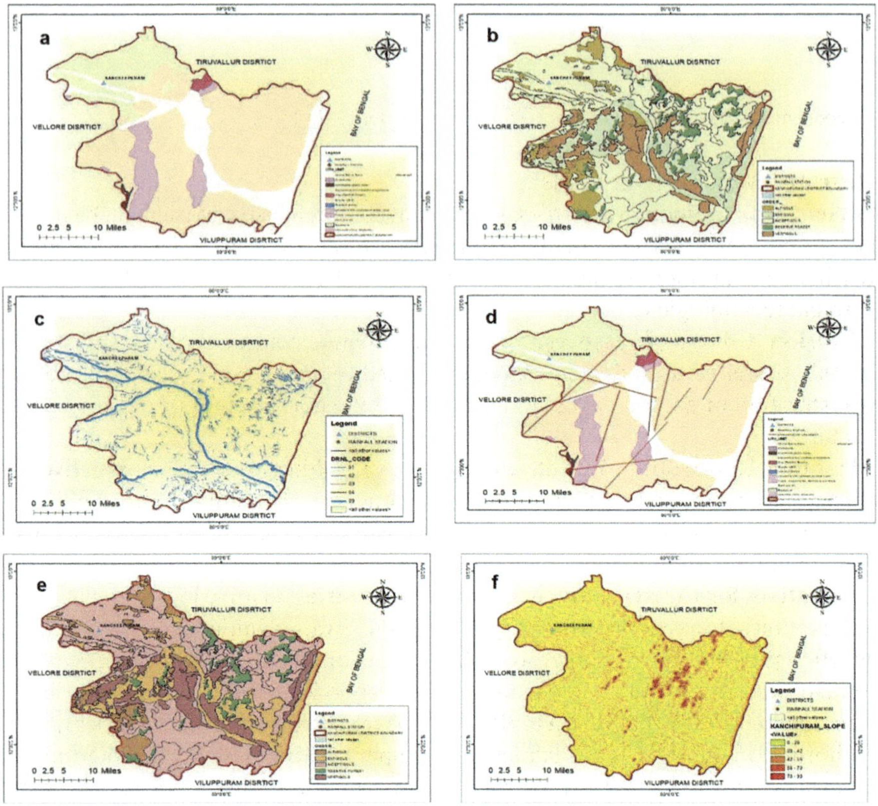

Fig. 6 **a** Geological evidence, **b** geomorphological evidence, **c** drainage evidence, **d** lineament evidence, **e** soil evidence, **f** slope evidence

Also, pediplains, pediments, and deltaic planes are mostly found in the southeast of the study area. In the Pankatucheri areas, these landforms are very porous and allow water to pass through them easily. It is noticed from the result that in the Kanchipuram district, spatially distributed rainfall stations are influencing water level stations by means of dug wells, irrigation wells, public wells, and monitoring wells. These wells are sunk in pediplains, pediments, plateaus, alluvial planes, and residual hills. Besides, it is understood from the results that the areas of Pankatucheri, Kanchipuram, Utharamerur, Klambakkam, and Sriperumdur have landforms that are pediplains, pediments, and alluvial planes. In this area, rainfall is normal to excess, while in other locations, it is very low. Because of excess rainfall, the geomorphological landforms of the area have an excellent-to- good fit, namely good groundwater recharge zones.

Drainage Evidence

Drainage is geomorphologically critical to evaluate groundwater recharge. It helps to assess the subsurface characteristics of the region's runoff, infiltration, porosity, and permeability. The study's drainage patterns depend on the topography and subsurface characteristics of the area (Fig. 6c). The region with a closeness of drainage lines is represented as high infiltration, and very low drainage lines are represented as less infiltration. The study area results reveal that 1st, 2nd, 3rd, 4th, and 9th order streams predominantly drain the Kanchipuram district. In this region, spatially distributed rainfall stations influence water level stations through the use of dug wells, irrigation wells, public wells, and monitoring wells. These wells are drilled in 1st, 2nd, 3rd, 4th, and 9th order streams. In general, drainage patterns control groundwater movement and are excellent indicators of favorable groundwater recharge zones.

Lineament Evidence

The lineaments are geologically/hydro-geologically critical; the structures controlled by linear or curvilinear features are developed by the tectonic activity. The study areas (Fig. 6d) lineaments indicate a low infiltration rate, and high lineament areas indicate a high infiltration rate. However, lineaments vary from a few meters to kilometers. The Kanchipuram location traverses additional major regional faults and fractures in the N-S, E-W, NE-SW, and NE-SW trending directions (Fig. 3). Therewith, some of the lineaments are ancient, being Precambrian Eon (Jurassic to Quarternary), whereas others are comparatively younger (Quarternary). In the study area, lineaments control groundwater movement with higher lineaments being an excellent fit to a good fit, namely good groundwater potential zones, and lower lineaments being a poor fit to no fit, namely poor groundwater potential zones.

Soil Evidence

Soil is also one of the most important factors in evaluating groundwater recharge. The alkaline and Endisols have a high infiltration rate, while vertisols, alkaline, and Alfisols have a low infiltration rate. It is noticed from the study area (Fig. 6e) that Kanchipuram district's spatially distributed rainfall stations are influencing water level stations by employing dug wells, irrigation wells, and monitoring wells. Entisols, Inceptisols, Alfisols, and Vertisols sink these wells. While soil character variations allow Inceptisols to infiltrate at high levels, other soil types significantly influence hydrological behavior. Besides, it is understood from the results that excess rainfall with Entisols and Inceptisols of the area has an excellent fit to a good fit, which means good groundwater recharge zones and some other locations. Alfisols and vertisols with less rainfall, with the soil character of the area having a poor to no fit, say poor groundwater recharge zones. Due to the excess rainfall, the soil characteristics (Entisols, Inceptisols) of the area exhibit an excellent to good fit.

Slope Evidence

The slope is one of the important factors in evaluating groundwater recharge. It defines the variation in elevation in a particular area that influences the runoff. Generally, slopes that control the runoff will have a remarkable effect on the infiltration of groundwater. The study area (Fig. 6f) has a high degree of slope in an area that has low groundwater potential due to high runoff; the low degree of slope retains the water flow, thus having a low runoff. Gentle slopes (56°–70°) have low runoff and are located in excellent groundwater resource zones; moderate slopes (42°–56°) have moderate runoff and are located in good groundwater resource zones; steep slopes (28°–42°) have high runoff and are located in poor groundwater resource zones; and very steep slopes (0°–28°) have very high runoff and are located in very poor groundwater resource zones. Because of excess rainfall, the geological setup of the area has an excellent fit to a good fit, namely good groundwater recharge zones, and some other locations with less rainfall have a poor fit to no fit with the geological setup of the area.

5 Conclusion

Groundwater is a valuable natural resource in the Palar River Basin. Furthermore, groundwater is the primary source of drinking water for people in the region, as well as an important source of water for irrigation and industry. It also provides a relatively consistent supply of water in some ecologically sensitive areas, allowing plant and animal species to grow. As the world's population increases and moves, so does the demand for these limited resources. Groundwater is frequently overexploited in a variety of ways, which must be recognized and measured if water resources are to be managed responsibly and sustainably. We statistically model the spatially distributed rainfall stations and water level variables in the Palar River Basin as an average rainfall versus an average water level. In the prevailing study area, there are two different types of monsoons: NEM and SWM. This season's maximum rainfall is 1395 mm from the Panpakkam rainfall station, and the minimum rainfall is 460 mm from the Arcot station. The study area's average annual rainfall is 958 mm. About 50% of the total annual average rainfall is received during the northeast monsoon, about 40% is received during the southeast monsoon, and the remaining 10% is split between both monsoon seasons. Kanchipuram district receives more rainfall from the northeast monsoon because of its proximity to the Bay of Bengal. It also has a good water table because most of the area is covered by sedimentary terrain. Many applications are used in the study area to constrain effective recharge, including the Theissen polygon method, time series analysis, the ARIMA model, and ArcGIS techniques. 11 rainfall stations and 53 water level stations are established. These wells result in a goodness of fit of excellent fit—40, good fit—108, poor fit—58, and no fit—106. Furthermore, geological evidence, geomorphological evidence, lineament evidence, drainage evidence, slope evidence, and soil evidence superimpose the goodness of fit results. It is identified that about 12.82% of the total area falls under the 'Excellent Fit,' 34.62% of the total area falls under the 'Good Fit,' 18.59% of the total area falls under the 'Poor Fit,' and 33.97% of the total area falls under the 'No Fit' groundwater recharge zones. The proximity analysis of wells to rain gauge stations is based on the assumption that each well is equally influenced by the amount of rainfall in a particular area marked by the Thiessen polygon and definitely controlled by geoparameters like lithology, geomorphology, slope, fractures, drainage patterns, structures, soil, etc. The proximal wells get more water than the distant ones. The upstream one gets less water than the downstream one. The fractures drain more water to wells where it falls on the fractures. If these conditions are not met, other parameters such as slope, lithology, geomorphology, and drainage pattern will have an impact, either individually or collectively.

References

Aflatooni, M., & Mehdi, M. (2011). Time series analysis of ground water table fluctuations due to temperature and rainfall change in Shiraz plain. *International Journal of Water Resources and Environmental Engineering, 3*(9), 176–188.

Box, G. E. P., Jenkins, G. M., & Reinsel, G. C. (1976). *Time series analysis, forecasting and control* (3rd ed., p. 592). Prentice Hall.

Box, G. E. P., Jenkins, G. M., Reinsel, G. C., & Ljung, G. M. (2015). *Time series analysis: Forecasting and control* (5th ed., p. 712). Wiley Publication.

Brockwell, P. J., & Davis, R. A. (1987). *Time series: Theory and method* (1st edn., p. 519). Springer Publication.

CGWB (2014). Dynamic Ground Water Resources of India (March 31, 2011) Ministry of Water Resources River Development & Ganga Rajuvenation. Government of India. Retrieved from: http://www.cgwb.gov.in/Documents-GW-Resourcces-2011.pdf, 11th April, 2017.

Chambers, J. M., Cleveland, W. S., Kleiner, B., & Tukey, P. A. (1983). *Graphical methods for data analysis* (1st edn., p. 336), Chapman and Hall/CRC.

Chitra, N., Shanmathi, R., & Rajesh, R. (2015). Application of ARIMA model using SPSS software—A case study in supply chain management. *International Journal of Science, Technology and Management, 4*(1), 206–215.

Cimbala, J. M. (2013). *Regression analysis.* Penn State University. https://online.stat.psu.edu/stat501/lesson/14

DPAP. (2008). Report of the technical committee on drought prone areas programme and desert development programme. *Annual Report by Ministry of Rural Development*. https://dolr.gov.in/en/drought-prone-areas-programme-dpap

Haan, C. T. (1977). *Statistical methods in hydrology* (pp. 378–391). Iowa State University Press.

IWMI. (2007). International Water Management Institute (IWMI). Technical reports on Salinity Management in Pakistan: multiple issues, IWMI, Colombo, 65.

IWMI. (2007). *Helping the world adapt to water scarcity*, Annual Report, 1–28.

Krishnamurthy, J., Kumar, N. V., Jayraman, V., & Manivel, M. (1996). An approach to demarcate groundwater potential zones through remote sensing and GIS. *International Journal of Remote Sensing, 17*, 10–12.

Manjunathan, N., & Thirukumaran. V. (2019). An impact of rain fall in ground water level fluctuation and prediction of rainfall at Kanchipuram district, Tamil Nadu, India. *International Journal for Research in Applied Science & Engineering Technology (IJRASET), 7*(3), 573–581.

Manjunathan, N., Thirukumaran, V., & Thirumurugan, V. (2021). Identification of groundwater potential zones using time series analysis and GIS applications in Palar river basin, Tamil Nadu, India. *NOVYI MIR Research Journal, 6*(4), 88–109.

Pankratz, A. (1983). Forecasting with univariate box Jenkins models: Concepts and cases. *Wiley Series in Probability and Statistics*, 562–568.

Preeja, K. R., Joseph, S., Thomas, J., & Vijith, H. (2011). Identification of groundwater potential zones of a tropical river basin (Kerala India) using remote sensing and GIS techniques. *Journal of the Indian Society of Remote Sensing, 39*(1), 83–94.

Ragunath, H. M. (1987). *Groundwater* (p. 563). Wiley Eastern Ltd.

Scott Armstrong, J. (2001). *Extrapolation for time-series and cross-sectional data, principles of forecasting: A handbook for researchers and practitioners* (pp. 217–243). Springer.

Singh, A., Srivastav, S. K., Kumar, S., & Chakrapani, G. J. (2015). A modified DRASTIC model (DRASTICA) for assessment of groundwater vulnerability to pollution in an urbanized environment in Lucknow, India. *Environmental Earth Science, 74*(7), 547–5490. https://doi.org/10.1007/s12665-015-4558-5

Srivastava, P. K., & Bhattacharya, A. K. (2006). Groundwater assessment through an integrated approach using remote sensing, GIS and resistivity techniques: A case study from a hard rock terrain. *International Journal of Remote Sensing, 27*(20), 4599–4620.

Thiessen, A. H. (1991). Precipitation averages for large areas. *Monthly Weather Review, 39*, 1082–1084.

Thirumurugan, V., Thirukumaran, V., Sridharan, V., & Manivannan, V. (2016a). Morphometric analysis for geohydrological studies of upper Amaravathi basin using geo spatial technology. *International Journal of Current Advanced Research, 5*, 1427–1436.

Thirumurugan, V., Thirukumaran, V., Sridharan, V., & Manivannan, V. (2016b). ARIMA modelling of water table fluctuations due to rainfall variations in Shanmuga Nadhi sub-basin of upper Amaravathi Basin, Tamil Nadu, India. *International Journal of Current Advanced Research, 7*, 14149–14164.

UNDP GWF. (2014). *Regional capacity building programme, promoting and developing water integrity in sub-Saharan Africa.* Interim Programme Report. UNDP, 1–128.

UNDP, Annual Report. (2023). *Human development report 2006: Beyond scarcity: Power, poverty and the global water crisis*, New York, 440.

UNEP. (2006). Annual Report, 1–87.

UNEP. (2008). *Vital water graphics: an overview of the state of the world's fresh and marine waters* (2nd edn., pp. 1–84) UNEP.

Simulation of Groundwater Flow in Coastal Aquifer of Ramanathapuram, Tamil Nadu, India

P. Thambidurai, Mathiazhagan Mookiah, and Anil Kumar Dikshit

1 Introduction

Groundwater is a vital resource that supports the livelihood of human communities. The coastal aquifer is a critical water resource for coastal regions, providing water to over a billion people globally (Antonellini et al., 2008; Ferguson & Gleeson, 2012; Brian, 2016), and serves as a link between the oceanic and hydrologic systems (Moore, 1996; Wetzelhuetter, 2013). The hydrogeological characteristics of aquifers are shaped by geological factors (Kumar & Thambidurai, 2022, 2023); in addition, the coastal aquifers are characterised by geological conditions, mixing zone, sea-level fluctuations (both short and long-term), and salinity-driven density gradients (Han et al., 2017). Groundwater is a valuable and accessible source of fresh water (Trenberth et al., 2007; Mishra, 2023). However, coastal groundwater is highly susceptible to degradation due to climate change, sea level rise, rapid urbanization, economic activities, and limited surface water availability (Jelgersma et al., 1993; Arnell, 1999; Alfarrah et al., 2017). In coastal aquifer systems with minimal topographic gradients, the impact of groundwater withdrawal is often more significant than that of sea-level rise or changes in groundwater recharge, due to the dynamic interaction between surface water and groundwater (Ferguson & Gleeson, 2012; Khairy & Janardhana, 2013). The over-extraction of groundwater leads to increased salinity through saltwater intrusion, rendering the aquifer unsuitable for use without advanced treatment (Oude Essink et al., 2010; Yin et al., 2022). Thus, implementing robust monitoring

P. Thambidurai (✉) · M. Mookiah
Department of Coastal Disaster Management, School of Physical, Chemical and Applied Sciences, Pondicherry University - Port Blair Campus, Sri Vijaya Puram, Andaman and Nicobar Islands, India
e-mail: thambiduraip@pondiuni.ac.in

M. Mookiah · A. K. Dikshit
Environmental Science and Engineering Department, Indian Institute of Technology Bombay, Mumbai, Maharashtra, India

P. Thambidurai and A. K. Dikshit (eds.), *Coastal Environments of India*,
https://doi.org/10.1007/978-3-031-97341-3_7

systems, such as numerical modelling, is crucial to determining sustainable pumping rates and mitigating seawater intrusion in coastal aquifers (Singh & Panda, 2013; Ayinippully Nalarajan et al., 2022). Various methods and approaches, such as hydrochemical evaluations, formulation of water quality indices, statistical techniques, and hydrological simulations, are employed to assess and track water resources within coastal aquifers. Groundwater models serve as a conceptual framework for representing physical systems through mathematical equations (Wang & Anderson, 1995; Roy & Datta, 2017). These models typically rely on three main techniques: the finite element method (Gelinas et al., 1981; Voss, 1984; Istok, 1989), the analytical element method (Freeze & Witherspoon, 1966; Jin et al., 2020), and the finite difference method (Trescott et al., 1976; McDonald & Harbaugh, 1986; Alam et al., 2021). For instance, Essink (2001) developed a density-dependent groundwater flow model to study seawater intrusion in the coastal aquifers. Likewise, the MOCDENS3D numerical model has been employed to analyse variations in coastal groundwater flow (Cheng & Chen, 2001; Bakker et al., 2004; Vandenbohede et al., 2009). Numerical models are extensively used to assess regional groundwater budgets across various aquifer systems (Elango & Sivakumar, 2008; Majumdar et al., 2009; Rani & Chen, 2010; Varalakshmi et al., 2014; Lin et al., 2023), assess the impacts of proposed developments, establish links between aquifer boundaries and specific locations, and further evaluate the groundwater quality and natural recharge rates (Ali et al., 2012; Mathiazhagan et al., 2021).

The Visual MODFLOW software package is globally popular for groundwater flow simulations due to its user-friendly and robust design (Rejani et al., 2008; Chen & Yang, 2011). The SEAWAT model, extensively applied in hydrogeological studies, enables the simulation of groundwater flow influenced by density variations and solute transport, making it effective for investigating saltwater intrusion dynamics (Rao et al., 2004; Badaruddin et al., 2022). A three-dimensional finite element model offers a more sophisticated method for simulating saltwater intrusion within both individual and interconnected coastal aquifer systems (Huyakorn et al., 1987; Senthilkumar & Elango, 2004; Lal & Datta, 2017). Integrating groundwater models with management models proves valuable for optimizing groundwater resource utilization and making informed decisions (Nalbantis et al., 2002; Yeh, 2015). These models have the capacity to simplify aquifer systems, albeit with some limitations, allowing for predictions of groundwater conditions and strategic planning for future use. Several decades ago , a sharp decline in water levels was observed across various regions of India, primarily driven by the intensified extraction of groundwater. The coastal region of Ramanathapuram in Tamil Nadu is one of the districts that has been facing water scarcity every year during the summer season. Government officials view postings in this district as punishment during their service period, which is the reason behind the water crisis. This study focused on developing a numerical model aimed at analyzing groundwater flow dynamics and the characteristics of shallow aquifers within the coastal area. The MODFLOW software was utilized for this study due to its flexible modular design, which effectively represents hydrogeological conditions. Moreover, MODFLOW, an open source and widely available software, is highly effective for computing water balance in small-scale aquifer systems. This technique supports the efficient use of shallow groundwater to help curb seawater intrusion. Being a groundbreaking effort in the area, the outcomes of this study are likely to play a vital role in guiding authorities toward more effective water management strategies.

2 Study Area

The research focuses on the coastal region of Ramanathapuram district, covering 2838 km^2 (Fig. 1); it covers between 9°09′15″ to 9°50′45″ N latitude and 78°10′ 30″ to 79°10′ 15″ E longitude. The area mapped falls within the Survey of India toposheets numbered 58 K/11, K/12, K/14, K/15, K/16, 58 O/3, and 58 O/4 at a 1:50,000 scale. The district features a coastline of 271 km, with 141 km falling under the Gulf of Mannar and the remaining portion under Palk Bay. The region experiences average temperatures ranging from 25.9 to 30.6 °C, with relative humidity hovering around 79%. The peak heat occurs between April and June, when temperatures typically climb by about 4 °C, coupled with elevated humidity. Owing to its geographic location, the cold season is short-lived, spanning only December and January. The area records an average yearly rainfall of approximately 827 mm. The majority of the area is covered by black clayey soil, which is predominantly black or blackish-brown in colour. This soil type is found in parts of Kamuthi, Ramanathapuram, Tiruvadanai, Mudukulathur, and Paramakudi blocks. Sandy soils are present in plain terrains along the Kadaladi and Rameshwaram blocks, while alluvial soils are found along the courses of the Vaigai and Gundar rivers, covering areas in Paramakudi, Tiruvadanai, and Muthukulathur blocks. Additionally, red ferruginous soil, characteristic of the Chettinad plains, appears in small pockets around the Paramakudi and Tiruvadanai blocks.

2.1 Geology and Hydrogeology

Geologically, the district is predominantly composed of sedimentary formations, including alluvium and Tertiary deposits, with only about 1% consisting of hard rock formations such as quartzite and gneiss. The study region exhibits varied physiographic features, with elevation levels spanning from 1 to 250 m above mean sea level (MSL). The entire region is covered by sedimentary formations from the Quaternary period, including fluvial, fluvio-marine, aeolian, and marine deposits composed of sand, silt, clay, calcareous sandstone, and fossiliferous sandstone. Tertiary deposits mainly consist of sandstone and claystone, while Cretaceous formations, primarily limestone, underlie them. Additionally, Archaean basement rock is reported at varying depths in the area. The aquifer system extends from 3 to 116 m in depth, with hydraulic conductivity ranging between 20 and 40 m per day (CGWB, 2009). The region's permeable geological formations are classified

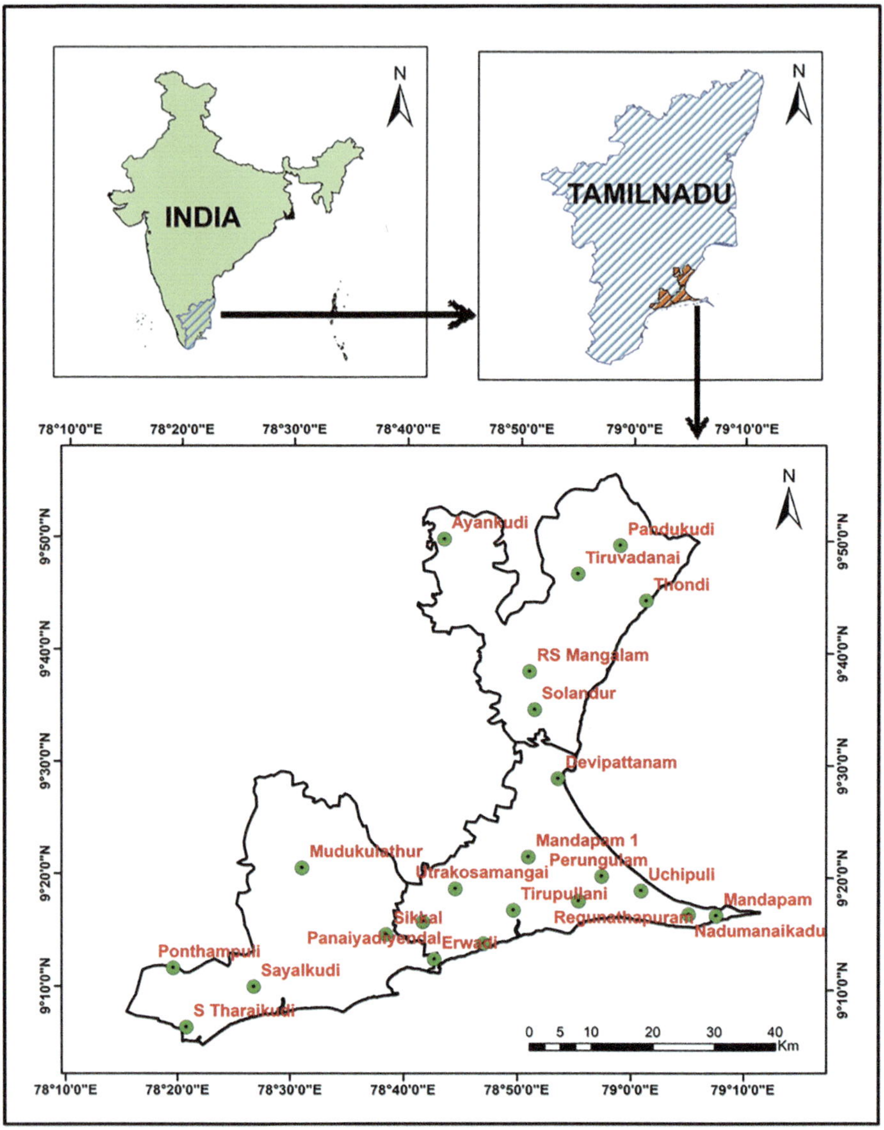

Fig. 1 Base map depicting study area and secondary observation well locations

into three distinct aquifer systems: Cretaceous, Tertiary, and Quaternary deposits. The Cretaceous aquifer, which includes both semi-confined and confined zones, comprises an upper stratum of compact, red, fossil-bearing sandstone and a lower layer characterized by pink or grey sandstone interlayered with shale. Owing to its dense sandstone makeup, this aquifer exhibits limited groundwater potential. In this area, groundwater occurs in two types: an unconfined layer with a thickness ranging from 15 to 20 m, and a confined layer exceeding 20 m in thickness. Unconfined aquifers can be developed through prospecting and drilled wells (Karthikeyan et al., 2022).

3 Groundwater Modelling

Visual MODFLOW, which incorporates the modular 3D finite-difference groundwater flow code developed by McDonald and Harbaugh (1986), is employed to simulate groundwater movement. A variety of numerical modelling techniques are utilized to represent groundwater flow patterns at both local and regional scales. In this study, the groundwater flow equation (Eq. 1) implemented in MODFLOW is applied to model three-dimensional flow (Anderson & Woessner, 1992):

$$\frac{\partial}{\partial x}\left(k_x\frac{\partial h}{\partial x}\right)+\frac{\partial}{\partial y}\left(k_y\frac{\partial h}{\partial y}\right)+\frac{\partial}{\partial z}\left(k_z\frac{\partial h}{\partial z}\right)=S_s\frac{\partial h}{\partial t}-R \tag{1}$$

where

- k_x, k_y, and k_z denote hydraulic conductivity in the x, y, and z directions, respectively.
- S_s represents specific storage,
- h is the hydraulic head,
- R indicates recharge or discharge sources.

Visual MODFLOW applies the finite-difference method (Eq. 2), assuming constant groundwater viscosity and density under transient flow conditions (Fetter, 2001):

$$k_x\frac{\partial^2 h}{\partial x^2}+k_y\frac{\partial^2 h}{\partial y^2}+k_z\frac{\partial^2 h}{\partial z^2}\pm W=S_s\frac{\partial h}{\partial t} \tag{2}$$

To solve the governing flow equations, Visual MODFLOW offers several computational solvers, including the Preconditioned Conjugate Gradient (PCG), Strongly Implicit Procedure (SIP), Slice Successive Over-Relaxation (SSOR), the Geometric Multigrid (GMG), and the WHS solver specifically developed for Visual MODFLOW. In this research, the WHS solver, enhanced by the Bi-Conjugate Gradient Stabilized (Bi-CGSTAB) method, is used to iteratively solve the partial differential equations that describe groundwater flow.

3.1 Model Discretization

The groundwater flow model for the Ramanathapuram district coastal area is developed to represent its physical conditions. The model includes a simulation for calibration over five years (2014–2018) and validation for the subsequent five years (2019–2023). The study involves the conceptualization of the aquifer system with two layers, defined based on geological and lithological characteristics. The numerical model is developed by converting the conceptual model through a series of preparatory steps. The study area spans 2838.16 km^2 and is discretized into 11,788 grid cells organized in 200 rows and 200 columns. Each cell measures 490 m × 490 m. The model includes two layers: the first layer's thickness ranges between 10 and 18 m, while the second layer varies from 7 to 15 m. Excel files were used to import ground elevation and layer elevation data into Visual MODFLOW. For this study, hydraulic head data from 23 observation wells were obtained from the India WIRS website, focusing on the Ramanathapuram coastal aquifer system (Fig. 1). The measured hydraulic head varies between 0.5 m near the coast and up to 10 m further inland. Groundwater flow simulation was initiated from the base date of January 1, 2014. For model calibration and validation, four annual hydraulic head datasets covering the period from 2014 to 2023 were employed. The model domain comprises five distinct zones with varying hydraulic conductivity values and specific storage characteristics assigned to each layer, as illustrated in Fig. 2a and b. Horizontal hydraulic conductivity ranges from 0.6 to 8 m per day, while vertical conductivity lies between 0.06 and 0.8 m per day (Fig. 3a). Specific storage values range between 8×10^{-6} and 0.001, while specific yield varies between 0.05 and 0.28. Effective porosity and total porosity are set at 0.15 and 0.30, respectively, in Fig. 3a and b. Two types of boundary conditions are applied in the model: the Bay of Bengal is treated as a constant head boundary, while natural rainfall infiltration is modelled as a recharge boundary, as depicted in Fig. 4.

3.2 Calibration and Validation

Calibration fine-tunes the groundwater model to mirror real-world conditions by matching computed hydraulic head values with observed measurements. In this study, calibration was performed using head data collected between 2014 and 2018. The calibration was carried out using an iterative trial-and-error approach, in which uncertain hydrogeological parameters like hydraulic conductivity and storage coefficientswere gradually refined through systematic adjustments. This refinement sought to reduce discrepancies between measured and simulated hydraulic head values, enhancing model precision. Calibration was conducted under transient conditions, replicating groundwater flow dynamics over a five-year span from January 2014 to December 2018. A successful calibration was achieved when the simulated head

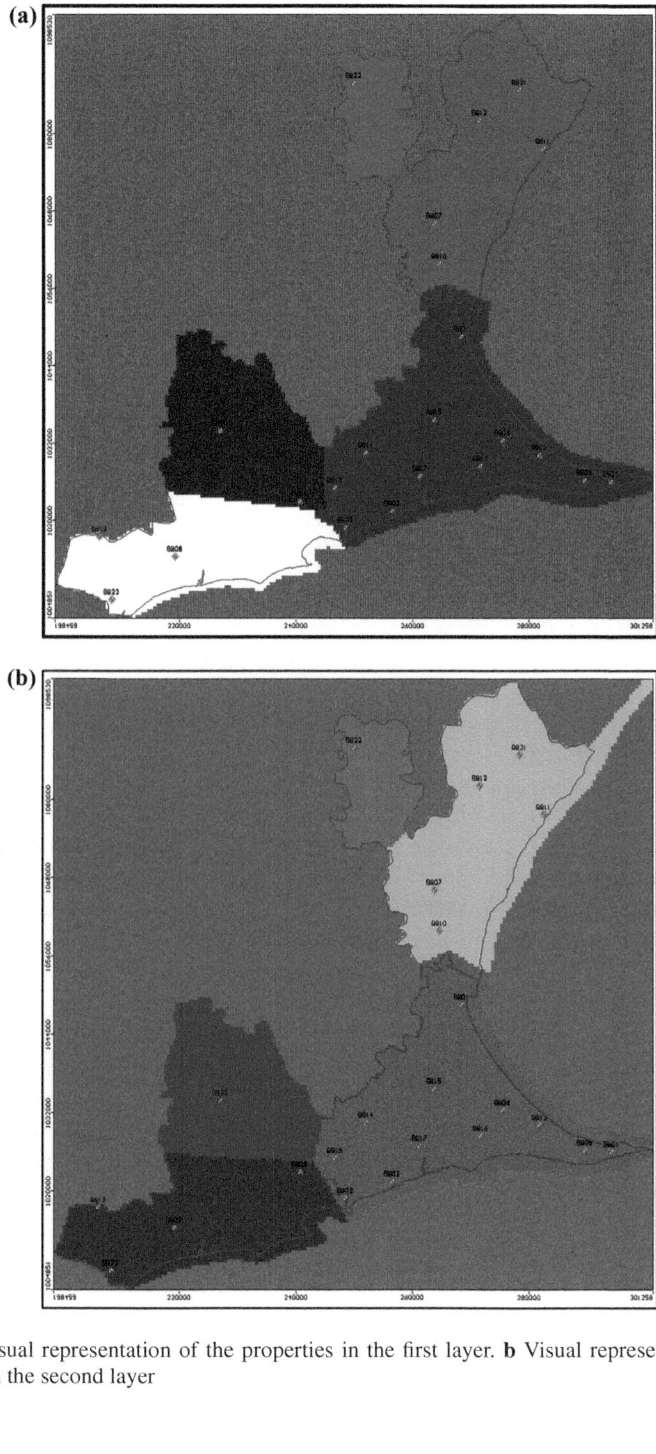

Fig. 2 **a** Visual representation of the properties in the first layer. **b** Visual representation of the properties in the second layer

(a)

Conductivity

Zone	Kx [m/d]	Ky [m/d]	Kz [m/d]	Active
1	8	8	0.8	☑
2	4.5	4.5	0.45	☑
3	1	1	0.1	☑
4	4.1	4.1	0.41	☑
5	2.9	2.9	0.29	☑
6	6	6	0.6	☑
7	2.8	2.8	0.28	☑
8	0.6	0.6	0.06	☑
9	0.7	0.7	0.07	☑
10	1.6	1.6	0.16	☑

(b)

Storage

Zone	Ss [1/m]	Sy []	Eff. Por. []	Tot. Por. []	Active
1	7E-5	0.28	0.15	0.3	☑
2	0.0003	0.18	0.15	0.3	☑
3	0.001	0.09	0.15	0.3	☑
4	0.0008	0.22	0.15	0.3	☑
5	3E-5	0.15	0.15	0.3	☑
6	8E-6	0.19	0.15	0.3	☑
7	0.0005	0.11	0.15	0.3	☑
8	1E-6	0.08	0.15	0.3	☑
9	0.0001	0.05	0.15	0.3	☑
10	9E-5	0.19	0.15	0.3	☑

Fig. 3 **a** Hydraulic conductivity properties: values derived from this study. **b** Storage properties: values derived from this study

Fig. 4 Rainfall zones classified using Thiessen polygons and constant head boundary assignment in the eastern coastal zone

values closely aligned with the observed data (Fig. 5a). This agreement was evaluated quantitatively through statistical indicators, including mean error, mean absolute error, and root mean square error. These indicators offered a precise evaluation of the model's effectiveness in replicating real groundwater behaviour.

After completing the calibration, the model underwent validation to verify its reliability and predictive capability. Validation involved simulating groundwater flow using hydraulic head data from January 2019 to December 2023. This step tested the model's ability to reproduce observed data in a different time frame, using the calibrated parameters without further adjustment. The validation process confirmed whether the model could predict future conditions accurately, thereby establishing its utility for groundwater management in this region (Fig. 5b). For the calibrated models, the correlation coefficient ranges from 0.957 to 0.999, whereas for the validated models, it falls between 0.966 and 0.978.

4 Results and Discussion

4.1 Qualitative Analysis

Groundwater level fluctuations serve as a clear indicator of head equipotential, which reflects the variation in pressure and elevation of the groundwater table across the study area. Figure 6a and b illustrate the spatial distribution of head equipotential lines, which map the contours of equal groundwater pressure at two distinct time points. Figure 6a corresponds to the final stage of the model calibration process, while Figure 6b represents the conclusion of the validation phase, capturing the changes in groundwater levels as the model's accuracy is refined. In both figures, the northwest region of the study area exhibits groundwater levels greater than 25 m, as indicated by brown and red colours. These areas with higher groundwater elevations suggest the presence of aquifers with more significant groundwater storage or recharge, possibly influenced by topographic features, geological formations, or hydrological conditions conducive to water accumulation. Such elevated water tables in the northwest may also correlate with natural recharge zones or areas with less abstraction or disturbance. Conversely, the eastern portion of the study area shows groundwater levels below 5 m, represented by blue shades. These lower water table elevations could be indicative of areas with limited groundwater storage, potentially influenced by factors such as drought conditions, lower precipitation, or over-extraction of groundwater resources. The blue-shaded regions could also correspond to more permeable zones where water quickly drains away, resulting in a shallower groundwater table. These outputs provide a detailed visual representation of how groundwater levels vary spatially across the study area. They reflect the underlying geological and hydrological differences, such as variations in rock formations, soil permeability, and recharge rates, as well as anthropogenic influences like water usage and land development. This spatial information is essential for understanding the

(a)

(b)

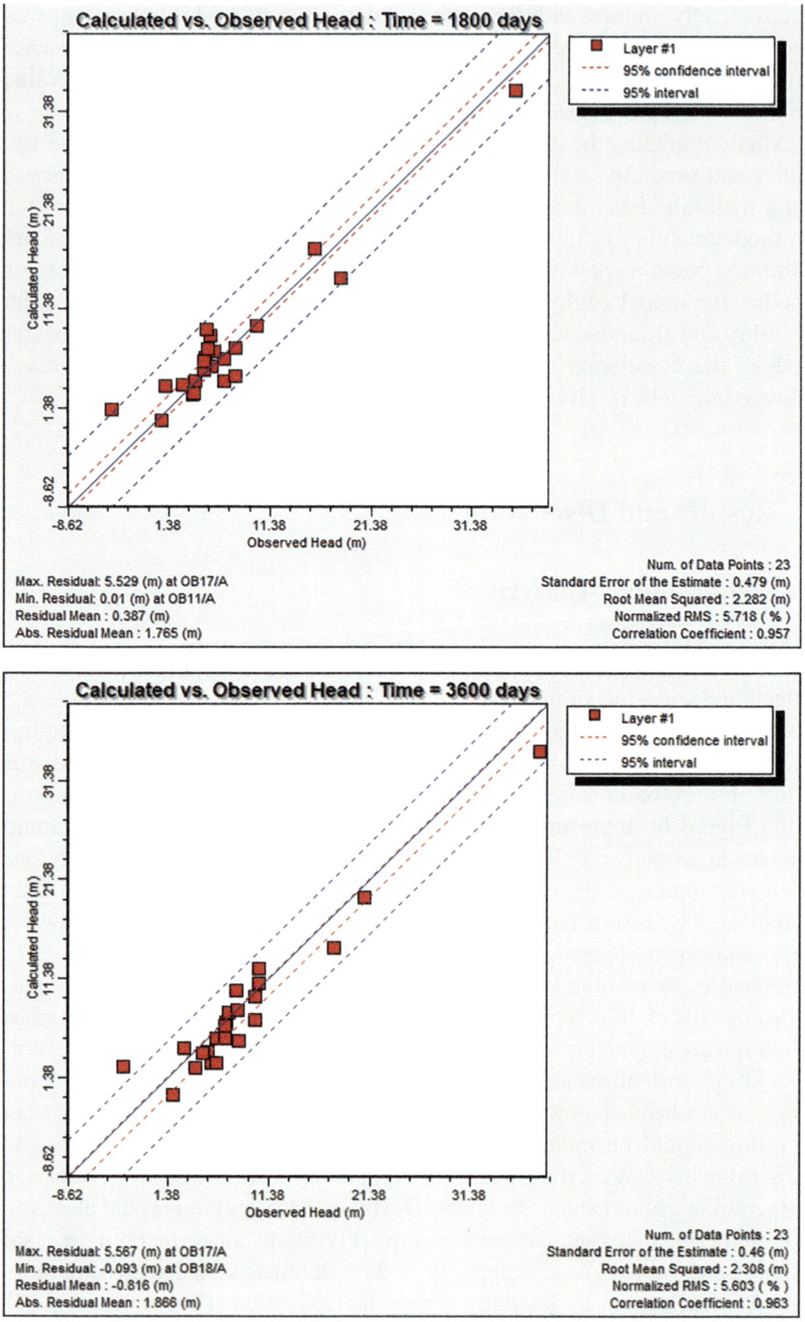

Fig. 5 **a** Comparison of computed head and observed head under transient state conditions during model calibration. **b** Assessment of observed versus calculated head under transient state conditions during model validation

dynamics of groundwater flow, predicting future water table changes, and managing groundwater resources more effectively. The clear differentiation between the high and low groundwater levels helps in identifying areas of concern for water resource management, conservation, and planning.

The selected wells, depicted in Fig. 7a and b, provide a detailed time series analysis of groundwater level fluctuations observed across various time intervals. The data illustrate how the groundwater levels in these wells change over time, with measurements taken at different stages of the study. The time series analysis reveals a significant correlation between the observed groundwater levels (measured directly in the field) and the modelled groundwater levels (predicted by the simulation model). This strong correlation suggests that the model accurately reflects the actual fluctuations in groundwater levels over the study period. The consistency observed between the observed and modelled data is particularly notable, as it indicates the success of the calibration process in aligning the model's predictions with real-world observations. This alignment is a critical indicator of the model's accuracy and reliability. It reinforces the fact that the model has been calibrated effectively to simulate the natural dynamics of groundwater flow and storage, with minimal discrepancies between the observed and predicted values. Furthermore, this level of accuracy highlights the model's robustness and its potential for real-world applications. By accurately replicating the behaviour of groundwater systems, the model serves as an effective instrument for forecasting future groundwater levels, understanding long-term trends, and making informed decisions in groundwater resource management. This high degree of model performance instils confidence in its ability to simulate real-world conditions across different wells, ensuring that it can be reliably used for various hydrological assessments and water resource planning in similar contexts.

Figure 8a, b display two separate maps representing the velocity directions of groundwater flow, specifically inward and outward directions, which are key to understanding the dynamics of groundwater movement through MODFLOW models. These visualizations are vital for hydrological modelling as they help illustrate how groundwater flows within the modelled area. The maps use a colour-coded system to distinguish between the different flow directions: blue represents outward flow (moving upward when viewed by layer), maroon indicates inward flow (moving downward by layer), and green indicates plane flow (Waterloo Hydrogeologic Inc., 2002). This colour-coding system enhances the user's ability to interpret the direction of groundwater movement quickly. By visually differentiating the flow patterns, these maps provide a clearer understanding of groundwater behaviour, which is crucial for modelling groundwater systems, predicting flow paths, and assessing the impact of various hydrological conditions on water movement within the study area. The eastern section of the coastal boundary, particularly near the areas marked by the blue and green arrows, indicates a potential risk of seawater intrusion. To assess this, a detailed groundwater quality analysis shall be conducted to confirm the extent of seawater intrusion. Based on the findings, appropriate recharge structures may be identify and implemented to mitigate the impact and enhance groundwater sustainability in the study area.

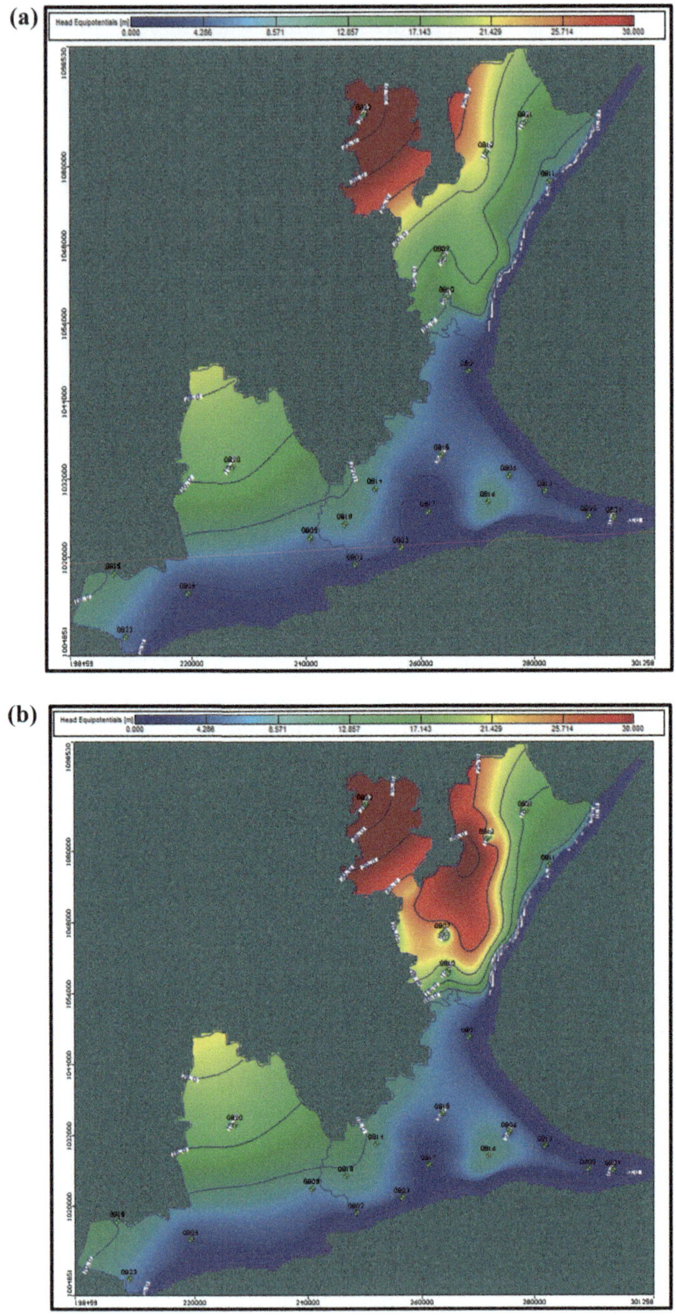

Fig. 6 **a** Spatial distribution of head equipotential map 1800 days. **b** Spatial distribution of head equipotential map 3600 days

(a)
(b)

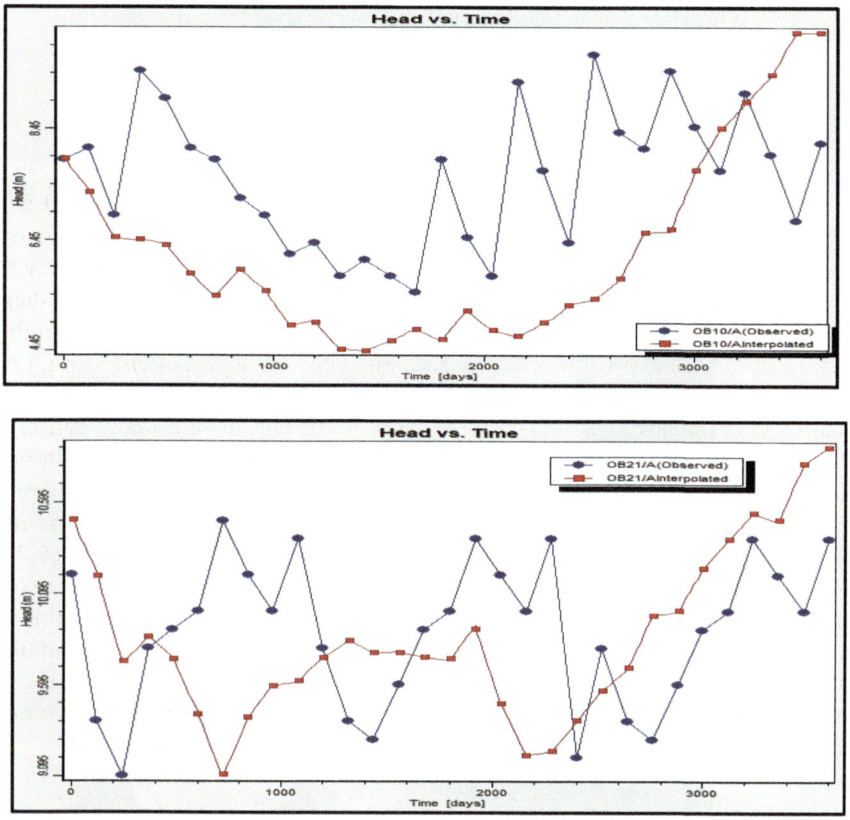

Fig. 7 Groundwater level fluctuations in observation wells **a** no. 10 and **b** no. 21 over the period of study

4.2 Quantitative Analysis

Interpreting the mass balance budget in Visual MODFLOW involves examining the quantitative aspects of water movement within a groundwater system. By analyzing this data, one can gain a deeper understanding of groundwater flow dynamics and make more informed decisions regarding water resource management. The mass balance provides a clear picture of the water inputs, outputs, and storage changes within a defined system. As shown in Table 1, the mass balance includes two inflow parameters, three outflow parameters, and the total amount of water available during both calibration and validation phases. Visual MODFLOW aids in this process by incorporating mass balance calculations, which are crucial for verifying that the model accurately represents real-world hydrological conditions. This ensures the model's reliability and its ability to simulate groundwater movement in a way that aligns with observed data.

(a)

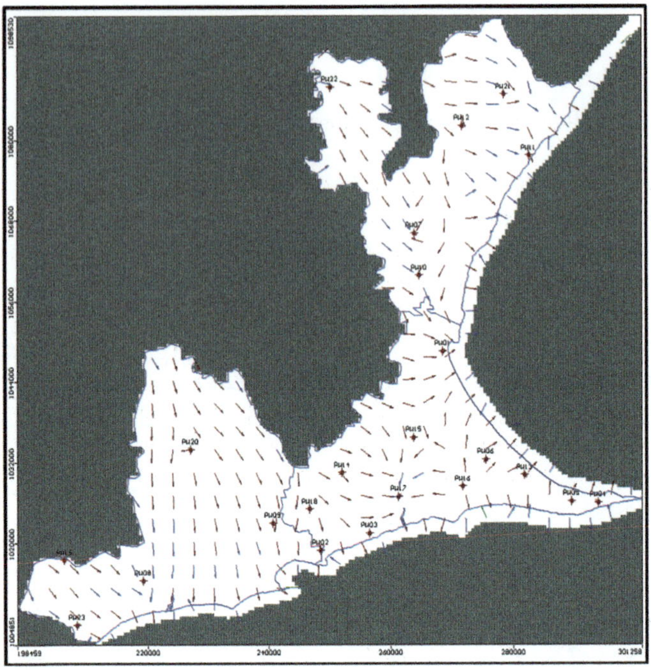

(b)

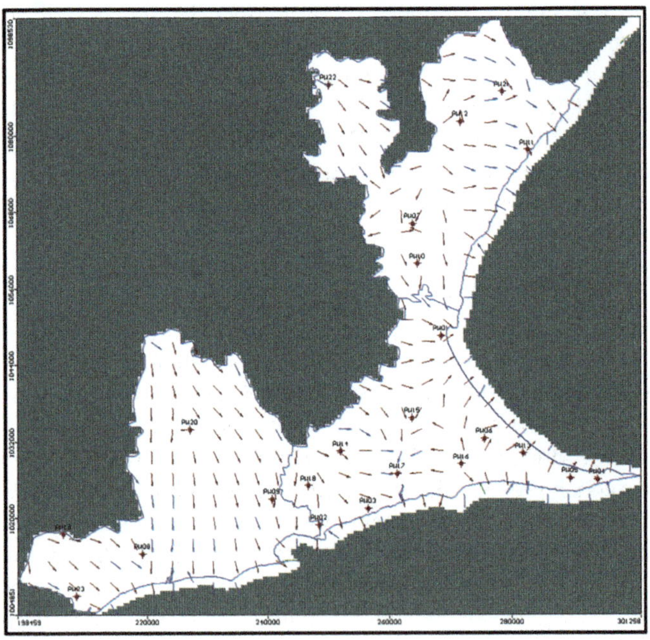

Fig. 8 **a** Velocity directions of end of the calibration period. **b** Velocity directions of end of the validation period

Table 1 Mass balance of the calibration and validation water inflow and outflow (2014–2023)

Inflow	Calibration (Mm^3)	Validation (Mm^3)	Outflow	Calibration (Mm^3)	Validation (Mm^3)
Storage	45.49	23.84	Storage	516.53	380.88
Recharge	587.12	456.15	Wells	12.44	13.56
			Constant head	103.64	85.55

5 Conclusion

The simulation of groundwater flow using a 3D model clearly indicates a west-to-east flow pattern within the study area aquifer. Elevated groundwater levels observed in the northwest suggest significant recharge zones and substantial aquifer storage, whereas lower levels in the eastern regions may point to over-extraction or limited recharge. The direction of groundwater flow velocity helps identify both freshwater potential zones and areas susceptible to seawater intrusion. The indication of groundwater flow from sea to landward could be seawater intruding into the freshwater aquifer. Additionally, mass balance calculations provide a quantitative analysis of the system's water inputs, outputs, and storage changes, ensuring that the model accurately represents observed hydrological conditions. These findings highlight the critical role of groundwater modelling in resource management, supporting sustainable water use strategies and informed decision-making for future water resource planning.

References

Alam, M. J., Loskor, W. Z., & Mohiuddin, M. (2021). Time dependent wave propagation modeling using finite difference scheme of 2D wave equation based on absorbing and reflecting boundaries. *Journal of Applied Mathematics and Physics, 09*(09), 2334–2344. https://doi.org/10.4236/jamp.2021.99148

Alfarrah, N., Berhane, G., Bakundukize, C., & Walraevens, K. (2017). Degradation of groundwater quality in coastal aquifer of Sabratah area, NW Libya. *Environmental Earth Sciences, 76*(19), 1–23. https://doi.org/10.1007/s12665-017-6999-5

Ali, R., McFarlane, D., Varma, S., Dawes, W., Emelyanova, I., Hodgson, G., & Charles, S. (2012). Potential climate change impacts on groundwater resources of south-western Australia. *Journal of Hydrology, 475*, 456–472. https://doi.org/10.1016/j.jhydrol.2012.04.043

Anderson, M. P., & Woessner, W. W. (1992). *Applied groundwater modeling: Simulation of flow and advective transport* (pp. 70–80). Gulf Professional Publishing.

Antonellini, M., Mollema, P., Giambastiani, B., Bishop, K., Caruso, L., Minchio, A., Pellegrini, L., Sabia, M., Ulazzi, E., & Gabbianelli, G. (2008). Salt water intrusion in the coastal aquifer of the southern Po Plain, Italy. *Hydrogeology Journal, 16*, 1541–1556.

Arnell, N. (1999). Climate change and global water resources. *Global Environmental Change, 9*, 31–49.

Ayinippully Nalarajan, N., Nambi, I. M., & Govindarajan, S. K. (2022). Numerical investigations on the reclaimable aquifer recharge from injection wells: A case study. *Sustainable Water Resources Management, 8*(5), 133. https://doi.org/10.1007/s40899-022-00730-0

Badaruddin, S., Suleman, A. R., Faisal, Z., Halim, H., & Saing, Z. (2022). Analytical and numerical model of saltwater intrusion in Jeneponto Regency, South Sulawesi province Indonesia. *Water Practice and Technology, 17*(5), 1146–1154. https://doi.org/10.2166/wpt.2022.052

Bakker, M., Essink, G., & Langevin, C. (2004). The rotating movement of three immiscible fluids—A benchmark problem. *Journal of Hydrology, 287*, 270–278.

Brian. (2016). *Coastal aquifers; Groundwater at sea*. Geological Digressions. https://www.geological-digressions.com/coastal-aquifers-groundwater-at-sea/

CGWB. (2009). District groundwater brochure Ramanathapuram district, Tamil Nadu, Technical report. 1–22.

Chen, S., & Yang, G. (2011). Eco-environmental quality appraisal of Yellow River delta wetland using the improved AHP method. *South-to-North Water Diversion and Water Science & Technology, 9*(1), 99–101. https://doi.org/10.3724/sp.j.1201.2011.01099

Cheng, J. M., & Chen, C. X. (2001). Three-dimensional modeling of density-dependent salt water intrusion in multilayered coastal aquifers in Jahe river Baisn, Shandong Province, China. *Groundwater, 39*(1), 137–143. https://doi.org/10.1111/j.1745-6584.2001.tb00359.x

Elango, L., & Sivakumar, C. (2008). Regional simulation of a groundwater flow in coastal aquifer, Tamil Nadu, India. In S. Ahmed, R. Jayakumar, A. Salih, (Eds.), *Groundwater dynamics in hard rock aquifers* (pp. 234–242). Springer, Dordrecht. https://doi.org/10.1007/978-1-4020-6540-8_18

Essink, G. H. O. (2001). Salt water intrusion in a three-dimensional groundwater system in the Netherlands: A numerical study. *Transport in Porous Media, 43*, 137–158.

Ferguson, G., & Gleeson, T. (2012). Vulnerability of coastal aquifers to groundwater use and climate change. *Nature Climate Change, 2*, 342–434.

Fetter, C. W. (2001). *Applied hydrogeology* (pp. 113–147). Prentice-Hall.

Freeze, R. A., & Witherspoon, P. A. (1966). Theoretical analysis of regional groundwater flow: 1. Analytical and numerical solutions to the mathematical model. *Water Resources Research, 2*, 641–656.

Gelinas, R., Doss, S., & Miller, K. (1981). The moving finite element method: Applications to general partial differential equations with multiple large gradients. *Journal of Computational Physics, 40*(1), 202–249. https://doi.org/10.1016/0021-9991(81)90207-2

Han, W., Meehl, G. A., Stammer, D., Hu, A., Hamlington, B., Kenigson, J., Palanisamy, H., & Thompson, P. (2017). *Spatial patterns of sea level variability associated with natural internal climate modes. Space sciences series of ISSI* (pp. 221–254). https://doi.org/10.1007/978-3-319-56490-6_10

Huyakorn, P. S., Andersen, P. F., Mercer, J. W., & White, H. O. (1987). Saltwater intrusion in aquifers: Development and testing of a three-dimensional finite element model. *Water Resources Research, 23*, 293–312.

Istok, J. (1989). *Groundwater modeling by the finite element method; water resources monograph series, 13* (pp. 255–283). American Geophysical Union.

Jelgersma, S., Zijp, V. M., & Brinkman, R. (1993). Sea level rise and the coastal lowlands in the developing world. *Journal of Coastal Research, 9*, 958–972.

Jin, Y., Yuan, W., Yin, Z., & Cheng, Y. (2020). An edge-based strain smoothing particle finite element method for large deformation problems in geotechnical engineering. *International Journal for Numerical and Analytical Methods in Geomechanics, 44*(7), 923–941. https://doi.org/10.1002/nag.3016

Karthikeyan, S., Kulandaisamy, P., Senapathi, V., Chung, S. Y., Thangaraj, K., Arumugam, M., Sugumaran, S., & Ho-Na, S. (2022). Hydrogeochemical survey along the northern coastal region of Ramanathapuram district, Tamilnadu, India. *Applied Sciences, 12*(11), 5595. https://doi.org/10.3390/app12115595

Khairy, H., & Janardhana, M. R. (2013). Hydrogeochemical features of groundwater of semi-confined coastal aquifer in amol–ghaemshahr plain, Mazandaran province, Northern Iran. *Environmental Monitoring and Assessment, 185*(11), 9237–9264. https://doi.org/10.1007/s10661-013-3248-6

Kumar, A., & Thambidurai, P. (2022). Relationship of topography, weathering, and water table in Usri watershed of Chota Nagpur Plateau, India. *Arabian Journal of Geosciences, 5–504*, 1–16. https://doi.org/10.1007/s12517-022-09624-z

Kumar, A., & Thambidurai, P. (2023). Impact of urbanization and river morphology on groundwater system in Patna Urban Area, Bihar, India. In P. Thambidurai, & A. K. Dikshit (Eds.), *Impacts of urbanization on hydrological systems in India* (pp. 167–185). Springer, Cham. https://doi.org/10.1007/978-3-031-21618-3_9

Lal, A., & Datta, B. (2017). Modelling saltwater intrusion processes and development of a multi-objective strategy for management of coastal aquifers utilizing planned artificial freshwater recharge. *Modeling Earth Systems and Environment, 4*(1), 111–126. https://doi.org/10.1007/s40808-017-0405-x

Lin, J., Ma, R., Sun, Z., & Tang, L. (2023). Assessing the connectivity of a regional fractured aquifer based on a hydraulic conductivity Field reversed by multi-well pumping tests and numerical groundwater flow modeling. *Journal of Earth Science, 34*(6), 1926–1939. https://doi.org/10.1007/s12583-022-1674-5

Majumdar, P. K., Ram, S., & Rao, P. R. (2009). Artificial recharge in multi aquifers of a mountainous watershed. *Journal of Hydrologic Engineering, 14*, 215–222.

Mathiazhagan, M., Jha, S. K., & Biswas, A. (2021). Assessment of groundwater mass balance and zone budget in the semi-arid region: A case study of Palar sub-basin, Tamil Nadu, India. *Journal of Earth System Science, 130*(4), 187. https://doi.org/10.1007/s12040-021-01691-2

McDonald, M. G., & Harbaugh, A. W. (1986). *A modular three-dimensional finite-difference ground-water flow model* (pp. 116–125). US Geological Survey Technical Manual of Water Resources Investigation, Book 6; USGS: Reston, VA, USA.

Mishra, R. K. (2023). Fresh water availability and its global challenge. *Journal of Marine Science and Research, 2*(1): 01–03. https://doi.org/10.58489/2836-5933/004

Moore, W. (1996). Large groundwater inputs to coastal waters revealed by 226Ra enrichments. *Nature, 380*, 612–614.

Nalbantis, I., Rozos, E., Tentes, G., Efstratiadis, A., & Koutsoyiannis, D. (2002). Integrating groundwater models within a decision support system. In G. Tsakiris (Ed.), *Proceedings of the 5th international conference of European water resources association: "water resources management in the era of transition"* (pp. 4–8), Athens.

Oude Essink, G. H. P., Van Baaren, E. S., & De Louw, P. G. (2010). Effects of climate change on coastal groundwater systems: A modeling study in the Netherlands. *Water Resources Research, 46*, 1–16.

Rani, F. M., & Chen, Z. H. (2010). Numerical Modeling of Groundwater Flow in Karst Aquifer, Makeng Mining Area. *American Journal of Environmental Sciences, 6*, 78–82.

Rao, S. V. N., Bhallamudi, S. M., Thandaveswara, B. S., & Mishra, G. C. (2004). Conjunctive use of surface and groundwater for coastal and deltaic systems. *Journal of Water Resources Planning and Management, 130*, 255–267.

Rejani, R., Jha, M. K., Panda, S. N., & Mull, R. (2008). Simulation modeling for efficient groundwater management in Balasore coastal basin, India. *Water Resources Management, 22*, 23–50.

Roy, D. K., & Datta, B. (2017). Saltwater intrusion processes in coastal aquifers—modelling and management: A review. *Desalination and Water Treatment, 78*, 57–89. https://doi.org/10.5004/dwt.2017.20381

Senthilkumar, M., & Elango, L. (2004). Three-dimensional mathematical model to simulate groundwater flow in the lower Palar River basin, southern India. *Hydrogeology Journal, 12*, 197–208.

Singh, A., & Panda, S. N. (2013). Optimization and simulation modelling for managing the problems of water resources. *Water Resources Management, 27*, 3421–3431.

Trenberth, K., Smith, L., Qian, T., Dai, A., & Fasullo, J. (2007). Estimates of the global water budget and its annual cycle using observational and model data. *Journal of Hydrometeorology, 8*, 758–769.

Trescott, P. C., Pinder, G. F., & Larson, S. P. (1976). *Finite-difference model for aquifer simulation in two dimensions with results of numerical experiments* (pp. 2–10), US Department of the Interior, Geological Survey: Reston, VA, USA.
Vandenbohede, A., Houtte, E., & Lebbe, L. (2009). Sustainable groundwater extraction in coastal areas: A Belgian example. *Environmental Geology, 57*, 735–774.
Varalakshmi, V., Venkateswara Rao, L. B., SuriNaidu, L., & Tejaswini, M. (2014). Groundwater flow modeling of a hard rock aquifer: Case study. *Journal of Hydrologic Engineering, 19*, 877–886.
Voss, C. I. (1984). *A finite-element simulation model for saturated-unsaturated, fluid-density-dependent ground-water flow with energy transport or chemically-reactive single-species solute transport* (pp. 65–80). US Geological Survey: Reston, VA, USA.
Wang, H. F., & Anderson, M. P. (1995). *Introduction to groundwater modeling: Finite difference and finite element methods* (pp. 47–85). Academic Press.
Waterloo Hydrogeologic Inc. (2002). *User's manual. Software-consulting-training* (pp. 386–390), Waterloo Hydrogeologic Inc. Waterloo, Ontario, Canada.
Wetzelhuetter, C. (2013). *Groundwater in the coastal zones of Asia-Pacific* (pp. 49–65). Springer Science & Business Media.
Yeh, W. W. (2015). Review: Optimization methods for groundwater modeling and management. *Hydrogeology Journal, 23*, 1051–1065.
Yin, J., Tsai, F. T., & Lu, C. (2022). Bi-objective extraction-injection optimization modeling for salt-water intrusion control considering surrogate model uncertainty. *Water Resources Management, 36*(15), 6017–6042. https://doi.org/10.1007/s11269-022-03340-9

Estimation of Submarine Groundwater Discharge in Coastal Aquifer of South Andaman, India

P. M. Mohan and P. Thambidurai

1 Introduction

SGD denotes the movement of groundwater from an aquifer to the seabed (Burnett et al., 2003; Moore, 2010). The boundary between freshwater discharge and saline ocean water creates a salinity gradient from land to sea (Cloern et al., 2017). SGD may stem from terrestrial groundwater, recirculated seawater, or a mix of both, influenced by multiple factors. The extent and distribution of terrestrial SGD are shaped by soil hydrological characteristics, including hydraulic conductivity, anisotropy, and preferential flow channels (Ruiz-González et al., 2021; Sajeev et al., 2023). Moreover, real-world complexities such as geological heterogeneities, shifting alongshore and cross-shore morphology (Robinson et al., 2018; Mathiazhagan et al., 2021; Kumar & Thambidurai, 2022, 2023), topographic variations, and human-made structures significantly influence the system. Globally, the discharge of groundwater into coastal regions is largely governed by an aquifer's flow capacity, which depends on factors such as permeability, topographic slope, and aquifer thickness (Senthilkumar & Elango, 2004; Luijendijk et al., 2020). Conversely, recirculated seawater, referred to as recirculated SGD, is primarily influenced by oceanic forces, including waves, tides, currents, and density variations (Taniguchi et al., 2019).The SGD also integrates recirculated seawater and nutrient fluxes from the sea into coastal aquifers (Swarzenski & Baskaran, 2007). SGD is widely acknowledged as a crucial contributor of nutrients (Garcia-Solsona et al., 2010), dissolved inorganic carbon

P. M. Mohan
Department of Ocean Studies and Marine Biology, Pondicherry University - Port Blair Campus, Sri Vijaya Puram, Andaman and Nicobar Islands, India

P. Thambidurai (✉)
Department of Coastal Disaster Management, School of Physical, Chemical and Applied Sciences, Pondicherry University - Port Blair Campus, Sri Vijaya Puram, Andaman and Nicobar Islands, India
e-mail: thambiduraip@pondiuni.ac.in

P. Thambidurai and A. K. Dikshit (eds.), *Coastal Environments of India*,
https://doi.org/10.1007/978-3-031-97341-3_8

(Jaffé et al., 2003), and trace metals to coastal waters, all of which play a pivotal role in driving ecological transformations in coastal ecosystems. The persistent input of these elements via SGD impacts water quality, often resulting in environmental degradation in coastal areas (LaRoche et al., 1997; Black et al., 2009; Lee et al., 2011; Rodellas et al., 2015; Trezzi et al., 2016). Due to its environmental importance, measuring SGD is essential; however, its quantification is challenging because of the discharge's slow, diffuse, and spatially variable nature beneath the water's surface (Hajati et al., 2019).

Estimating SGD and analyzing its spatiotemporal variability is essential. On a local scale, commonly used field methods for evaluating freshwater SGD include seepage meters (Shaw & Prepas, 1990; Hata et al., 2016; Das et al., 2020), piezometers (Freeze & Cherry, 1979), and natural tracers like radium and radon (Burnett et al., 2001; Porcelli & Swarzenski, 2003; Hwang et al., 2005). However, the limited spatial coverage of direct measurements poses challenges in precisely quantifying fresh SGD, given its high variability. Computational approaches provide an effective alternative for estimating freshwater SGD across different spatial scales, ranging from global (Sawyer et al., 2016; Zhou et al., 2019) to regional levels (Destouni et al., 2008; Zhou et al., 2018; Russo et al., 2023). Baseflow plays a crucial role in the exchange of water between streams and groundwater systems, directly impacting stream water quality (Miller et al., 2016). In small island watersheds, this interaction also extends to coastal water quality. Blanco et al. (2011) utilized radon measurements to investigate nearshore groundwater discharge, revealing that subsurface geology, groundwater recharge, hydraulic gradients, and tidal pumping significantly shape nearshore hydrodynamics. Additionally, Zhu et al. (2019) employed a mass balance method combined with a topographic model based on empirical relationships to estimate SGD flux in an island setting. Their findings indicated that elevated SGD rates were strongly associated with steep terrain.

Growing awareness of SGD's impact on coastal environments has led to an increasing number of studies focused on estimating SGD rates and associated fluxes using various methods (Peterson et al., 2008; Gleeson et al., 2013; Tait et al., 2013; Krall et al., 2017; Szymczycha et al., 2012, 2016; Idczak et al., 2020). Among various methodologies, numerical groundwater flow models have been designed to evaluate and forecast SGD and its fluxes under both present and future scenarios (Anderson & Woessner, 1992; Kaleris et al., 2002; Langevin, 2003). Three-dimensional (3D) geological models, which depict the structure and stratigraphy of depositional sequences, serve as the foundational framework for groundwater flow simulations (Pasanen & Okkonen, 2017). Precise estimation of SGD and its fluxes through numerical modelling necessitates a comprehensive understanding of aquifer three-dimensional geometry, porosity distribution within host sediments, and groundwater flow pathways (Elango & Sivakumar, 2008; Bishop et al., 2017; Luijendijk et al., 2020; Unnikrishnan et al., 2021). This study explores the effectiveness of numerical modelling in assessing submarine groundwater discharge (SGD) under island conditions. Through numerical simulations, it aims to improve the understanding of SGD dynamics by accounting for aquifer characteristics, topography, and hydrological interactions. The results will offer valuable insights into SGD quantity in the target region, aiding informed decision-making for sustainable groundwater management and coastal resource conservation.

2 Study Area

The study area is situated in the eastern part of the Bay of Bengal, between latitudes 11°30' to 11°45' N and longitudes 92°30' to 94°50' E. The South Andaman Island, a part of the Great Andaman group, exhibits an undulating topography characterized by small terrain tracts, coastal low-lying areas, and hilly landscapes with escarpments, rolling terrain, and scattered hillocks/mounds. Sri Vijaya Puram (Port Blair) serves as the district and sub-divisional headquarters of South Andaman, also functioning as the capital of the Andaman and Nicobar Islands. Positioned in the equatorial belt, the islands experience significant marine influences and a tropical climate that is warm, moist, and equable. The temperatures typically range from 22 to 32 °C, while annual rainfall varies between 3000 and 3500 mm and the humidity levels fluctuate between 66 and 85%. Most rainfall occurs from May to December, with approximately 2250 mm received during the southwest monsoon (May–September) and 685 mm during the northeast monsoon (October–December). The period from January to April is relatively dry, with minimal rainfall, typically limited to three rainy days per month. The soils in the region are predominantly classified as entisols, inceptisols and alfisols. Agricultural activities are primarily conducted on alluvial and colluvial soils, while coastal areas prone to tidal flooding consist of sulfate-rich acidic soils. The soil thickness varies from a few meters to several tens of meters, but these soils are generally nutritionally deficient with low organic matter content.

2.1 Geology and Hydrogeology

The study area predominantly consists of thick Eocene sedimentary formations, accompanied by Pre-Tertiary sandstone, siltstone, and shale, interspersed with intrusions of basic and ultrabasic igneous rocks. The igneous rocks mainly belong to the ophiolitic suite. The Richie's Archipelago is predominantly made up of calcareous sandstones. The region's marine sedimentary formations consist of greywackes, graded sandstones, siltstones, black shale, and volcanogenic sediments. Clayey sediments, characteristic of fine-grained deep-sea deposits, are particularly dominant. Additionally, coralline formations can be observed along the coastlines and in transitional zones between sedimentary and igneous formations. The ophiolitic and marine sedimentary rocks in the study area have undergone varying degrees of tectonic activity, including folding, faulting, and upliftment. These processes have resulted in the development of fractures and fissures, facilitating rainwater infiltration and its subsequent percolation into deeper layers. The loose marine sedimentary formations contain a mix of sand and a high proportion of clay, making them poor water yielders and leading to low-potential aquifer systems at both shallow and deeper levels. In these formations, the water table depth varies between 2.0 and 4.4 m below ground level (bgl). According to the Central Ground Water Board (CGWB) data, boreholes drilled in these formations were largely unsuccessful, with the exception of one at Prothrapur, which yielded brackish water (5500 EC) at a discharge rate of 17,000 L per hour. Even in igneous rocks, potential aquifers are not well-developed and offer only moderate water yield compared to their sedimentary counterparts (CGWB 2016). The primary aquifers are formed within the weathered mantle and fractured basement rocks, but clayey residues often clog fractures, fully sealing the conduits. These fractures are estimated to extend up to 45 m in depth. Aquifers in this region are primarily utilized through dug wells, ponds, and private tube wells and also observed some coralline formations certainly contribute to potential aquifers. Recent CGWB studies indicate that valley-fill deposits, composed of boulders, pebbles, cobbles, and sand, provide promising near-surface aquifers (Fig. 1).

2.2 *Data Collection*

The surface elevation of the study area was represented using a Digital Elevation Model (DEM) acquired from OpenTopography. The DEM features a spatial resolution of 30 $\times$ 30 m, ensuring comprehensive coverage of the entire study region. The key hydrological parameters, including observed water levels, well depths, and lithological data, were sourced from the India Water Resources Information System (India WRIS). Furthermore, key aquifer characteristics, including hydraulic conductivity, specific yield, specific storage, effective porosity, and total porosity, were sourced from the CGWB. Meteorological data were gathered from various sources: rainfall records were retrieved from the Indian Meteorological Department's online portal, while humidity data were obtained from the Department of Agriculture, Andaman and Nicobar Administration, located in Sri Vijaya Puram.

3 Methodology

The groundwater model is developed using multiple packages, each designed to perform specific functions within the MODFLOW framework. MODFLOW provides a suite of packages that enable the simulation of groundwater flow and associated processes. The model domain is structured as a grid matrix consisting of 160 rows $\times$ 120 columns $\times$ 1 layer. In this model, a single-layer configuration incorporates spatially varying hydrogeological properties (Fig. 2) and exhibits variable thickness across the study area. Each model cell is uniformly shaped, maintaining a regular rectangular structure. For the entire island, the model identifies 9330 active cells, with each cell having a spatial resolution of 177 $\times$ 177 m. This grid-based representation allows for a detailed simulation of groundwater flow dynamics across the study area.

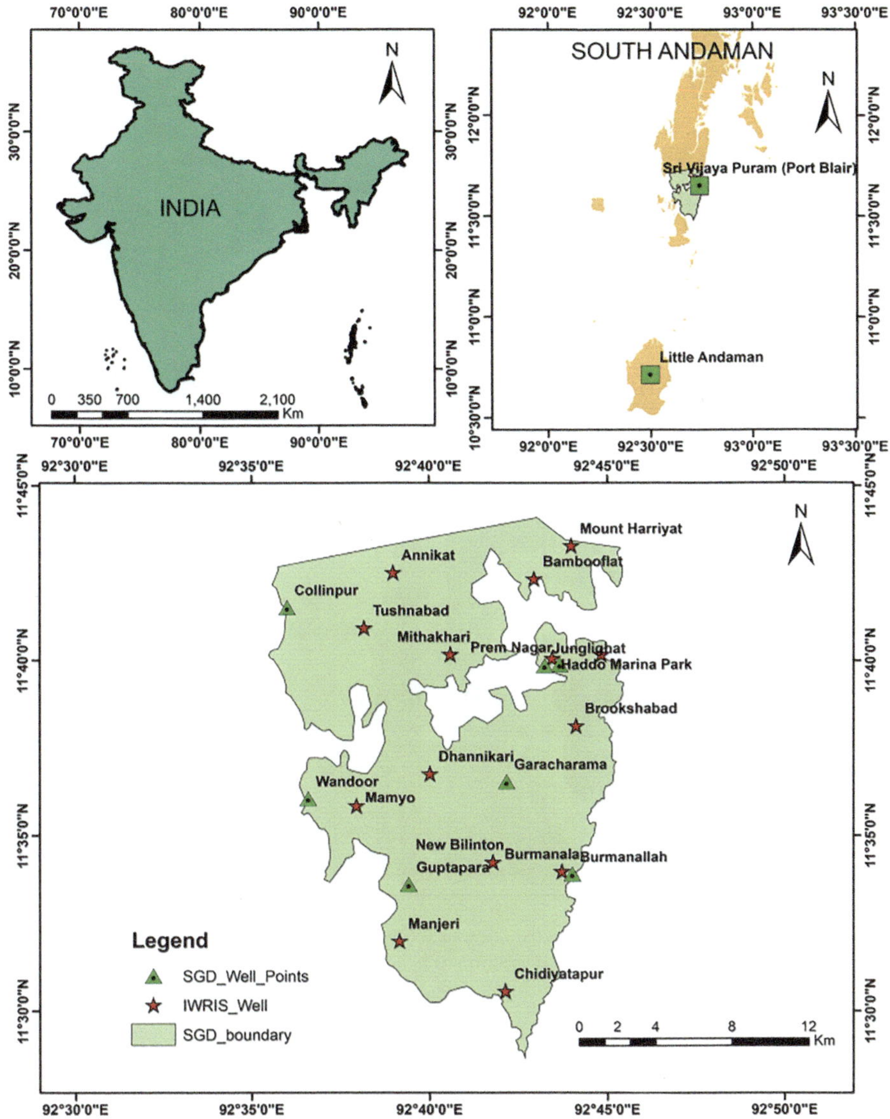

Fig. 1 Study area map with observation well locations

Figure 2a illustrates the distribution of active and inactive cells within the study area. The land portion of the island is designated as the active cell region, while a constant head boundary was applied to a few cells surrounding the study area boundary (Fig. 2d). The remaining portion is classified as inactive cells, indicating areas where groundwater flow is not simulated. Figure 2b depicts the hydraulic conductivity zones, which are assigned based on the geological variations across

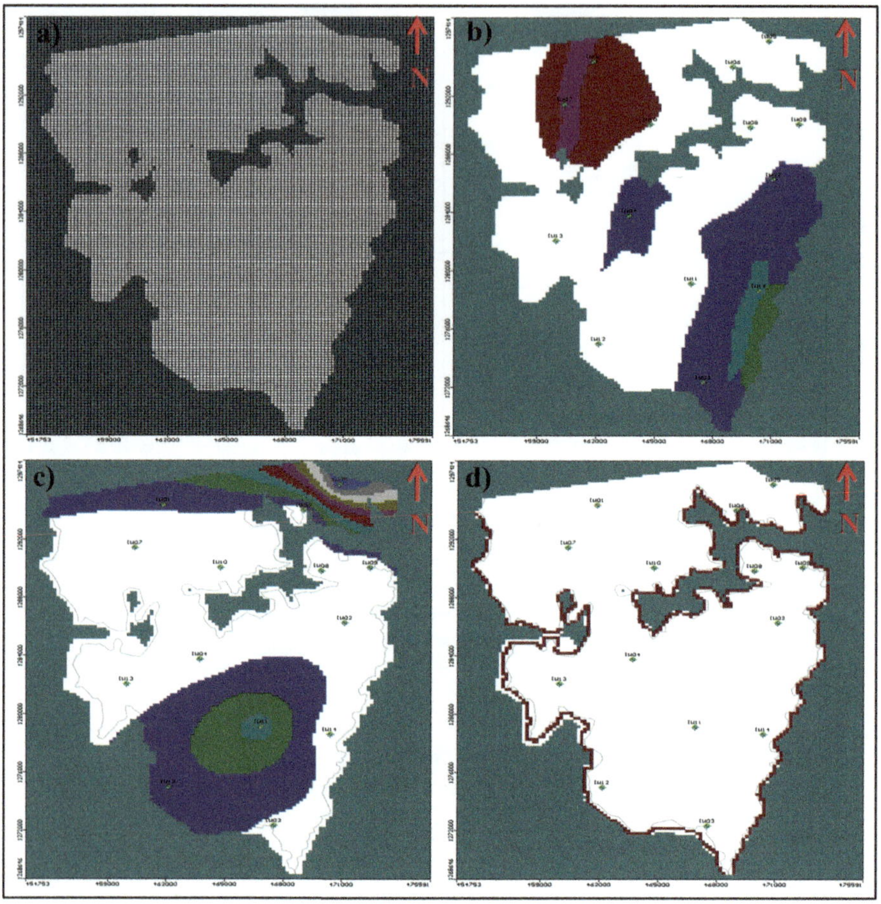

Fig. 2 Model simulation inputs, **a** distribution of active and inactive cells, **b** application of constant head boundary along the shoreline and location of head observation wells, **c** hydraulic conductivity zones categorized based on geological formations, **d** spatial variation of initial head

the study region. The distribution of hydraulic conductivity is crucial for accurately modelling subsurface flow dynamics. Fourteen (14) head observation well data (Fig. 1) were plotted and presented in Fig. 2b, c and d. These wells serve as reference points for monitoring groundwater levels. The initial head values for the model were derived from the starting measurements recorded at these observation wells. The spatial variation of the initial head, as shown in Fig. 2c, highlights differences in water levels across the study area. Regions with elevated water levels are represented using distinct colours, while areas with lower water levels are depicted in white, indicating zones of minimal groundwater elevation.

The Head-Observation Package (HOB) is utilized to define and monitor hydraulic head observations within the groundwater model. For each simulation, two essential error metrics, Mean Absolute Error (MAE) and Root Mean Square Error (RMSE), are calculated. The MAE quantifies the average absolute difference between observed and simulated hydraulic head values. A lower MAE indicates a closer match between observed and model-calculated heads, signifying a more accurate simulation. The MAE is determined using the following equation, as described by Aghlmand and Abbasi (2019),

$$MAE = \frac{1}{n}\sum_{i=1}^{n}|(h_o - h_c)_i|$$

Here, n represents the total number of observations, h_o denotes the observed hydraulic head (L), and h_c refers to the calculated hydraulic head (L). The RMSE is calculated by first averaging the squared differences between observed and simulated head values and then taking the square root of this mean. It acts as a measure of model accuracy, where lower RMSE values indicate a closer match between observed and simulated heads. The following equation, as presented by Aghlmand and Abbasi (2019), is used to determine RMSE

$$RMSE = \left[\frac{1}{n}\sum_{i=1}^{n}(h_o - h_c)_i^2\right]^{0.5}$$

The optimal hydraulic conductivity values were determined through model calibration by minimizing MAE and RMSE. The calibration process was designed to optimize the alignment between observed and simulated hydraulic head values. The final calibrated values indicate that horizontal hydraulic conductivity (kx = ky) varies between 1 and 3 m/day, while vertical hydraulic conductivity (kz) ranges from 0.1 to 0.3 m/day. These values capture the geological and hydrogeological traits of the study area, enhancing the accuracy of groundwater flow representation.

4 Results and Discussion

4.1 Aquifer Characterization

Aquifer characterization is the process of assessing an aquifer's permeability and its economic viability based on the available quantity of water. The study area features steep slopes, rugged terrain, and hills situated in close proximity to the coastline. Due to the low infiltration capacity of the island's sediments, approximately 70% of the total area is covered by these formations. Around 15% of the area consists of indigenous rock formations, while the remaining 15% is composed of coralline and limestone formations. These islands are located in an active tectonic zone and frequently experience shallow to deep-focus earthquakes, leading to fractures and fissures in the igneous and sedimentary rock formations. The igneous rocks, primarily of ophiolitic nature, are limited to a small area but provide moderate to high groundwater yields in shallow and deep tubewells, as well as dug wells, with yields ranging from 1 to 10 L per second (lps). A significant portion of the area is covered by marine sedimentary rocks, which yield only a minimal amount of groundwater (0.1–0.5 lps) due to the presence of fine marine clay interspersed within these formations. Areas covered with coralline limestone formations yield between 5 and 25 lps during the rainy season. However, in the dry season, the yield varies due to the lack of recharge in these formations.

4.2 Groundwater Modelling

In this study, three-dimensional modelling has been conducted to determine the direction of the SGD flow. Initially, a three-dimensional physiographic model was prepared to develop a representation of the study area, highlighting its topographical variations (Fig. 3a). The study area's highest elevation is located in the northeast, with another notable peak situated in the southwest. Figure 3b illustrates the simulated groundwater levels with respect to mean sea level (MSL). The results highlight two prominent groundwater bulges, one in the northeast rising above 140 m and another in the southwest, reaching below 80 m. The overall groundwater flow direction is from the central part of the island towards the coastline, driven by the natural hydraulic gradient, where water moves from regions of higher to lower hydraulic head. This flow pattern is primarily influenced by the specified head boundary conditions along the coastal bays. Moreover, the groundwater flow budget is controlled by recharge and outflow through boundary conditions. Table 1 presents a detailed breakdown of this flow budget, as depicted in the regional plan view model.

The quantification of outflow storage and discharge across the constant head boundary reveals a substantial volume of water being released into the sea, representing the potential SGD in the study area. Based on this assessment, several SGD

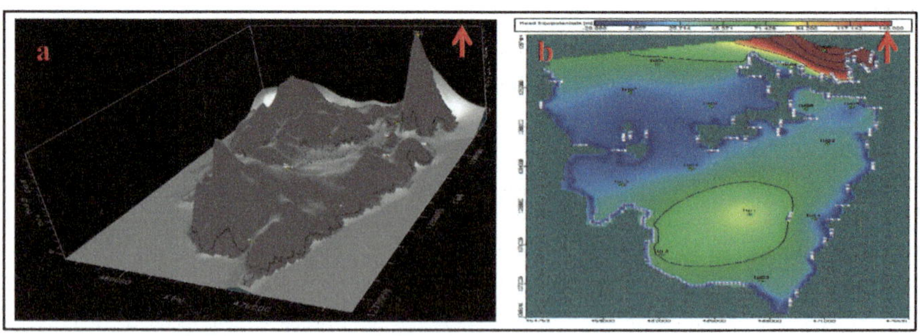

Fig. 3 **a** 3D Physiographic representation of the study area, **b** spatial distribution of groundwater head equipotential contours

Table 1 Regional plan view model—flow budget

	Inflow (Mm^3/day)	Outflow (Mm^3/day)
Storage	1.21	0.158
Constant Head	0.001	0.82
Recharge	0.322	0
Wells	0	0.015
ET	0	0.541

locations have been identified and displayed in the row view of the groundwater flow model. These selected sites were further analyzed and validated using model output data, including water level and Total Dissolved Solids (TDS) concentrations. The outflow across the boundary predominantly regulates the SGD. To provide a clearer understanding of this process, a detailed cross-sectional analysis has been conducted, with a zoomed-in visualization of the boundary conditions presented in Fig. 4a–g. The velocity vectors in these figures indicate horizontal groundwater movement until they encounter the freshwater-saltwater interface, where flow dynamics change. Below the red line and within the circled area, the velocity vectors clearly indicate groundwater movement from the land towards the sea, identifying potential SGD points.

Location 1—Prem Nagar

In the Prem Nagar area (Centre of the city at Sri Vijaya Puram), four wells were identified, and water samples were collected. Among the four, two wells are situated approximately 50 m from the coastline at an elevation of around 2.5 m. The depth of groundwater in these wells varies between 0.76 and 1.70 m. The underlying lithology in this area is composed of flysch formations. The measured TDS in the groundwater samples ranges from 164 to 486 mg/L, and the highest TDS concentration was observed near the shore, and lower concentrations were recorded at greater distances from the shoreline. Notably, the water level remains approximately 2.5 m above the high tide line at a distance of 50 m from the shore. It strongly indicates that a potential groundwater discharge in a circle-marked area, velocity direction land area into the sea (Fig. 4a), is likely facilitated by existing fractures or cleavage planes within the flysch formation. In the selected row view of the model, three potential SGD points were identified from left to right, with the third one being the Prem Nagar SGD point.

Location 2—Junglighat

At this site, four wells were identified near the shoreline, distributed over a distance of approximately 1 km. These wells are located slightly above the high tide line. The lithology of all these studied wells is composed of flysch formations. Of the four analyzed wells, two were located in the northern part and exhibited slightly higher TDS levels, ranging from 216 to 941 mg/L, but the other two were still used for potable water. In contrast, these two wells on the southern side contained relatively better-quality water. The groundwater levels in these four wells ranged from 0.58 to 1.35 m, which is nearly at the same level as the high tide line. In the selected row view of the model, two potential SGD points were identified from left to right, with the second one (marked in a circle) being the Junglighat SGD point (Fig. 4b).

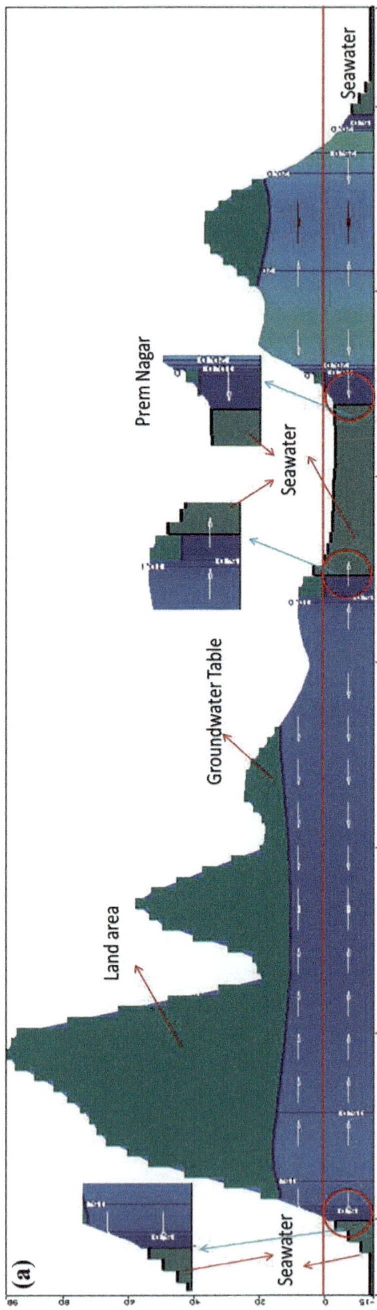

Fig. 4 SGD point of the study area. **a** Prem Nagar. **b** Junglighat. **c** Garacharma. **d** Guptapara. **e** Wandoor. **f** Burmanallah. **g** Colinpur

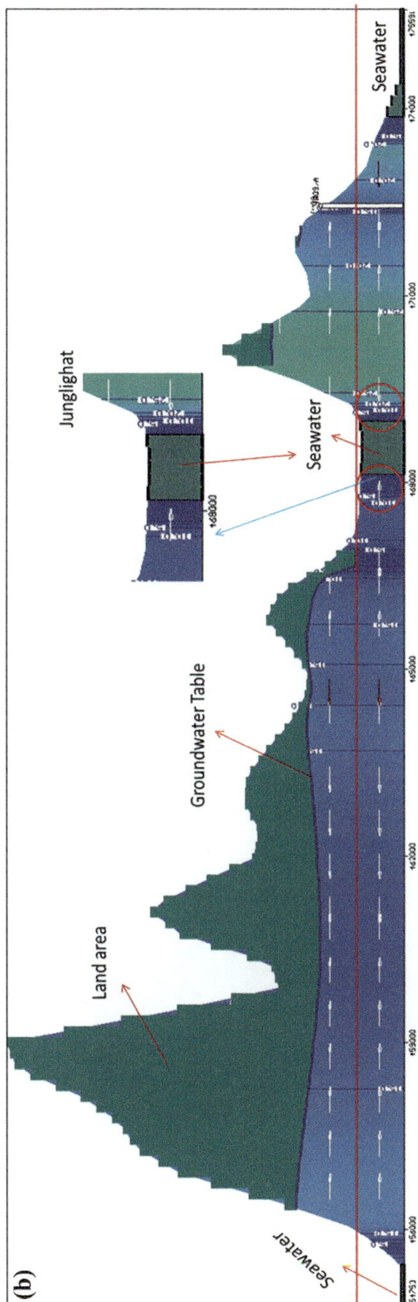

Fig. 4 (continued)

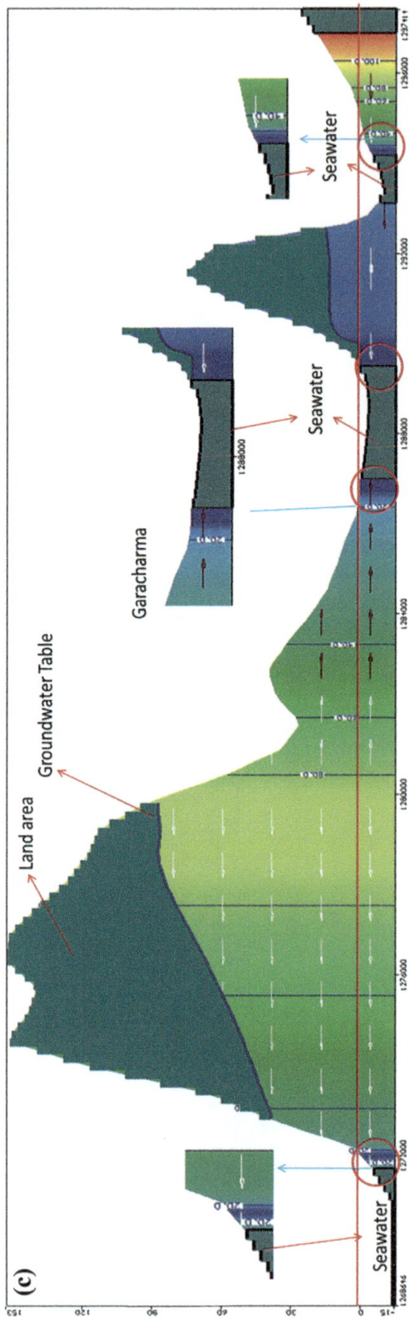

Fig. 4 (continued)

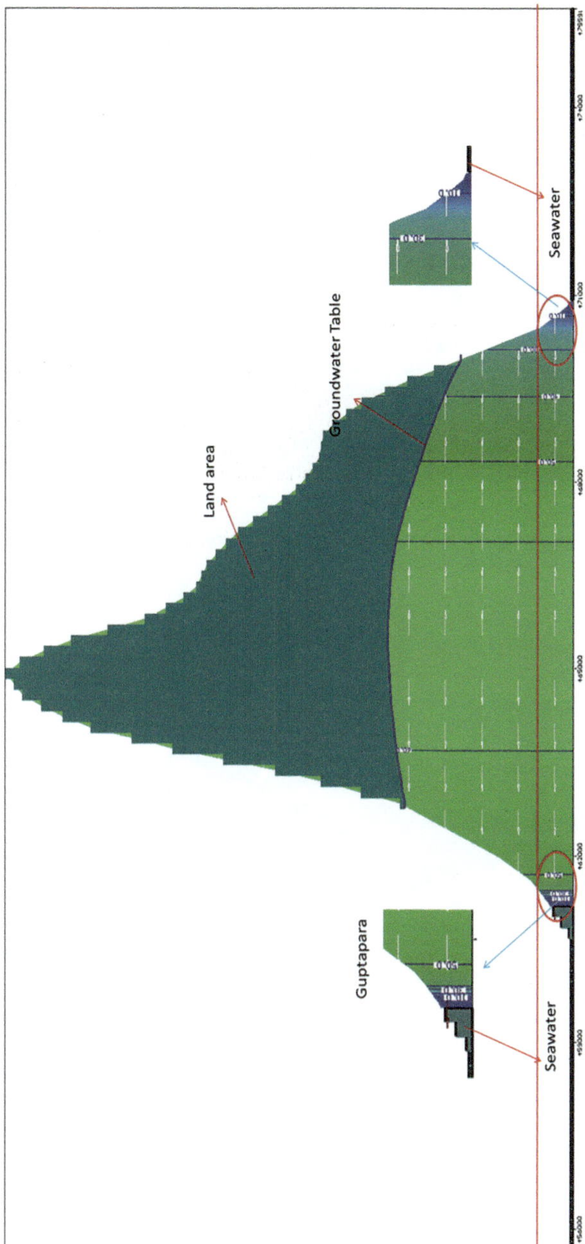

Fig. 4 (continued)

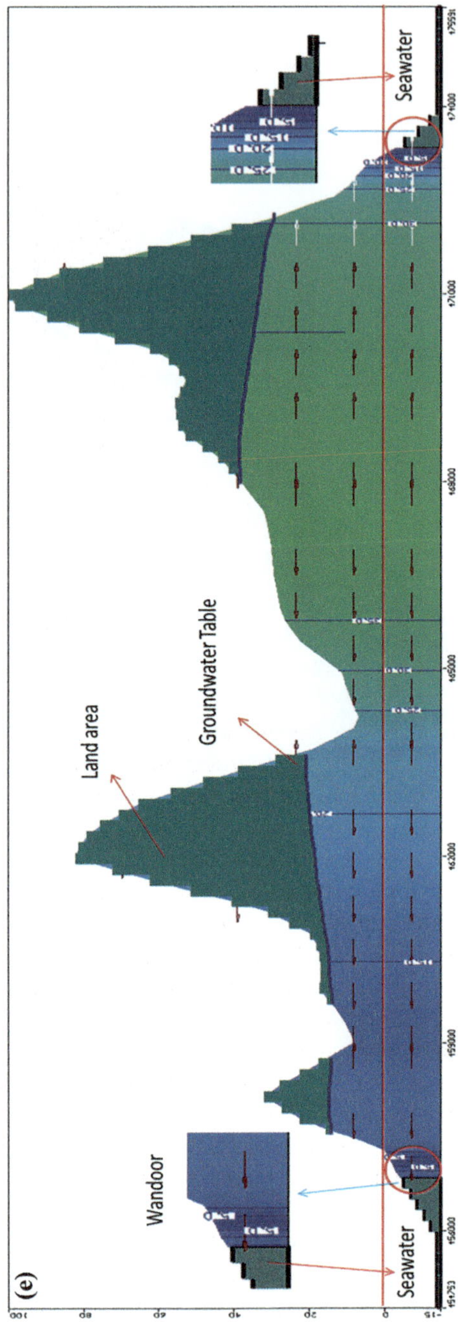

Fig. 4 (continued)

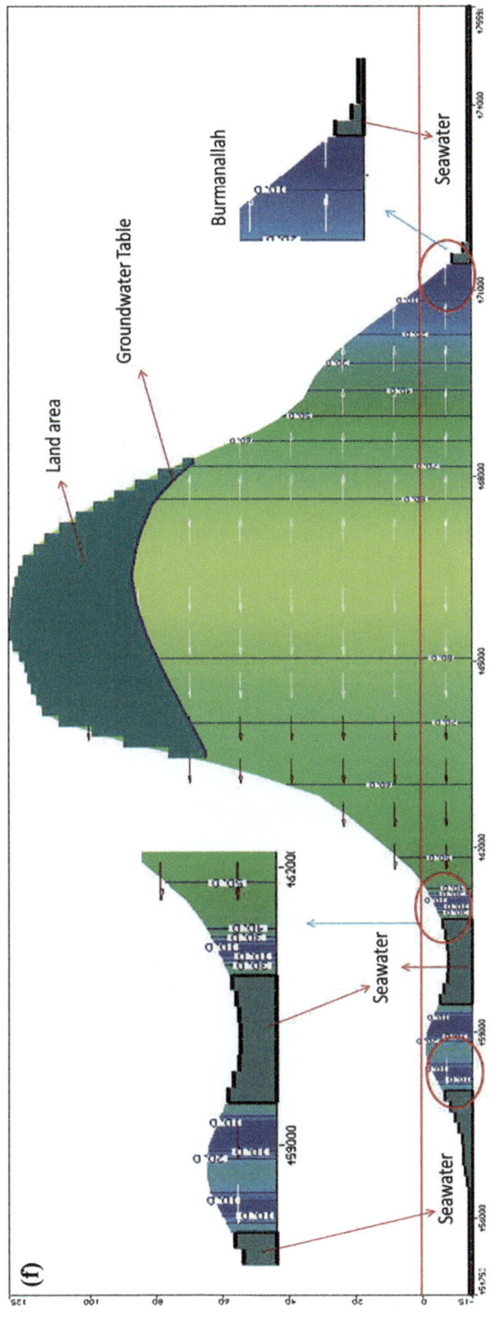

Fig. 4 (continued)

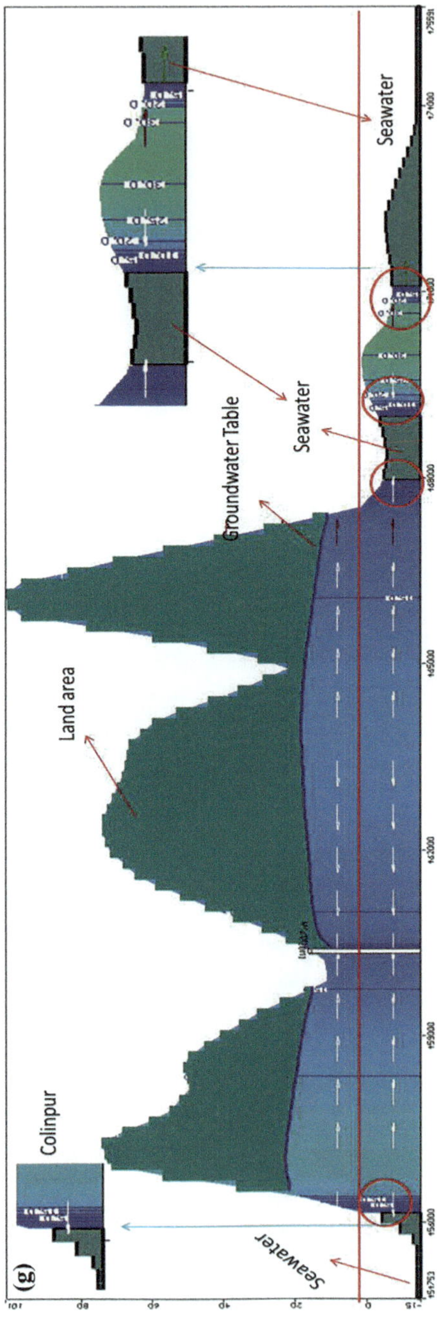

Fig. 4 (continued)

Location 3—Garacharma

In this location, only one well was identified within the Tsunami-Created Wetland (TCW). This area falls within a reclaimed zone that was submerged during the 2004 mega earthquake. The groundwater in this well exhibits good quality, with TDS levels falling within the potable water (188 mg/L). Offshore from this location, there may be an influence of freshwater discharge near the tidal seawater fluctuation zone. Since this area was originally land before being submerged under tidal influences, the presence of groundwater at a depth of approximately 3.3 m suggests that freshwater may extend below the seawater in the tidal fluctuation region. In the selected column view of the model, four potential SGD points were identified from left to right, with the second one (marked in the circle) being the Garacharma SGD point (Fig. 4c).

Location 4—Guptapara

The well investigated at Guptapara for the study is located within a tidal influence zone, approximately 500 m from the freshwater channel. The water in the well showed a TDS value of 129 mg/L, and the groundwater level was around 7 m deep. This suggests the possibility of groundwater flow occurring at a significant depth, potentially contributing to SGD in this region. Additionally, in the selected row view of the model, two potential SGD points were identified from left to right, with the first one (marked in the circle) being the Guptapara SGD point (Fig. 4d).

Location 5—Wandoor

The Wandoor is located on the western side of South Andaman. A total of 9 wells were identified and fell within 500 m of shorelines. The lithology of these wells consists of shelly limestone. The TDS values in the sampled wells ranged from 115 to 1047 mg/L. The depth of groundwater levels in these wells varied between 1 and 6 m. Notably, one well, located very close to the inundation channel and at the base of a 20-m foothill, exhibited a significantly higher TDS level of 4118 mg/L. In the selected row view of the model, two potential SGD points were identified from left to right, with the first one (marked in the circle) being the Wandoor SGD point (Fig. 4e).

Location 6—Burmanallah

The location exhibits a feature of three wells situated close to the shoreline at an elevation of approximately 2.5 m above the high tide line. The lithology of two wells consists of shelly limestone, while the third well is composed of alluvial soil. The water in these wells has TDS levels ranging from 199 to 474 mg/L. Interestingly, despite being around 3.5 m above the low tide level, no visible water seepage was observed on the surface. This could be attributed to the presence of a thin layer of loose sediment and the construction of a 5-m seawall, which may restrict the interaction between land and sea. Consequently, in the selected row view of the model, three potential SGD points (marked in the circle) were identified from left to right, with the third one being the Burmanallah SGD point (Fig. 4f).

Location 7—Colinpur

Although, at this point, the well was not officially marked by the PBMC records, it was identified during the survey as a domestic well constructed in the backyard of a house. The high tide line was observed approximately 100 m from the well. The groundwater depth at this well is 2.12 m, and the elevation of the location is nearly at the same level as the high tide line. The water quality has a TDS value of 305 mg/

L. In the selected row view of the model, four potential SGD points were identified from left to right, with the first one (marked in the circle) being the Colinpur SGD point (Fig. 4g).

5 Conclusion

For the first time, an effort has been made to estimate the SGD in coastal aquifers in the Andaman Islands. The present study provides a comprehensive analysis of aquifer characteristics and groundwater flow dynamics in the study area, focusing on identifying and quantifying the SGD. The region's hydrogeology is shaped by topography, lithology, and tectonics, influencing fractures and permeability, with igneous and limestone formations yielding more groundwater than marine sedimentary rocks. The groundwater modelling results depict the 3D physiography and simulated water levels, highlighting groundwater bulges in the northeast and southwest. The general groundwater flow follows a natural hydraulic gradient from elevated inland areas toward the coastline, with specified head boundary conditions playing a crucial role in shaping the flow regime. The flow budget analysis further emphasizes the significant contribution of recharge and outflow processes in regulating groundwater dynamics. The identification of SGD locations was achieved through field investigations, groundwater level measurements, and the TDS analysis. The selected sites exhibit varying degrees of groundwater interaction with coastal and marine environments, with some locations showing potential SGD through fractures in flysch formations. In contrast, others display deeper groundwater flow influenced by geological constraints and human interventions such as seawalls. Among the studied locations, Premnagar, Junglighat, and Wandoor exhibited clear indicators of SGD, with groundwater levels above the high tide line, suggesting a direct discharge into the sea. The Burmanala and Colinpur sites also present potential SGD occurrences, albeit at subsurface levels, influenced by sedimentary barriers. Garacharma and Guptapara, located in reclaimed and tidally influenced zones, suggest that freshwater may extend offshore, particularly during monsoonal recharge periods.

References

Aghlmand, R., & Abbasi, A. (2019). Application of MODFLOW with boundary conditions analyses based on limited available observations: A case study of Birjand plain in east Iran. *Water, 11*(9), 1904. https://doi.org/10.3390/w11091904

Anderson, M. P., & Woessner, W. W. (1992). Applied groundwater modeling. In *Simulation of flow and advective transport* (p. 381). Academic Press.

Bishop, J. M., Glenn, C. R., Amato, D. W., & Dulai, H. (2017). Effect of land use and groundwater flow path on submarine groundwater discharge nutrient flux. *Journal of Hydrology: Regional Studies, 11*, 194–218. https://doi.org/10.1016/j.ejrh.2015.10.008

Black, F. J., Paytan, A., Knee, K. L., De Sieyes, N. R., Ganguli, P. M., Gray, E., & Flegal, A. R. (2009). Submarine groundwater discharge of total mercury and Monomethylmercury to central California coastal waters. *Environmental Science & Technology, 43*(15), 5652–5659. https://doi.org/10.1021/es900539c

Blanco, A. C., Watanabe, A., Nadaoka, K., Motooka, S., Herrera, E. C., & Yamamoto, T. (2011). Estimation of nearshore groundwater discharge and its potential effects on a fringing coral reef. *Marine Pollution Bulletin, 62*(4), 770–785. https://doi.org/10.1016/j.marpolbul.2011.01.005

Burnett, W. C., Bokuniewicz, H., Huettel, M., Moore, W. S., & Taniguchi, M. (2003). Groundwater and pore water inputs to the coastal zone. *Biogeochemistry, 66*(1/2), 3–33. https://doi.org/10.1023/b:biog.0000006066.21240.53

Burnett, W. C., Taniguchi, M., & Oberdorfer, J. (2001). Measurement and significance of the direct discharge of groundwater into the coastal zone. *Journal of Sea Research, 46*(2), 109–116. https://doi.org/10.1016/s1385-1101(01)00075-2

Cloern, J. E., Jassby, A. D., Schraga, T. S., Nejad, E., & Martin, C. (2017). Ecosystem variability along the estuarine salinity gradient: Examples from long-term study of San Francisco Bay. *Limnology and Oceanography, 62*(S1). https://doi.org/10.1002/lno.10537

Das, K., Debnath, P., Duttagupta, S., Sarkar, S., Agrahari, S., & Mukherjee, A. (2020). Implication of submarine groundwater discharge to coastal ecology of the Bay of Bengal. *Journal of Earth System Science, 129*(1). https://doi.org/10.1007/s12040-019-1317-0

Destouni, G., Hannerz, F., Prieto, C., Jarsjö, J., & Shibuo, Y. (2008). Small unmonitored near-coastal catchment areas yielding large mass loading to the sea. *Global Biogeochemical Cycles, 22*(4). https://doi.org/10.1029/2008gb003287

Elango, L., & Sivakumar, C. (2008). Regional simulation of a groundwater flow in coastal aquifer, Tamil Nadu, India. In S. Ahmed, R. Jayakumar, & A. Salih (Eds.), *Groundwater dynamics in hard rock aquifers* (pp. 234–242). Springer, Dordrecht. https://doi.org/10.1007/978-1-4020-6540-8_18

Freeze, R. A., & Cherry, J. A. (1979). *Groundwater*. Prentice Hall.

Garcia-Solsona, E., Garcia-Orellana, J., Masqué, P., Rodellas, V., Mejías, M., Ballesteros, B., & Domínguez, J. A. (2010). Groundwater and nutrient discharge through karstic coastal springs (*Castelló*, Spain). *Biogeosciences, 7*(9), 2625–2638. https://doi.org/10.5194/bg-7-2625-2010

Gleeson, J., Santos, I. R., Maher, D. T., & Golsby-Smith, L. (2013). Groundwater–surface water exchange in a mangrove tidal Creek: Evidence from natural geochemical tracers and implications for nutrient budgets. *Marine Chemistry, 156*, 27–37. https://doi.org/10.1016/j.marchem.2013.02.001

Hajati, M., Sutanudjaja, E., & Moosdorf, N. (2019). Quantifying regional fresh submarine groundwater discharge with the lumped modeling approach CoCa-RFSGD. *Water Resources Research, 55*(7), 5321–5341. https://doi.org/10.1029/2018wr024248

Hata, M., Sugimoto, R., Hori, M., Tomiyama, T., & Shoji, J. (2016). Occurrence, distribution and prey items of juvenile marbled sole Pseudopleuronectes yokohamae around a submarine groundwater seepage on a tidal flat in southwestern Japan. *Journal of Sea Research, 111*, 47–53. https://doi.org/10.1016/j.seares.2016.01.009

Hwang, D., Kim, G., Lee, Y., & Yang, H. (2005). Estimating submarine inputs of groundwater and nutrients to a coastal Bay using radium isotopes. *Marine Chemistry, 96*(1–2), 61–71. https://doi.org/10.1016/j.marchem.2004.11.002

Idczak, J., Brodecka-Goluch, A., Łukawska-Matuszewska, K., Graca, B., Gorska, N., Klusek, Z., Pezacki, P. D., & Bolałek, J. (2020). A geophysical, geochemical and microbiological study of a newly discovered pockmark with active gas seepage and submarine groundwater discharge (MET1-BH, central Gulf of Gdansk, southern Baltic Sea). *Science of the Total Environment, 742*, Article 140306. https://doi.org/10.1016/j.scitotenv.2020.140306

Jaffé, R., Gardinali, P. R., Cai, Y., Sudburry, A., Fernandez, A., & Hay, B. J. (2003). Organic compounds and trace metals of anthropogenic origin in sediments from Montego Bay, Jamaica: Assessment of sources and distribution pathways. *Environmental Pollution, 123*(2), 291–299. https://doi.org/10.1016/s0269-7491(02)00368-8

Kaleris, V., Lagas, G., Marczinek, S., & Piotrowski, J. A. (2002). Modelling submarine groundwater discharge: An example from the western Baltic Sea. *Journal of Hydrology, 265*(1–4), 76–99.

Krall, L., Trezzi, G., Garcia-Orellana, J., Rodellas, V., Mörth, C., & Andersson, P. (2017). Submarine groundwater discharge at Forsmark, Gulf of Bothnia, provided by Ra isotopes. *Marine Chemistry, 196*, 162–172. https://doi.org/10.1016/j.marchem.2017.09.003

Kumar, A., & Thambidurai, P. (2022). Relationship of topography, weathering, and water table in Usri watershed of Chota Nagpur Plateau, India. *Arabian Journal of Geosciences, 5–504*, 1–16. https://doi.org/10.1007/s12517-022-09624-z

Kumar, A. & Thambidurai, P. (2023). Impact of urbanization and river morphology on groundwater system in Patna Urban Area, Bihar, India. In P. Thambidurai, & A. K. Dikshit (Eds.), *Impacts of urbanization on hydrological systems in India* (pp. 167–185). Springer, Cham. https://doi.org/10.1007/978-3-031-21618-3_9

Langevin, C. D. (2003). Simulation of submarine ground water discharge to a marine estuary: Biscayne Bay, Florida. *Ground Water, 41*(6), 758–771. https://doi.org/10.1111/j.1745-6584.2003.tb02417.x

Laroche, J., Nuzzi, R., Waters, R., Wyman, K., Falkowski, P., & Wallace, D. (1997). Brown tide blooms in Long Island's coastal waters linked to interannual variability in groundwater flow. *Global Change Biology, 3*(5), 397–410. https://doi.org/10.1046/j.1365-2486.1997.00117.x

Lee, Y., Rahman, M. M., Kim, G., & Han, S. (2011). Mass balance of total mercury and Monomethylmercury in coastal embayments of a volcanic island: Significance of submarine groundwater discharge. *Environmental Science & Technology, 45*(23), 9891–9900. https://doi.org/10.1021/es202093z

Luijendijk, E., Gleeson, T., & Moosdorf, N. (2020). Fresh groundwater discharge insignificant for the world's oceans but important for coastal ecosystems. *Nature Communications, 11*(1). https://doi.org/10.1038/s41467-020-15064-8

Mathiazhagan, M., Jha, S. K., & Biswas, A. (2021). Assessment of groundwater mass balance and zone budget in the semi-arid region: A case study of Palar sub-basin, Tamil Nadu, India. *Journal of Earth System Science, 130*(4), 187. https://doi.org/10.1007/s12040-021-01691-2

Miller, M. P., Buto, S. G., Susong, D. D., & Rumsey, C. A. (2016). The importance of base flow in sustaining surface water flow in the upper Colorado River Basin. *Water Resources Research, 52*(5), 3547–3562. https://doi.org/10.1002/2015wr017963

Moore, W. S. (2010). The effect of submarine groundwater discharge on the ocean. *Annual Review of Marine Science, 2*(1), 59–88. https://doi.org/10.1146/annurev-marine-120308-081019

Pasanen, A. H., & Okkonen, J. S. (2017). 3D geological models to groundwater flow models: Data integration between GSI3D and groundwater flow modelling software GMS and FeFlow®. *Geological Society, London, Special Publications, 408*(1), 71–87. https://doi.org/10.1144/sp408.15

Peterson, R. N., Burnett, W. C., Taniguchi, M., Chen, J., Santos, I. R., & Ishitobi, T. (2008). Radon and radium isotope assessment of submarine groundwater discharge in the Yellow River delta, China. *Journal of Geophysical Research: Oceans, 113*(C9). https://doi.org/10.1029/2008jc004776

Porcelli, D., & Swarzenski, P. W. (2003). 8. The behavior of U- and Th-series nuclides in groundwater. *Uranium-series Geochemistry*, 317–362. https://doi.org/10.1515/9781501509308-013

Robinson, C. E., Xin, P., Santos, I. R., Charette, M. A., Li, L., & Barry, D. (2018). Groundwater dynamics in subterranean estuaries of coastal unconfined aquifers: Controls on submarine groundwater discharge and chemical inputs to the ocean. *Advances in Water Resources, 115*, 315–331. https://doi.org/10.1016/j.advwatres.2017.10.041

Rodellas, V., Garcia-Orellana, J., Masqué, P., Feldman, M., & Weinstein, Y. (2015). Submarine groundwater discharge as a major source of nutrients to the Mediterranean Sea. *Proceedings of the National Academy of Sciences, 112*(13), 3926–3930. https://doi.org/10.1073/pnas.1419049112

Ruiz-González, C., Rodellas, V., & Garcia-Orellana, J. (2021). The microbial dimension of submarine groundwater discharge: Current challenges and future directions. *FEMS Microbiology Reviews, 45*(5). https://doi.org/10.1093/femsre/fuab010

Russo, A., Boutt, D. F., Munk, L. A., & Jenckes, J. (2023). Contribution of fresh submarine groundwater discharge to the Gulf of Alaska. https://doi.org/10.22541/essoar.168121396.66122111/v1

Sajeev, S., Muthukumar, P., & Selvam, S. (2023). Submarine groundwater discharge: An Asian overview. *Chemosphere, 325*, Article 138261. https://doi.org/10.1016/j.chemosphere.2023.138261

Sawyer, A. H., David, C. H., & Famiglietti, J. S. (2016). Continental patterns of submarine groundwater discharge reveal coastal vulnerabilities. *Science, 353*(6300), 705–707. https://doi.org/10.1126/science.aag1058

Senthilkumar, M., & Elango, L. (2004). Three-dimensional mathematical model to simulate groundwater flow in the lower Palar River basin, southern India. *Hydrogeology Journal, 12*, 197–208.

Shaw, R., & Prepas, E. (1990). Groundwater-lake interactions: I. Accuracy of seepage meter estimates of lake seepage. *Journal of Hydrology, 119*(1–4), 105–120. https://doi.org/10.1016/0022-1694(90)90037-x

Swarzenski, P. W., & Baskaran, M. (2007). Uranium distribution in the coastal waters and pore waters of Tampa Bay, Florida. *Marine Chemistry, 104*(1–2), 43–57. https://doi.org/10.1016/j.marchem.2006.05.002

Szymczycha, B., Kroeger, K. D., & Pempkowiak, J. (2016). Significance of groundwater discharge along the coast of Poland as a source of dissolved metals to the southern Baltic Sea. *Marine Pollution Bulletin, 109*(1), 151–162. https://doi.org/10.1016/j.marpolbul.2016.06.008

Szymczycha, B., Vogler, S., & Pempkowiak, J. (2012). Nutrient fluxes via submarine groundwater discharge to the Bay of puck, southern Baltic Sea. *Science of the Total Environment, 438*, 86–93. https://doi.org/10.1016/j.scitotenv.2012.08.058

Tait, D. R., Santos, I. R., Erler, D. V., Befus, K. M., Cardenas, M. B., & Eyre, B. D. (2013). Estimating submarine groundwater discharge in a South Pacific coral reef lagoon using different radioisotope and geophysical approaches. *Marine Chemistry, 156*, 49–60. https://doi.org/10.1016/j.marchem.2013.03.004

Taniguchi, M., Dulai, H., Burnett, K. M., Santos, I. R., Sugimoto, R., Stieglitz, T., Kim, G., Moosdorf, N., & Burnett, W. C. (2019). Submarine groundwater discharge: Updates on its measurement techniques, geophysical drivers, magnitudes, and effects. *Frontiers in Environmental Science, 7*. https://doi.org/10.3389/fenvs.2019.00141

Trezzi, G., Garcia-Orellana, J., Rodellas, V., Santos-Echeandia, J., Tovar-Sánchez, A., Garcia-Solsona, E., & Masqué, P. (2016). Submarine groundwater discharge: A significant source of dissolved trace metals to the north western Mediterranean Sea. *Marine Chemistry, 186*, 90–100. https://doi.org/10.1016/j.marchem.2016.08.004

Unnikrishnan, P., Srinivas, R., Ramasamy, M., & Suresh Babu, D. (2021). Computation of submarine groundwater discharge from geomorphologically different coastal catchments of SW India using numerical modeling. *Regional Studies in Marine Science, 47*, Article 101963. https://doi.org/10.1016/j.rsma.2021.101963

Walker, D. (n.d.). *Nutrient pollution of coastal rivers, bays, and seas*. U.S. Geological Survey Publications Warehouse. https://pubs.usgs.gov/publication/70174406

Zhou, Y., Befus, K. M., Sawyer, A. H., & David, C. H. (2018). Opportunities and challenges in computing fresh groundwater discharge to continental coastlines: A Multimodel comparison for the United States Gulf and Atlantic coasts. *Water Resources Research, 54*(10), 8363–8380. https://doi.org/10.1029/2018wr023126

Zhou, Y., Sawyer, A. H., David, C. H., & Famiglietti, J. S. (2019). Fresh submarine groundwater discharge to the near-global coast. *Geophysical Research Letters, 46*(11), 5855–5863. https://doi.org/10.1029/2019gl082749

Zhu, A., Saito, M., Onodera, S., Shimizu, Y., Jin, G., Ohta, T., & Chen, J. (2019). Evaluation of the spatial distribution of submarine groundwater discharge in a small island scale using the 222Rn tracer method and comparative modeling. *Marine Chemistry, 209*, 25–35. https://doi.org/10.1016/j.marchem.2018.12.003

Heavy Metals Eminence in Coastal Groundwater of Manali Industrial Region in Chennai, India

S. Senthilkumar, K. Srinivasamoorthy, A. Rajesh Kanna, and C. Babu

1 Introduction

Groundwater is an essential and vital component of our life support system. Groundwater resources are utilized for industrial, agricultural, and potable reasons. Groundwater resources are strained due to rapid population growth, urbanization, industrialization, and agricultural practices (Ramesh et al., 2023; Visvalingam et al., 2024). Concerns over the deterioration of groundwater quality caused by anthropogenic and geogenic activities are emerging (Ayotte et al., 2011; Senthilkumar et al., 2017a). The kind and stratification of soils and subsurface geological formations through which groundwater traverses is significantly influence its natural chemical makeup. The atmosphere and surface water bodies influence groundwater quality (Srinivasamoorthy et al., 2011; Gopinath & Srinivasamoorthy, 2015; Vazhacharickal & Thambidurai, 2021; Senthilkumar et al., 2023). Moreover, human activities substantially influence groundwater quality. Excessive groundwater extraction in coastal regions can result in seawater intrusion and elevated groundwater salinity. The widespread application of fertilizers and pesticides in agriculture, along with the inadequate disposal of urban and industrial waste, might result in the contamination of groundwater resources (Ayotte et al., 2011; Senthilkumar et al., 2017a). Groundwater consists of many dissolved inorganic chemical elements at varying concentrations, resulting from the chemical and biological interactions between water and geological minerals. Inorganic contaminants, including salinity, chloride, fluoride, nitrate, iron, and arsenic, are essential for evaluating the potability of groundwater, which is acknowledged as the most critical natural resource underpinning the ecological system.

Trace elements are referred to by various names, including potentially toxic elements, trace metals, heavy metals, micro nutrients, and minor elements. The term

S. Senthilkumar (✉) · K. Srinivasamoorthy · A. Rajesh Kanna · C. Babu
Department of Earth Sciences, Pondicherry University, Puducherry, India
e-mail: siva.senthil.geo@gmail.com

P. Thambidurai and A. K. Dikshit (eds.), *Coastal Environments of India*,
https://doi.org/10.1007/978-3-031-97341-3_9

"potentially toxic elements" refers to those elements that can be harmful to humans and plants, but not all of them exhibit toxicity (Yang et al., 2023). Trace metals such as Fe, Mn, Cu, Zn, Co, and Ni serve as micronutrients for living organisms; their deficiency or excess can result in various health issues in humans. Typically, trace elements in groundwater exist in very low concentrations, often below 1 mg/L (USGS, 1993). Certain trace elements are classified as micronutrients in living organisms, and they are required in daily amounts ranging from 50 to 18 mg (USEPA, 2011). A collection of 15 trace elements, comprising As, Cu, Cd, Fe, Zn, Cr, B, Pb, Sr, Ba, Mn, Co, Ni, Se, and Al, has been classified according to their position in the periodic table and toxicity levels into the following categories: toxic elements (Pb, As), alkaline earth metals (Sr, Ba), transition metals (Mn, Co, Ni), metallic elements (Cu, Cd, Fe, Zn, Cr), and non-metals (B, Se, Al). The concentrations of trace elements in groundwater were analyzed in aquifers from eight principal geological material categories: unconsolidated sand and gravel; glacial unconsolidated sand and gravel; semi-consolidated sand; sandstone; sandstone and carbonate rock; basaltic and other volcanic rock; and crystalline rock (Ayotte et al., 2011). Diverse methodologies exist for assessing heavy metal pollution. A key approach entails evaluating the spatial distribution of heavy metals in industrial areas and mapping their locations using geographic information systems (GIS) (Hou et al., 2017). Pollution indices offer an effective method for evaluating heavy metal contamination. This study examines important ion chemistry and the variation of trace elements as examined by Ravichandran and Jayaprakash (2011), Senthilkumar et al. (2017a, (2023), and Syed Ahamudul Rafeek et al. (2024). This study seeks to ascertain the spatial distribution of heavy metal concentrations and assess the environmental risk linked to these metals through the Heavy Metal Pollution Index (HIP).

Study Area

The Manali, Ennore region and its vicinity are located in the Thiruvallur District, with an area of 1402.79 km^2. This study region is geographically situated between 12°15′ N and 13°15′ N latitude and 79°15′ E and 80°20′ E longitude, as seen on Survey of India Toposheets No. 66C3, 66C4, 66C7, and 66C8. It is located in the northeastern region of Tamil Nadu, India (Fig. 1). The Thiruvallur district is delineated by the Bay of Bengal to the east, Andhra Pradesh to the north, Chennai city to the southeast, Kanchipuram to the south, and Vellore to the west. The region experiences an annual precipitation of approximately 1200 mm, with roughly 60% derived from the northeast monsoon occurring from October to December, and the other 40% resulting from the southwest monsoon occurring from July to September (CGWB, 2007).The interval from April to June is typically characterized by high temperatures and aridity (CGWB, 2007; Senthilkumar et al., 2021).

The yearly average lowest and maximum temperatures are 24.3 °C and 32.9 °C, respectively. The primary drainage systems within the research region include Tanks, the Korattalayar River, and the Buckingham Canal. The Buckingham Canal extends in a north–south orientation, reaching the Korattalayar River and ultimately connecting with the Bay of Bengal, exhibiting a non-perennial character. The Korattalaiyar River has unconsolidated coarse-grained sands, gravels, pebbles, clay, and sandstones, flowing in an easterly direction (Senthilkumar et al., 2017b). The area

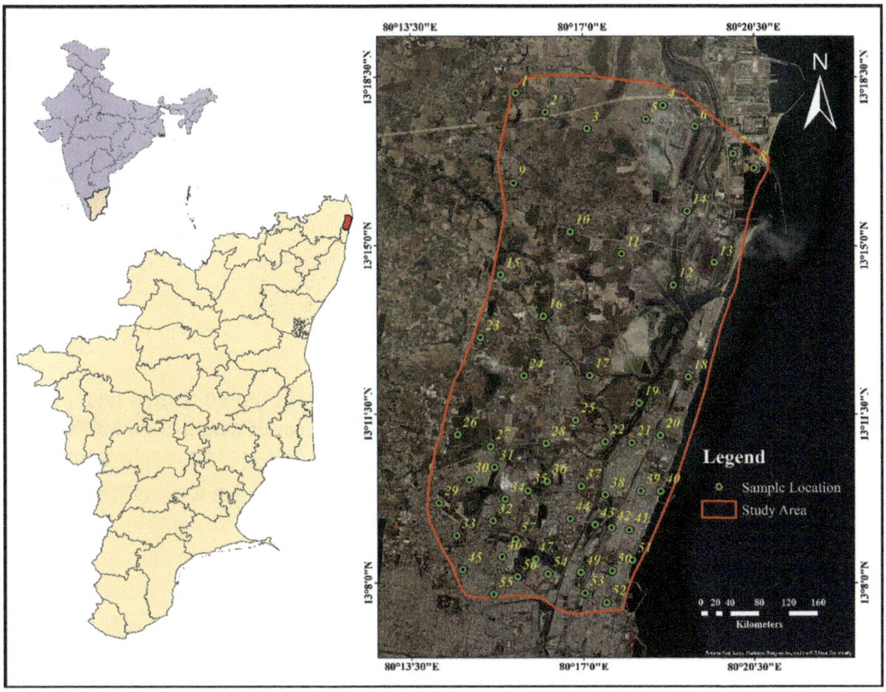

Fig. 1 Sampling location map of the study area

experiences precipitation from both the southwest and northeast monsoons. The northeast monsoon mostly influences the precipitation in the study area (CGWB, 2007). The geological map of the research region is derived from the district resource map created by the Geological Survey of India. The geology of the study area is characterized by the Upper Gondwana formation. The geological formations of the research region comprise Jurassic-age sandstones and clays, Cretaceous-age marine sediments, Tertiary-age sandstones, shales, and clays, as well as recent coastal alluvial deposits. Quaternary deposits consisting of sand, silt, and clay are restricted to the principal units of the study region. Groundwater exists under phreatic to semi-confined conditions within the inter-granular pore spaces of sands and sandstones, as well as in the bedding planes and minor fractures of shale (CGWB, 2007). The western section of the study area is comprised of agricultural fields. Another 579.27 km^2 of the total area is designated for agricultural use, whereas another 388.3608 km^2 is classified as barren ground, commonly referred to as wasteland. Extensive water bodies cover 187.43 km^2, whereas shallow water bodies, including ponds and small tanks, occupy 74.18 km^2 of the studied area. Vegetated settlements encompass 164.83 km^2 of the research area. The primary geomorphological components of the research area consist of a plain terrain that gently slopes eastward in the northern and central regions, and southward in the southern region. The significant geomorphic units discerned in the district by satellite imagery analysis include alluvial plains, built-up land, coastal plains, flood plains, and sedimentary highlands. The southern region of the district features sedimentary plains characterized by diverse landforms, including natural levees, swales, marshy areas, lagoon/backwater coastal plains, and beach ridges (CGWB, 2007). The predominant soil types in the study area are clay and loamy clay, with minor areas of loamy sand located in the coastline region.

2 Material and Methods

A field research was conducted, and water samples were obtained from 47 accessible tubewells and dug wells. Physicochemical parameters such as Electronic Conductivity (EC), Total Dissolved Solids (TDS), Salinity, pH, Dissolved Oxygen (DO), and temperature are assessed at the sampling location. Water samples were collected in 1-L polyethylene bottles for hydrogeochemical analysis, and 250 ml samples were acidified with 1 ml of pure HNO_3 to achieve a pH of 2.0. All samples were thereafter stored in the laboratory for further analysis. The analysis of trace elements was conducted using inductively coupled plasma mass spectrometry (ICP-MS Agilent 7700X) at the Department of Earth Sciences, Pondicherry University, following established protocols (APHA, 2017; Vinnarasi et al., 2022) to ascertain the concentrations of trace metals including As, Pb, Sr, Ba, Mn, Co, Ni, Cu, Cd, Fe, Zn, Cr, B, Se, and Al. The outcomes have been rigorously evaluated employing diverse geochemical methodologies for the assessment of heavy metal pollution indices. Spatial distribution maps for various trace element concentrations have been generated using the ArcGIS platform, encompassing diverse physicochemical and geochemical parameters for the study area and subsequently analyzed.

3 Results and Discussion

The analyzed water sample data from the research region indicates that the pH value ranges from 6.04 to 8.3, with an average of 6.7. The pH values of the groundwater samples range from acidic to basic. The electrical conductivity (EC) varies from 692 to 19,093 μs/cm, with a mean of 3977 μs/cm, while the total dissolved solids (TDS) range from 443 mg/l to 12,220 mg/l, averaging 2545 mg/l. All groundwater samples collected from the Ennore area are classified as fresh, brackish, or saline water types. The heightened levels of EC, TDS, and salinity indicate the potential for seawater intrusion resulting from excessive groundwater extraction and tidal effects (Rajesh Kanna et al., 2023). The trace elements identified in the groundwater analysis of the study area have been categorized into distinct groups: toxic elements (Pb and As), alkaline earth metals (Sr and Ba), transition metals (Mn, Co, and Ni), metallic elements (Cu, Cd, Fe, Zn, and Cr), and other nonmetallic elements (B, Se, and Al), as detailed in Table 1. Jinwal et al. (2009) estimated that trace metals like Fe, Mn, Cu, Zn, Co, and Ni are micronutrients for the living system. Their deficiency or excess can lead to a number of disorders in the human body. Some trace metals like Cd, Pb and Cr can be lethal to human beings even at low concentrations because of their tendency to accumulate in the body (Khan et al., 2008) (Table 2).

Arsenic (As) naturally occurs in ores and can be concentrated in sulfide-bearing minerals, such as arsenopyrite (FeAsS). Arsenic infiltrates the environment by natural and anthropogenic mechanisms (Shukla et al., 2010). Arsenic contaminates groundwater through industrial effluents, atmospheric deposition, and the application of pesticides, insecticides, and herbicides; the permitted concentration of arsenic is 0.05 mg/L, as stipulated by the Bureau of Indian Standards (BIS). The concentration of arsenic in the research region varied from 0.32 to 42.36 μg/L, with a mean value of 6.79 μg/L (Fig. 2a), which is below the permitted standard limit. Possible sources of lead in groundwater encompass diesel fuel widely utilized in agricultural regions, discarded batteries, paint, and leaded gasoline. Lead is utilized in certain pesticides, including lead arsenate. Lead is detrimental to both the central and peripheral nervous systems, resulting in sub-encephalopathic neurological and behavioral consequences. Excessive intake may lead to hearing loss, hematological problems, hypertension, and potentially deadly outcomes (ISI, 1991; WHO, 1993). The acceptable lead concentration in drinking water was established at 0.01 mg/L (Fig. 2b), as per ISI (1991). The concentration of Pb in the studied area varied from Not Detectable (ND) to 0.46 μg/L, with an average of 0.07 μg/L, which is below

Table 1 Statistical measurement of trace elements of the groundwater samples during Pre-monsoon period (values are in μg/L except for EC in μs/cm and TDS mg/l)

Variables	Max	Min	Avg	SD
pH	8.30	6.04	6.70	0.40
EC	1909.30	692.10	39,772.70	3359.80
TDS	12,220.00	443.00	2545.40	2150.30
Ni	498.00	10.49	984.90	90.55
Cu	29.66	0.53	6.02	5.37
Zn	9031.00	0.26	296.09	1521.95
As	42.36	0.31	6.79	7.93
Ba	1435.00	5.46	170.00	288.05
Fe	757.00	6.56	132.45	191.59
Mn	5091.00	0.24	541.04	976.31
Cr	21.50	0.40	6.61	4.59
Sr	8976.00	183.64	1889.92	1759.25
Co	7.88	0.07	0.75	1.42
Al	1.44	0.12	0.69	0.36
Pb	0.46	0.00	0.07	0.12

Table 2 Risk-based drinking water criteria of trace elements (in mg/l)

Parameter	WHO	USEPA	BIS
Li	–	–	–
Al	0.2	0.2	0.2
Cr	0.05	0.1	0.05
Mn	0.4	0.05	0.5
Fe	–	0.3	1
Co	–	–	–
Ni	0.07	0.1	0.02
Cu	2	1.3	1.5
Zn	3	5	10
As	0.01	0.01	0.05
Se	0.04	0.05	0.01
Mo	0.07	–	0.07
Cd	0.003	0.005	0.01
Ba	0.7	2	0.7
Pb	0.01	0.015	0.1
pH	6.5–8.5	6.5–8.5	6.5–9.2
TDS	1000	500	2000
EC	1500	–	–

the allowed limit of 0.01 mg/L (WHO, 2017). Strontium (Sr) minerals are extensively disseminated across the planet and are introduced into groundwater through the natural re-crystallization and weathering of rocks and soils (Greve et al., 2007). Elevated quantities of strontium in water and soil, when ingested, can result in fractures and osteodystrophy. The concentration of strontium exceeded the acceptable limit of 0.07 mg/L (WHO, 1984) over the area, ranging from 183.64 to 8976 μg/L, with an average of 1889 μg/L (Fig. 3a). The elevated amounts primarily result from anthropogenic causes. Strontium concentrations in soil may also result from the disposal of industrial waste. Strontium in soil solubilizes in water, enabling it to percolate deeper into the groundwater. Barium (Ba) exists in many compounds within the earth's crust and is utilized in numerous industrial applications; nevertheless, it mostly enters groundwater from natural sources. The concentration of barium (Ba) in the groundwater at the chosen study location ranged from 5.46 to 1435 μg/L, with a mean value of 170 μg/L (Fig. 3b). Barium ions primarily originate from natural sources; however, due to industrial applications of barium, sample locations such as G1 and G21 exceeded the legal limit of 0.7 mg/L (WHO, 1984), while other areas in the research region remained under the acceptable level. Manganese (Mn) is among the most prevalent metals in the Earth's crust and typically coexists with iron. The presence of manganese in aerated water frequently results from industrial contamination. Regarding oral consumption, manganese (Mn) is frequently considered one of the least harmful elements; yet, various investigations have indicated a correlation between elevated Mn intake and neurological consequences. The WHO has established a health-based recommended value of 0.4 mg/l (WHO, 2004). The concentration range of the Manganese ion varies from 0.24 to 5091 μg/L (Fig. 4a). Cobalt (Co) is one of the three magnetic minerals, alongside iron (Fe) and nickel, commonly utilized in magnet alloys. It possesses significant industrial applications. Cobalt is utilized in the paint and dye industries; cobalt deficiency has been linked to disruptions in vitamin B12 production. The concentration of cobalt in the groundwater of the research area varies from 0.07 to 7.88 μg/L, with a mean value of 0.75 μg/L (Fig. 4b). Elevated levels of this metal in the human body may lead to hypothyroidism, erythrocytosis, pulmonary fibrosis, and asthma.

Soil, water, air, and biological components all contain nickel (Ni), a naturally occurring element. Natural sources of nickel are dust from volcanic emissions and the weathering of rocks and soils. The nickel concentration in the research area varies from 10.99 to 498 μg/L, with a mean value of 84.9 μg/L (Fig. 5a). Inorganic fertilizers, especially phosphate fertilizers, exhibit varying concentrations of nickel. The BIS (2012) specifies that the standard content of Nickel in groundwater is 0.02 mg/l. Copper (Cu) is a vital trace element for the bodies of plants and animals. Copper is a malleable metal frequently employed in substantial quantities as an electrical conductor, including in alloys such as brass and bronze. Industries such as steel production, plastic manufacturing, and blast furnaces contribute to the accumulation of copper in water. Furthermore, agrochemicals and waste disposal sites represent significant sources. The values of Cu in the study area ranged from 0.53 to 29.66 μg/L,

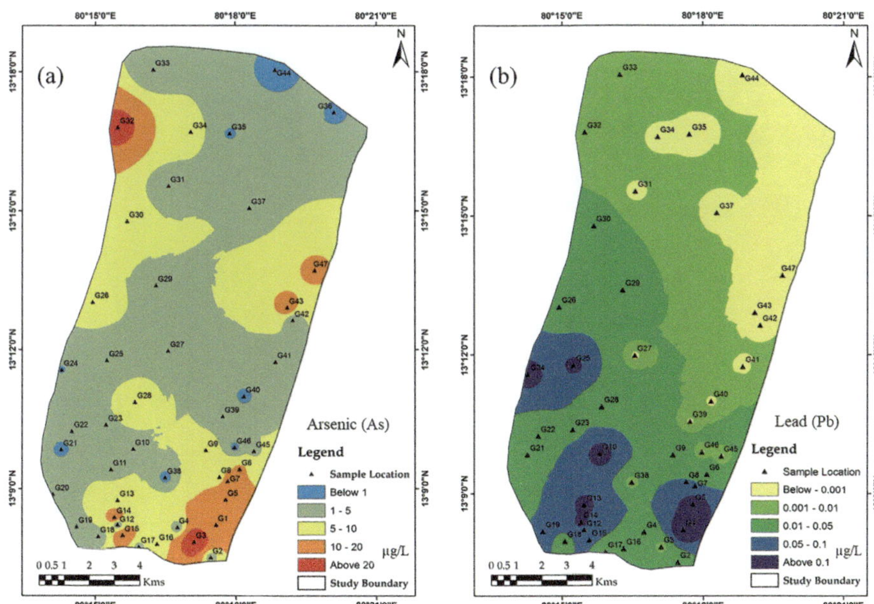

Fig. 2 Spatial distribution map of arsenic (**a**) and lead (**b**)

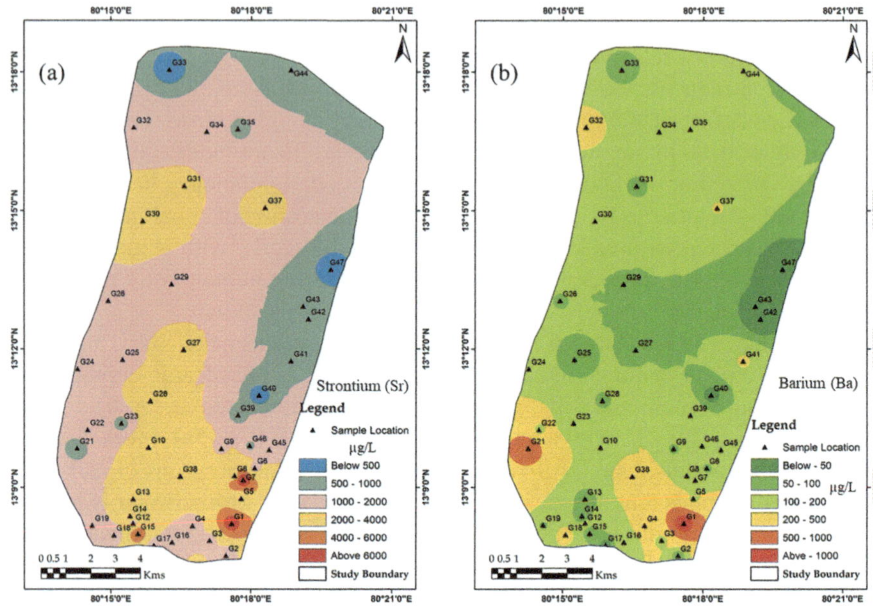

Fig. 3 Spatial distribution map of strontium (**a**) and barium (**b**)

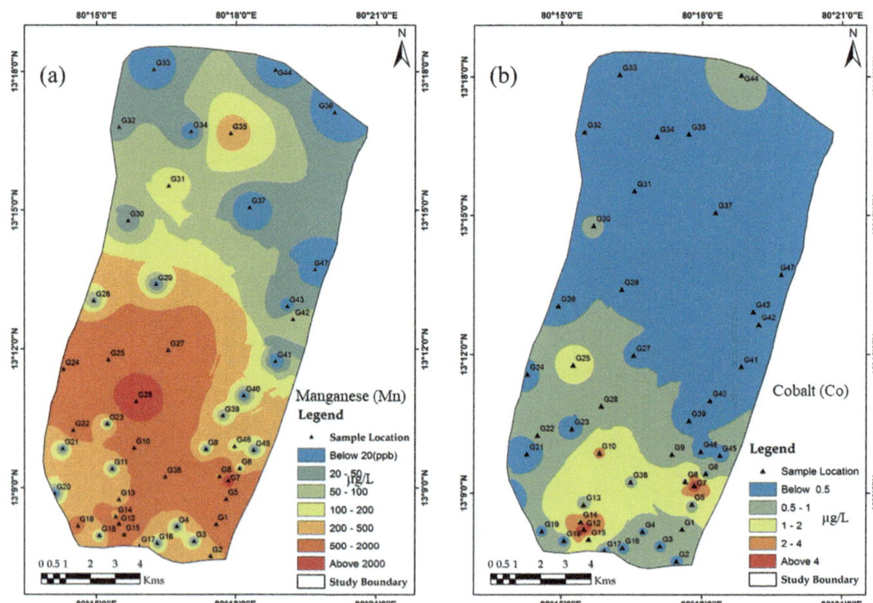

Fig. 4 Spatial distribution map of Manganese (**a**) and Cobalt (**b**)

with an average of 6.02 μg/L (Fig. 5b). According to BIS guidelines, certain samples fall within the maximum allowable amount of 1.5 mg/L. Copper (Cu) deficiency is rare in healthy persons, though it may manifest in babies. The principal symptoms linked to copper deficiency are fatigue, anemia, and diminished white blood cell count. Iron (Fe), the most abundant element by mass on Earth, is a substantial part of both the outer and inner core. It is the fourth most prevalent element, following oxygen, silicon, and aluminum. The body's iron content is approximately 3–4 g, equating to a concentration of about 40–50 mg of Fe per kilogram of body weight. The recommended dietary allowance (RDA) for iron (Fe) in healthy people is established at 8 mg per day for men and post-menopausal women, and 18 mg per day for menstruating women, owing to the considerable blood loss associated with their menstrual cycle. Zinc (Zn) is an essential trace element found in nearly all food sources and drinking water, either in the form of salts or chemical complexes. Zinc concentrations range from 0.26 to 9031 μg/L, with a mean value calculated at 296 μg/L. As per BIS standard specifications, the Zn concentration values were beneath the permissible threshold of 10 mg/L; however, the G22 sample locations approached the maximum allowable limit, potentially leading to health issues such as dwarfism, dermatitis, and impaired taste perception (Fig. 6).

Chromium (Cr) is a vital trace element required by humans in small amounts. Chromium is prevalent in the dietary supply, however most food sources contain very minimal quantities. It can be found in foods including egg yolks, whole-grain goods, high-bran morning cereals, coffee, almonds, green beans, broccoli, beef,

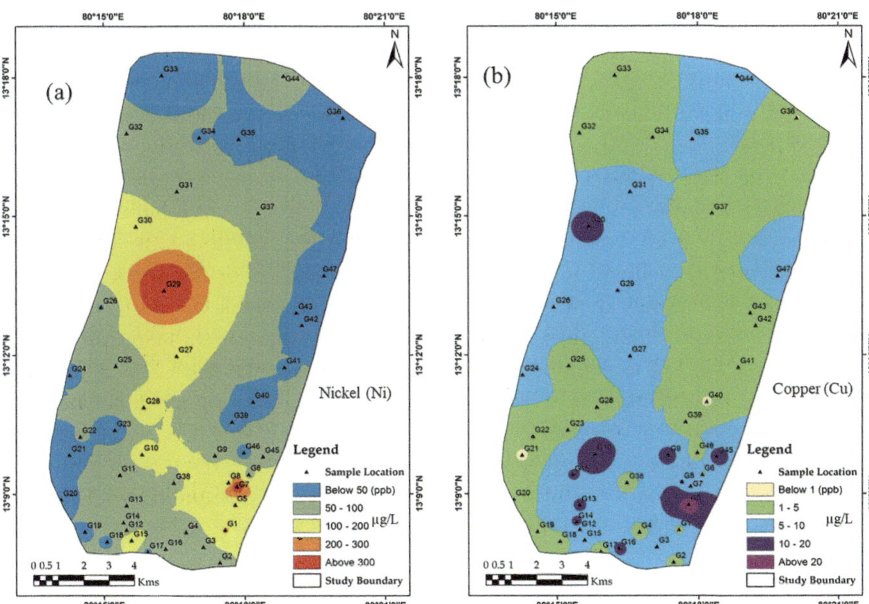

Fig. 5 Spatial distribution map of Nickel (a) and Copper (b)

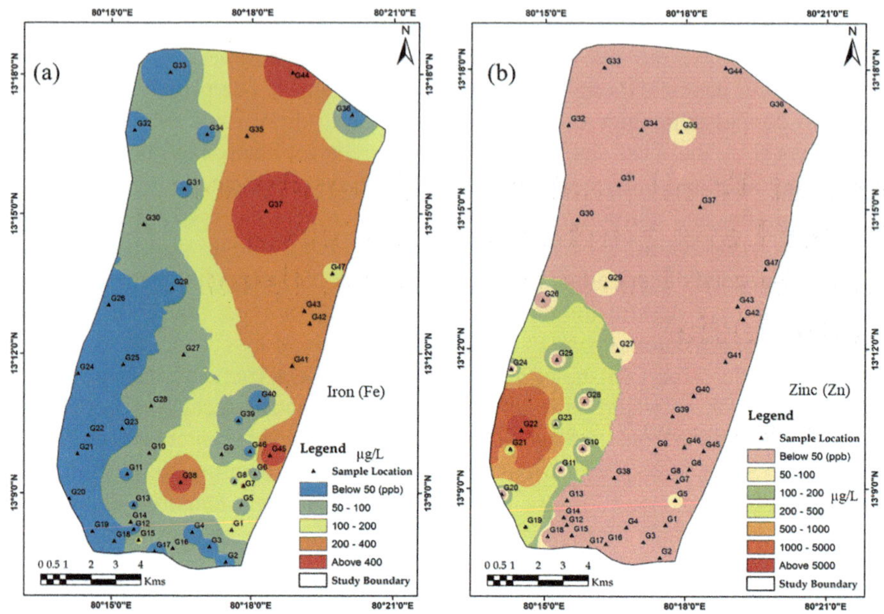

Fig. 6 Spatial distribution map of Iron (**a**) and Zinc (**b**)

and brewer's yeast. In the examined region, the concentration of Chromium varied from 0.40 to 21.50 µg/L, with a mean of 6.61 µg/L (Fig. 7a). Consequently, all samples exhibited low concentration levels, approaching the undesirable threshold of 0.05 mg/L according to the standard value (WHO, 2011). Increased concentrations of chromium in groundwater can result in nasal septum ulceration and dermatitis (Singh & Bhayana, 1986). Generally, elevated concentrations of Cr are expected in areas with alkaline rock formations; however, in this study location, the sole hypothesis is that the Cr present in groundwater is derived from anthropogenic sources. Aluminum (Al) concentrations in groundwater samples varied from 0.12 to 1.44 µg/L, with a mean concentration of 0.69 µg/L (Fig. 7b). The maximum allowable concentration of aluminum in the drinking water within the research area is considerably lower than the standard value of 0.2 mg/L (ISI, 1991). The origin of aluminum in groundwater may be ascribed to the weathering of bedrock and soil or to industrial operations that have contaminated the water supply with aluminum. Numerous investigations have established a correlation between neurological diseases and elevated aluminum levels in bodily tissues. A correlation has been observed between aluminum and Alzheimer's disease, as Alzheimer's patients exhibit elevated levels of aluminum in the grey matter of their brains.

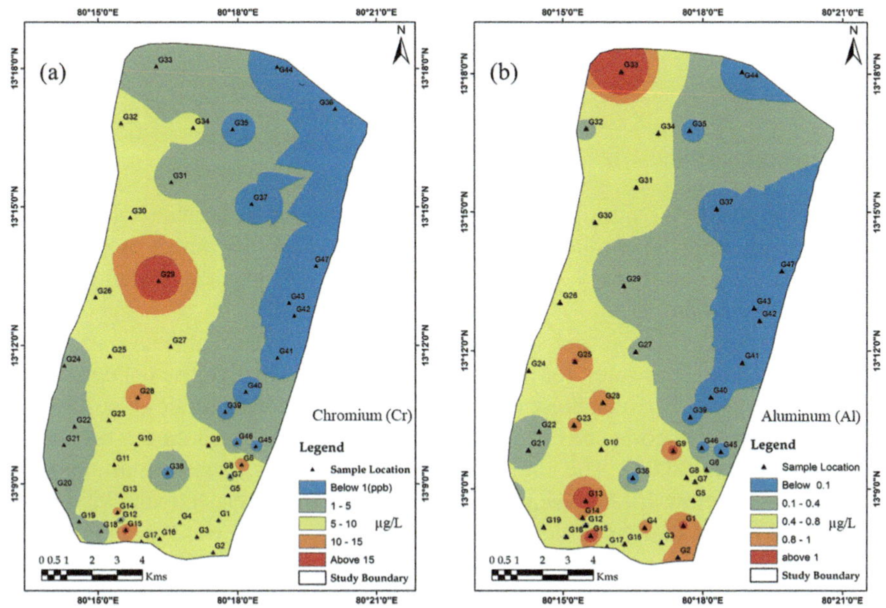

Fig. 7 Spatial distribution map of chromium (**a**) and aluminum (**b**)

3.1 Pollution Assessment Indices

The HPI methodology evaluates the total quantity of heavy metals in drinking water (Prasad & Sangita, 2008). This approach was established by allocating a weight (Wi) to each heavy metal. This weight is a numerical number between zero and one, inversely correlated with the suggested standard (Si) for each metal (Prasad & Jaiprakas, 1999). The highest Si denotes the maximum allowable content in potable water. The maximum allowable limits (li) denote the standard thresholds for the heavy metals present in drinking water (Table 3).

Table 3 Computation of Heavy metal Pollution Index (HPI) for G1 sample location

Heavy metals	Mean value (Mi) (µg/L)	Standard permissible value (Si) (µg/L)	Highest desirable value (Ii) (µg/L)	Sub index (Qi)	Unit weightage (Wi)	Qi*Wi
Pb	0.18	50	10	24.55	0.02	0.491
Cr	9.46	50	10	74	0.02	1.48
Cd	0.16	–	–	1.6	0.1	0.16
Zn	16.06	1500	50	3397.333	0.0002	0.6795
Cu	4.31	1500	50	4572.333	0.001	4.5723
∑Qi*Wi 7.3828			∑Wi = 0.1412		HPI = 52.28612	

The HPI was calculated using the equation outlined by Prasad and Jaiprakash in (1999).

$$HPI = \frac{\sum_{i=1}^{n} WiQi}{\sum_{i=1}^{n} Wi} \tag{1}$$

In this context, Wi signifies the unit weight of the (i) parameter, Qi represents the sub-index of the (i) parameter, and n symbolizes the total number of parameters considered. The sub-index value (Qi) for the parameter was calculated using the subsequent equation:

In this context, Mi, Si, and Ii denote the observed values of heavy metal, the standard, and the ideal value for the (i) parameter, respectively. The (–) denotes the numerical disparity between these two values, excluding the algebraic sign. A revised scale with three categories has been utilized in this study, based on the research of Edet and Offiong (2002). The categories are designated as low, medium, and high for HPI levels below 15, between 15 and 30, and over 30, respectively. Prasad and Bose (2001) assert that the critical pollution index for potable water is an HPI value of 100. The groundwater drinking standards established by the USEPA, WHO, and the Chinese drinking water standards (GB 5749–2006) are outlined in Table 4.

$$Qi = \sum_{i=1}^{n} \frac{\{Mi(-)li\}}{(Si - li)} \times 100 \tag{2}$$

To calculate the HPI of the groundwater in Ennore and Manali industrial area, the concentrations of Cr, Cd, Zn, Cu, Pb were considered. Calculated HPI values of the study area is range from 32.53 to 1423.75 with average of 156.79. Based on HPI value, all the groundwater location is high contamination stage. Spatial distribution map of the heavy metal pollution index is shown in Fig. 8.

According to the HPI value, all groundwater sites exhibit a high level of contamination. As per the critical pollution index for drinking water suggested by Prasad and Bose (2001), HPI values below 100 indicate critical contamination; in this study area, 68% of the samples fall into the category of critical contamination due to heavy metals. These findings suggest that, overall, the groundwater is severely contaminated with heavy metals.

Table 4 Drinking standards of groundwater for toxic metals in μg/L

Heavy metals	USEPA (2012)	WHO (2011)	GB 5749–2006 (2006)
Cr	100	50	50
Ni	NA	70	20
Zn	5000	NA	1000
As	10	10	10
Pb	15	10	10

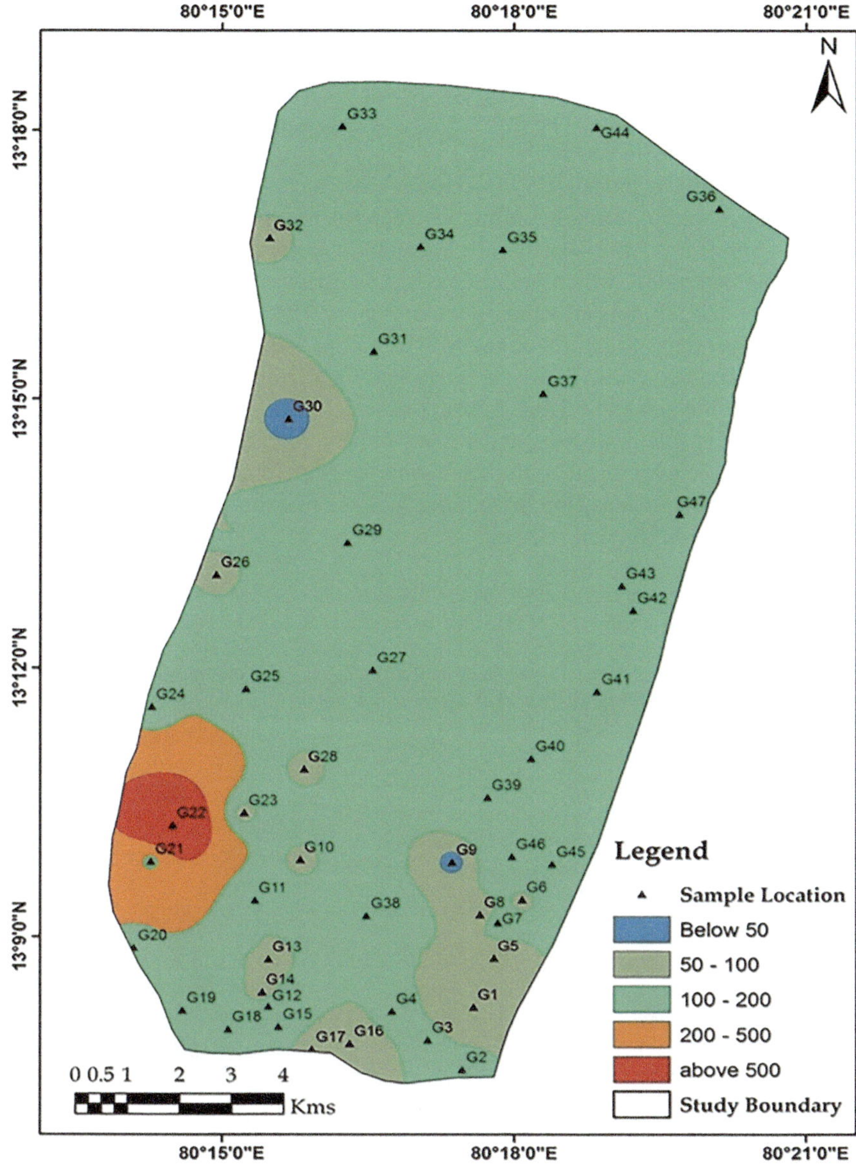

Fig. 8 Spatial distribution map of calculated heavy metal pollution index

4 Conclusion

The study region mainly hinge on on groundwater, surface and rainwater for drinking and domestic commitments. The groundwater exhibits a range from acidic to basic pH levels. According to the evaluations of Total Dissolved Solids (TDS) and the Heavy Metal Pollution Index (HPI), it is determined that the groundwater analyzed is generally unsuitable for drinking, with the exception of a few samples from western tubewells. Based on Trace metals, all are under the maximum permissible limit, except Ni concentration, which only exceeds the maximum permissible limits in all locations, which is strongly affected by contaminants. The strong positive correlation analysis results suggest that ions like Cr, Ni, Sr, Co, Mo, Zr, Y, and Nb are led to common pollution sources. Thus, the analytical data from the Ennore and Manali industrial area shows that the Trace metal concentration of groundwater is under the maximum permissible limits except for a few metals that exceed the limits to affect human health. The geological composition and landform characteristics, along with additional elements such as land utilization, indicate that the groundwater is mostly inappropriate for drinking and domestic use, with a few exclusions.

Acknowledgements The primary author acknowledges the University Grant Commission (UGC) in New Delhi for the financial support provided under the D.S. Kothari Postdoctoral Fellowship (DSKPDF), Grant No. (F. 4-2/2006 (BSR)/E.S./18-19/0006 dated 26/12/2018). Furthermore, heartfelt gratitude is expressed to the Department of Earth Sciences at Pondicherry University for permitting access to the ICPMS facility for the purpose of conducting analysis.

References

APHA. (2017). *Standard methods for the examination of water and wastewater* (23rd1st ed.). American Public Health Association.

Ayotte, J. D., Szabo, Z., Focazio, M. J., & Eberts, S. M. (2011). Effects of human-induced alteration of groundwater flow on concentrations of naturally occurring trace elements at water-supply wells. *Applied Geochemistry, 26*(5), 747–762.

BIS. (2012). Bureau of Indian Standards Specification for drinking water. IS: 10500:91. Revised 2012, Bureau of Indian Standards, New Delhi. BIS, 2.

CGWB. (2007). Central Ground Water Board, District groundwater brochure, Thiruvallur district, Tamil Nadu. CGWB, Chennai, 26.

Edet, A. E., & Offiong, O. E. (2002). Evaluation of water quality pollution indices for heavy metal contamination monitoring. a study case from Akpabuyo-Odukpani area, Lower Cross River Basin (southeastern Nigeria). *GeoJournal, 57*, 295–304.

Gopinath, S., & Srinivasamoorthy, K. (2015). Application of geophysical and hydrogeochemical tracers to investigate salinisation sources in Nagapatinam and Karaikal coastal aquifers, south India. *Aquatic Proc, 4*, 65–71.

Greve, K., Nielsen, E., & Ladefoged, O. (2007). Evaluation of health hazards by exposure to strontium in drinking water. *Toxicology Letters, 172*, 1–7.

Hou, D., O'Connor, D., Nathanail, P., Tian, L., & Ma, Y. (2017). Integrated GIS and multivariate statistical analysis for regional scale assessment of heavy metal soil contamination: A critical review. *Journal Environmental Pollution, 231*, 1188–1200.

ISI. (1991). Indian standard specifications for drinking water, I.S. 10500, New Delhi, 22.

Jinwal, A., Dixit, S., & Malik, S. (2009). Some trace elements investigation in ground water of Bhopal & Sehore district in Madhya Pradesh: India. *Journal of Applied Sciences and Environmental Management, 13*(4), 47–50.

Khan, S., Cao, Q., Zheng, Y. M., Huang, Y. Z., & Zhu, Y. G. (2008). Health risks of heavy metals in contaminated soils and food crops irrigated with wastewater in Beijing. *China. Environ. Pollut., 152*(3), 686–692.

Prasad, B., & Sangita, K. (2008). Heavy metal pollution index of ground water of an abandoned open cast mine filled with fly ash: A case study. *Mine Water and the Environment, 27*, 265–267.

Prasad, B., & Bose, J. M. (2001). Evaluation of heavy metal pollution index for surface and spring water near a limestone mining area of the lower Himalayas. *Environm. Geol., 41*, 183–188.

Prasad, B., & Jaiprakas, K. C. (1999). Evaluation of heavy metals in groundwater near mining area and development of heavy metal pollution index. *Journal of Environmental Science and Health, A34*(1), 91–102.

Rajesh Kanna, A., Srinivasamoorthy, K., Babu, C., & Ponnumani, G. (2023). Groundwater and pore water inputs to the Bay of Bengal supported by geochemical tracers: Example from Tamil Nadu and Pondicherry, East Coast of India. *Chemosphere, 336*, Article 139327. https://doi.org/10.1016/j.chemosphere.2023.139327

Ramesh, K., Srinivasamoorthy, K., Rajesh Kanna, A., Gopalakrishnan, V., Supriya Varshini, D., & Subramanian, S. (2023). Simulation of the impact of sea level rise groundwater flooding along the south-eastern coast of India. *Urban Climate, 52*, Article 101732.

Ravichandran, K., & Jayaprakash, M. (2011). Seasonal variation on physico-chemical parameters and trace metals in groundwater of an industrial area of north Chennai, India. *Indian Journal of Science and Technology, 4*(6), 646–649.

Senthilkumar, S., Balasubramanian, N., Gowtham, B., & Lawrence, J. F. (2017a). Geochemical signatures of groundwater in the coastal aquifers of Thiruvallur district, South India. *Applied Water Science, 7*, 263–274. https://doi.org/10.1007/s13201-014-0242-2

Senthilkumar, S., Gowtham, B., Srinivasamoorthy, K., & Gopinath, S. (2021). Hydrogeochemical delineation of groundwater fitness for drinking and agricultural utilities in Thiruvallur district, South India. *Arabian Journal of Geosciences, 14*, 1–20. https://doi.org/10.1007/s12517-021-06547-z

Senthilkumar, S., Gowtham, B., Sundararajan, M., Chidamparam., S, Lawrence, F. J., & Prasanna, M. V. (2017b). Impact of landuse on the groundwater quality along coastal aquifer of Thiruvallur district, South India. *Sustainable Water Resources Management, 4*(13), 1–25.

Senthilkumar, S., Srinivasamoorthy, K., & Gowtham, B. (2023). The effect of urbanization on groundwater quality and hydrochemical characteristics in Ennore coastal aquifers of Chennai, South India. In P. Thambidurai, & A. K. Dikshit, (Eds.), *Impacts of urbanization on hydrological systems in India* (pp. 273–291). Springer, Cham. https://doi.org/10.1007/978-3-031-21618-3_14.

Shukla, D. P., Dubey, C., Singh, N. P., Tajbakhsh, M., & Chaudhry, M. (2010). Sources and controls of arsenic contamination in groundwater of Rajnandgaon and Kanker District, Chattisgarh Central India. *Journal of Hydrology, 395*, 49–66.

Singh, K. P., & Bhayana, N. (1986). Geochemistry of groundwater of Ludhiana area, Punjab with special reference to its use for India. *Geol. Cong.*, 579–594.

Srinivasamoorthy, K., Nanthakumar, C., Vasanthavigar, M., Vijayaraghavan, K., Rajivgandhi, R., Chidambaram, S., Anandhan, P., Manivannan, R., & Vasudevan, S. (2011). Groundwater quality assessment from a hard rock terrain, Salem district of Tamil Nadu, India. *Arabian Journal of Geosciences, 4*, 91–102. https://doi.org/10.1007/s12517-009-0076-7

Syed Ahamudul Rafeek, S., Mohamed Basith, N., Mohamed Hanipha, M., Meera Moydeen Abdul Hameed, & Senthilkumar, S. (2024). Assessing the influence of hydrogeochemical characteristics and microbial communities on monsoon dynamics in groundwater quality in north Chennai city, India. *Physics and Chemistry of the Earth, Parts A/B/C, 136*, 103735, ISSN 1474-7065. https://doi.org/10.1016/j.pce.2024.103735

Vazhacharickal, P. J., Thambidurai, P. (2021). Wastewater irrigation and quality-practices, safe reuse and perspectives: Scenarios from urban agriculture production across railway gardens and farms across the Mumbai Metropolitan Region (MMR). In Tiwari, A. K., Kumar, A., Singh, A. K., Singh, T. N., Suozzi, E., Matta, G., & Russo, S. L. (Eds.), *India in water resources: Scarcity, contamination and management* (pp. 499–523).

USEPA. (2011). *Edition of the drinking water standards and health advisories* (pp. 731–732). Office of Water U.S. USEPA.

USGS. (1993). National water summary 190–1991 stream water quality. *USGS Water Supply, 59*, 2400.

Vinnarasi, F., Srinivasamoorthy, K., Saravanan, K., Rajesh Kanna, A., Gopinath, Prakash, R., Ponnumani, G., & Babu, C. (2022). Hydrogeochemical characteristics and risk evaluation of potential toxic elements in groundwater from Shanmughanadhi, Tamilnadu, India. *Environmental Research, 204*(8), 112199.

Visvalingam, G., Srinivasamoorthy, K., Andiyappan, A. K., Ramesh, K., Supriya, V. D., & Surinaidu, L. (2024). Understanding the impact of climate-induced sea level rise on groundwater inundation in a low-lying coastal area: A numerical simulation in Southeast India. *Regional Studies in Marine Science, 70*(4), Article 103401.

WHO. (1984). *Guidelines for drinking water quality V. 1, Recommendations* (p. 130). Switzerland: Geneva.

WHO. (1993). *Guidelines for drinking water quality* (Vol. 1, p. 188). WHO.

WHO. (2004). The World Health Report 2004: Changing History (p. 96).

WHO. (2011). Guidelines for Drinking-Water Quality, fourth ed. World Health Organisation, Geneva, Switzerland, 541.

WHO. (2017). *Guidelines for drinking water quality: Fourth edition incorporating the first addendum*. World Health Organization.

Yang, Z., Wang, M., Dong, Z., Tan, Z., & Guo, X. (2023). Potentially toxic elements contamination, risk and source analysis in sediments of Beiyun River supplied with reclaimed water, China. *Ecological Indicators, 154*, Article 110622.

Shoreline Dynamics in Thiruvananthapuram District of Kerala, India

P. Thambidurai, Venkatesan Shiva Shankar, and V. S. Harsha

1 Introduction

The coastal environment is very important for mankind, providing natural protection from storms and other natural disasters. The majority of countries around the world encompass 74.87% of the coastline (Bojang et al., 2023). The world's coastline length is estimated to be 3,56,000 km (Sekar et al., 2023) with more than two billion people (40%) dwelling within 100 km of coastal zone (Ferreira et al., 2021). Moreover, this coastal zone accounts for a land surface area of less than 20% (Dewi & Bijker, 2020). Since times immemorial coastal frontiers have preferably used for bountiful food resources, ease of transportation, recreation, human settlement, trade and commerce, agriculture, and potential sources of green energy like wave and tidal energy (Ferreira et al., 2021; Samra & Ali, 2021; Bojang et al., 2023; Kilar, 2023). Owing to the high human population density in the coastal frontiers more than 70% of the beaches around the world are imperilled by erosion (Kilar, 2023). Luijendijk et al. (2018) reported that globally, 24% of the coastal area has been lost to erosion at an alarming rate of 0.5 m/yr, while 28% of accretion and the rest have been stable in the past three decades. Further, they stated the gravity of the threat posed to the human communities dwelling on the coastal frontiers. Coastal shorelines are dynamic transitional zones that respond differently to various natural and anthropogenic variables (Al-Attar & Basheer, 2023; Elkotby et al., 2023; Mirdan et al., 2023). The natural variables are such as coastal morphology, climatology, and geology and while the anthropogenic variables are commonly linked with the establishment of infrastructures like ports, harbours, jetties, seawalls, etc., (Mageswaran et al., 2021; Santos et al., 2021; Rezaee et al., 2022; Kumar & Thambidurai, 2022, 2023; Kilar, 2023).

P. Thambidurai (✉) · V. S. Shankar · V. S. Harsha
Department of Coastal Disaster Management, School of Physical, Chemical and Applied Sciences, Pondicherry University-Port Blair Campus, Sri Vijaya Puram, Andaman and Nicobar Islands, India
e-mail: thambiduraip@pondiuni.ac.in

P. Thambidurai and A. K. Dikshit (eds.), *Coastal Environments of India*,
https://doi.org/10.1007/978-3-031-97341-3_10

The dynamics of shoreline are eventually associated with both erosion and accretion induced by anthropogenic and natural variables that imperial infrastructures, communities, and natural ecosystems along the coast (Raj et al., 2019; Sam & Gurugnanam, 2022; Mirdan et al., 2023). Coastal areas are increasingly vulnerable to natural and anthropogenic hazards that cause coastal erosion. Shoreline retreat is recognised as a growing potential threat as a result of global climate change that imperials sustainable beaches and coasts. Henceforth, quantification of shoreline changes that may be either short-term or long-term, their shape, and location are the need of the hour for effective coastal zone management and hazard assessment pertaining to the coast (Dewi & Bijker, 2020; Kilar, 2023; Mirdan et al., 2023). Coastal erosion is a major issue concerning the entire vast Indian peninsular shoreline (7516.6 km) that has an impending impact on the neighbouring maritime countries, with a broad array of coastal processes predominating the dynamics of the coastal area. Among coastal states in India, West Bengal (63%) experiences the highest coastal erosion, followed by Puducherry (57%) and Kerala (45%), being in the third position (Kankara et al., 2018). The Kerala coastline stretches for about 560 km in the NNW-SSE direction and suffers from severe erosion caused by natural processes, resulting in a change in the sediment budget that disrupts the natural equilibrium (Mallik et al., 1987).

Remote sensing data products derived from various airborne and spaceborne platforms were extensively used by various researchers to assess shoreline changes (Gens, 2010; Ferreira et al., 2021). Especially the infra-red band of the satellite imageries aids in discriminating the land–water interface clearly (Esmail et al., 2019; Ferreira et al., 2021). Further, Geographical Information Systems (GIS) is a candid embellishing tool to scrutinize and decipher more trustworthy and consistent information about the shoreline from satellite imagery (Louati et al., 2015; Esmail et al., 2019). Digital Shoreline Analysis System (DSAS) was extensively used by several researchers to quantify the short-term or long-term changes in shorelines (Samra & Ali, 2021; Sam & Gurugnanam, 2022; Sanhory et al., 2022; Al-Attar and Basheer, 2023; Bojang et al., 2023; Elkotby et al., 2023; Kilar, 2023; Mirdan et al., 2023; Sekar et al., 2023). The present study attempts to assess the recent spatio-temporal changes in shorelines over the past two decades (2003–2023) in detail using DSAS. The objective of the study encompasses village-wise quantification of End Point Rate (EPR) and Net Shoreline Movement (NSM) in an interval of a decade. Consequently, the spatio-temporal shoreline change (SLC) in Thiruvananthapuram district is vital in forecasting the impending dynamics of shoreline behaviour and evolving effective strategies for coastal zone management.

2 Study Area

Thiruvananthapuram is a coastal district of Kerala, located on the southwest coast of the Indian peninsular with a shoreline of 76 km long. It is bounded by the geographical coordinates 8°17′27″–8°46′57″ N latitudes and 76°40′25″–77°05′56″ E longitude (Fig. 1). On the west, the study area is sandwiched between the east coast of the

Lakshadweep archipelago in the Arabian Sea. In the north, it is bordered by Kollam, another coastal district of Kerala, while on the east and south, it is bounded by Tirunelveli and Kanyakumari districts of Tamil Nadu, respectively. Out of the four taluks, three of them viz., Thiruvananthapuram, Chirayinkeezhu, and Neyyatinkara are located on the coastal frontiers of the Arabian Sea while Nedumangad taluk is land-locked. The study is generally characterized by two formations viz., Vizhinjam (sedimentary) and Kovalam (Crystalline rocks). Coastal cliffs, coastal plains, channel bars, estuaries sand bars, and sandy beaches are the coastal geomorphic features that are encountered in the Thiruvananthapuram district.

The rocky coasts are creased by undulating lateritic terrain defined by either structural or residual hills. The study area consists of two lagoons namely Akkulam and Kadinamkulam. Furthermore, the study area has four non-perineal rivers namely Veli, Adimalathura, Punthura, and Poovar (Shaji, 2014; Princy et al., 2023). The coast is mainly composed of sand (80%), followed by 15%, and 5% of silt and clay (Kadavan & Joseph, 2017). Henceforth, its coastal area is susceptible to high rates of erosion (~ 84%) owing to the high percentage of sand (Kumar et al., 2006). The study area is endowed with a hot tropical climate with an average annual precipitation of 1760 mm and an annual temperature range between 20 and 35 °C. The Thiruvananthapuram district receives rainfall during both southwest monsoon (June to September) and the northeast monsoon (October to November). Comparatively, the study area receives a majority of rainfall during the southwest monsoon (46.3%),

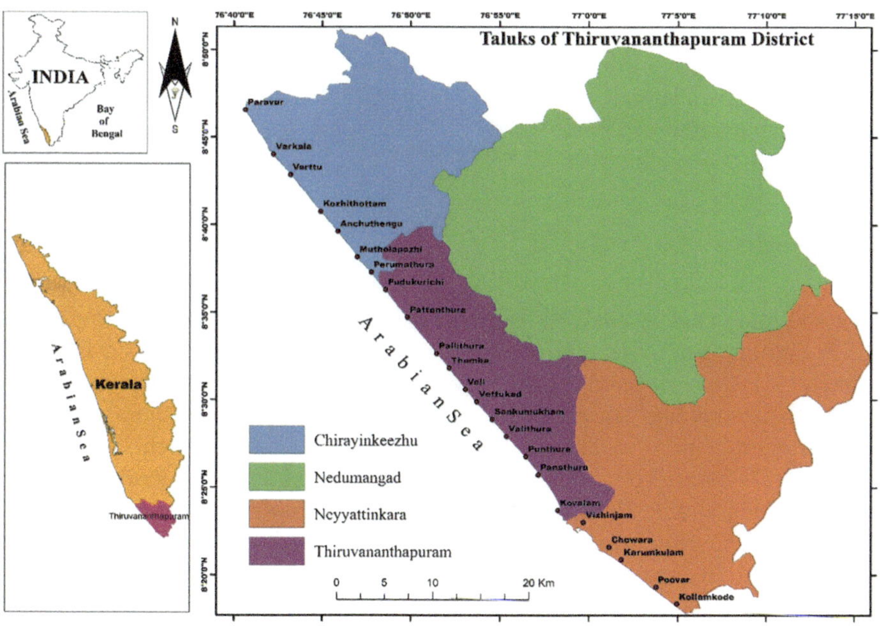

Fig. 1 Study area map

followed by northeast monsoon (28%), and the rest (25.7%) during the pre-monsoon/ dry season.

3 Materials and Methods

In order to comprehend the objective of the study, the data collection was taken from the USGS portal (www.earthexplorer.usgs.gov). There are two multi-temporal Landsat-7 (2003, 2014) and a Landsat-8 (2023) satellite images having the path/ row (143/54) for quantifying the SLC in the three coastal taluks of Thiruvananthapuram district. These three satellite images were clipped to the district boundary, and the multi-temporal shoreline of the three coastal taluks was digitised using High Tide Line (HTL) as a common shoreline proxy (Selvan et al., 2016; Mirdan et al., 2023; Sekar et al., 2023). Multi-temporal satellite images and their shoreline were re-projected to UTM zone 43 N in ArcMap 10.5. Further, two separate statistical methods, viz., End Point Rate (EPR) and Net Shoreline Movement (NSM), were employed to quantify the SLC rate using DSAS (Thieler et al., 2009). The former method requires two different shorelines (2003–2014 and 2014–2023) to quantify the rate in m/yr while the latter method is used to measure the overall change in meters (2003–2023) in the shoreline. The formula for the quantification of EPR and NSM in DSAS was pointed out by several researchers (Sam & Gurugnanam, 2022; Sekar et al., 2023). Negative and positive values of EPR signify erosion and accretion, respectively. Similarly, retreating and advancing shorelines are implied by negative and positive values of NSM. To appreciate the SLC, an onshore baseline was digitised parallel to the coast, and a total of 1440 transects were created at 50 m intervals with a transect length of 2500 m orthogonal to the coast using DSAS.

4 Results

The shoreline change analyses were carried out for two decades, from 2003 to 2023, and the results are discussed in the section. There are three coastal taluks, namely (1) *Chirayinkeezhu taluk* covers seven villages viz., Paravur, Varkala, Verttu, Kozhithottam, Anchuthengu, Mutholapozhi, and Perumathura. (2) *Thiruvananthapuram taluk* comprises eleven villages viz., Pudukurichi, Pattanthura, Pallithura, Thumba, Veli, Vettukad, Sankumukham, Valithura, Punthura, Panathura, and Kovalam, and *3) Neyyattinkara taluk* encompasses five coastal villages viz., Vizhinjam, Chowara, Karumkulam, Poovar, and Kollamkode, The EPR values were classified into five categories such as (1) High Erosion (< − 2 m/yr), (2) Erosion (− 2 to − 1 m/yr), (3) Stable Beach (− 1 to 1 m/yr), (4) Accretion (1–2 m/yr), and (5) High Accretion (> 2 m/yr). On the other hand, the NSM has been categorised into two, namely, Retrogradation (Erosion) and Progradation (Accretion). The EPR and NSM of 2003–2014 and 2014–2023 are *hitherto* referred to as past and recent spells, respectively. The village-wise results of EPR and NSM due to shoreline changes are presented below in detailed sections.

4.1 *Chirayinkeezhu*

The analysed data and different calculated values of the Chirayinkeehu taluk are presented in Figs. 2, 3, 4 and 5. The taluk covers a coastal length of 23.11 km and is bounded by 460 transects (1–460), encompassing the entire stretch of the taluk. The Paravur-Varkala has a coast length of 5.62 km, incorporated within 113 transects. Around 69.03% of the coast has been accreted, while the rest has eroded (30.97%) during the past spell. However, the EPR values showed the maximum erosion and accretion of the past spell are – 4.96 m/yr and 4.62 m/yr, respectively. The overall EPR value of 0.59 m/yr strongly indicates that it is a stable coast. The overall NSM value of 9.7 m during this spell suggests an advancing shoreline with a maximum of – 44.77 m. But, in the recent spell, a total reversal of the scenario was observed, with 74.33% of the coast between the villages experiencing erosion and 25.67% accretion. During the recent period, the maximum erosion and accretion were found to be – 6.12 m/yr and 4.27 m/yr, respectively. The result suggests that the coastal area within these villages is erosion-prone, with a mean erosion of – 1.51 m/yr. The overall NSM value of – 16.8 m articulates shoreline retreat with a maximum value of – 67.45 m. Comparative results (EPR and NSM) of the past and the recent spell are indicative of the transformation of a stable shoreline to a retreating shoreline (Fig. 2).

The coastal stretch of Varakala-Verttu is 2.89 km, with 56 transects (113–168). In the past spell, 53.58% of the coast exhibited accretion, while the rest was eroded (46.42%). The maximum EPR accretion and erosion are – 4.35 m/yr and 3.35 m/yr, respectively. This is articulating that the coast was stable with an overall EPR value of – 0.21 m/yr. The overall NSM value of 5.96 m indicates an advancing shoreline with a maximum value of 30.25 m. However, during the recent spell, the EPR of erosion and accretion was 57.14% and 42.86%, respectively, with an overall mean EPR of – 1.1 m/yr, enunciating the one set of erosion in this coastal stretch. The overall NSM value of – 11.86 m indicates shoreline retreat with a maximum of – 65.21 m. Furthermore, a comparison of the EPR of the past and present spells shows that the coast in this section of the village transformed from a stable to an erosional coast. Also, the contrasting values of NSM of past and recent spells suggest the metamorphosis of an advancing shoreline to shoreline retreat.

The next coastal stretch in this taluk is Verttu-Kozhithottam, with a shore length of 4.46 km covered by 91 transects (169–259). The past spell shows 83.52% of the shoreline was exhibiting erosion. A maximum EPR erosion and accretion value was observed to be – 27.9 m/yr and 0.91 m/yr, respectively. However, the mean EPR value of – 2.87 m/yr indicates that this stretch was experiencing very high erosion. On the other hand, the overall NSM value of 0.22 m reveals an advancing shoreline despite erosion predominance. Similarly, a very high erosion trend continued in the

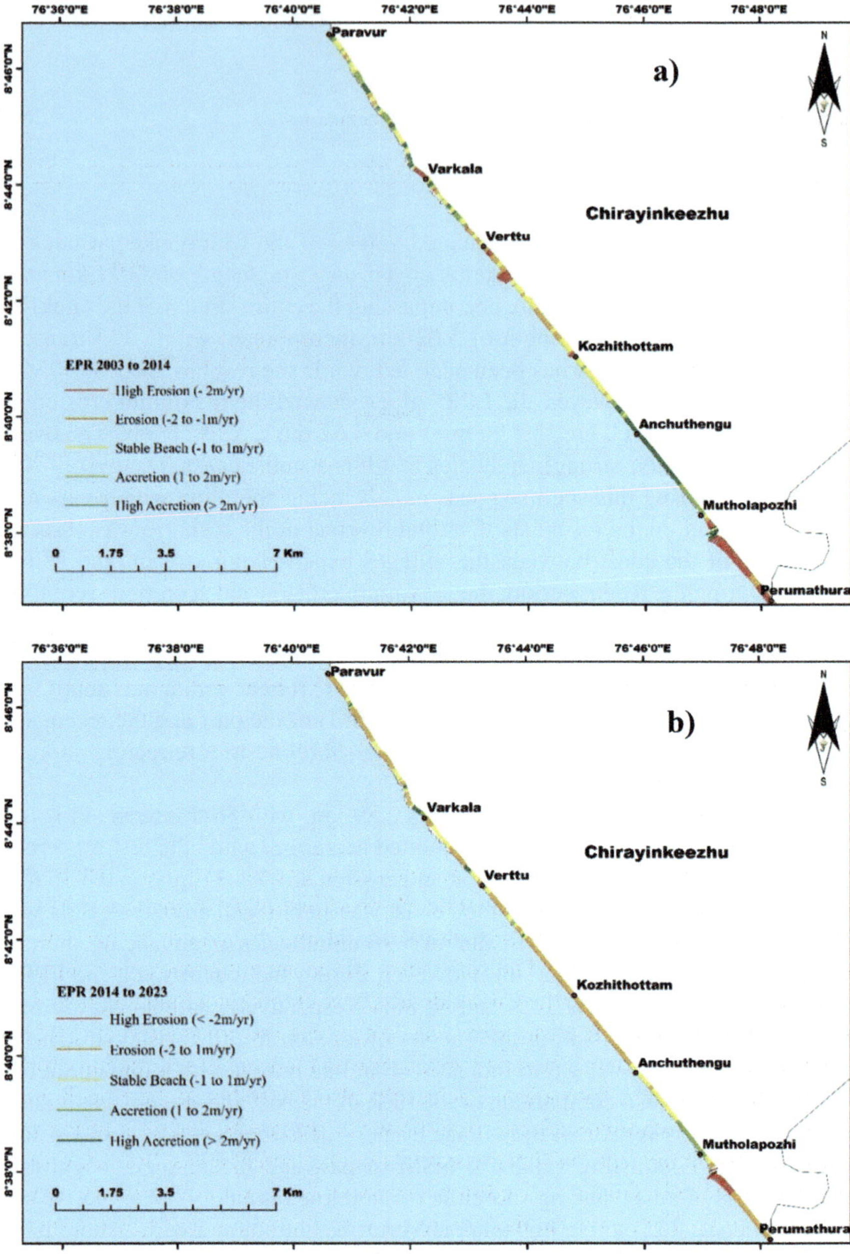

Fig. 2 Spatial distribution of EPR for Chirayinkeezhu Taluk **a** 2003–2014 and **b** 2014–2023

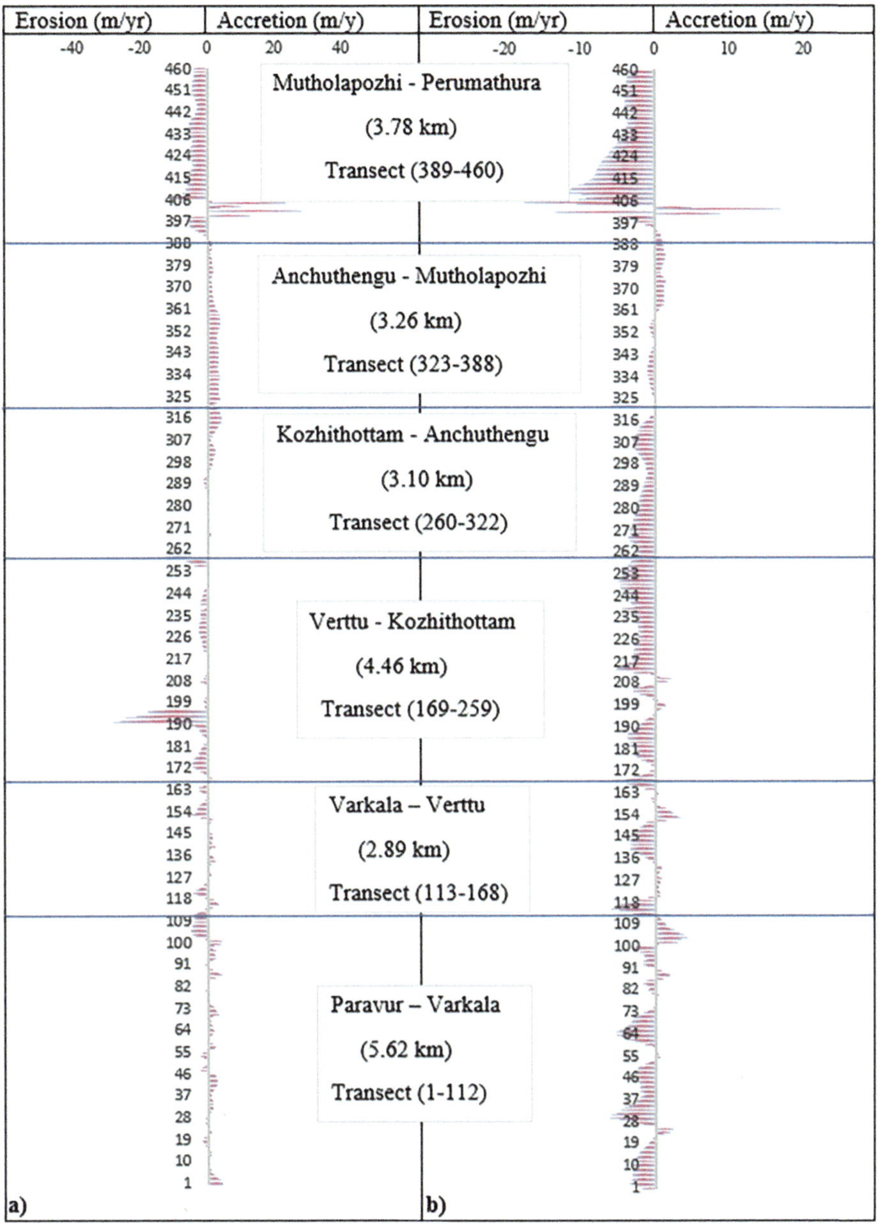

Fig. 3 Village wise estimation of EPR for Chirayinkeezhu Taluk **a** 2003–2014 and **b** 2014–2023

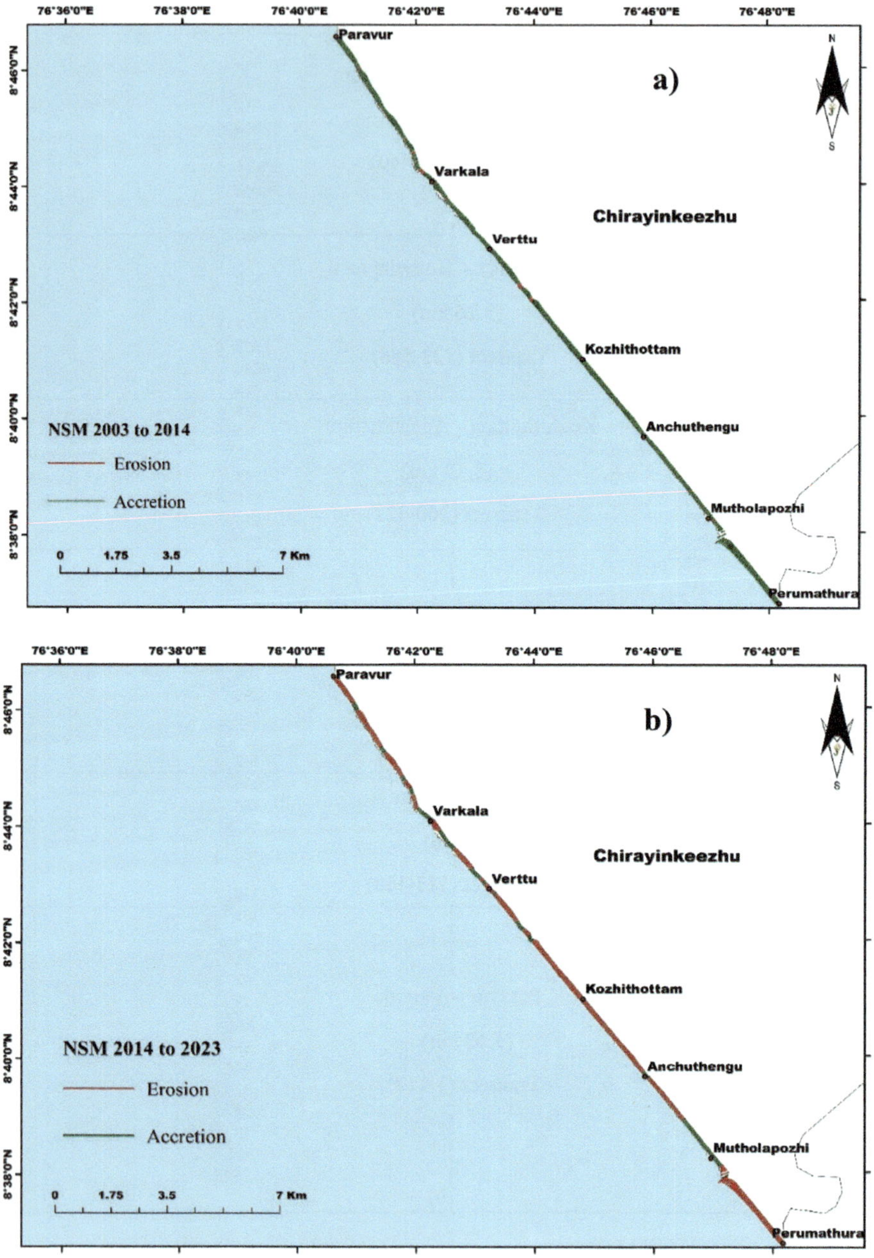

Fig. 4 Spatial distribution of NSM for Chirayinkeezhu Taluk **a** 2003–2014 and **b** 2014–2023

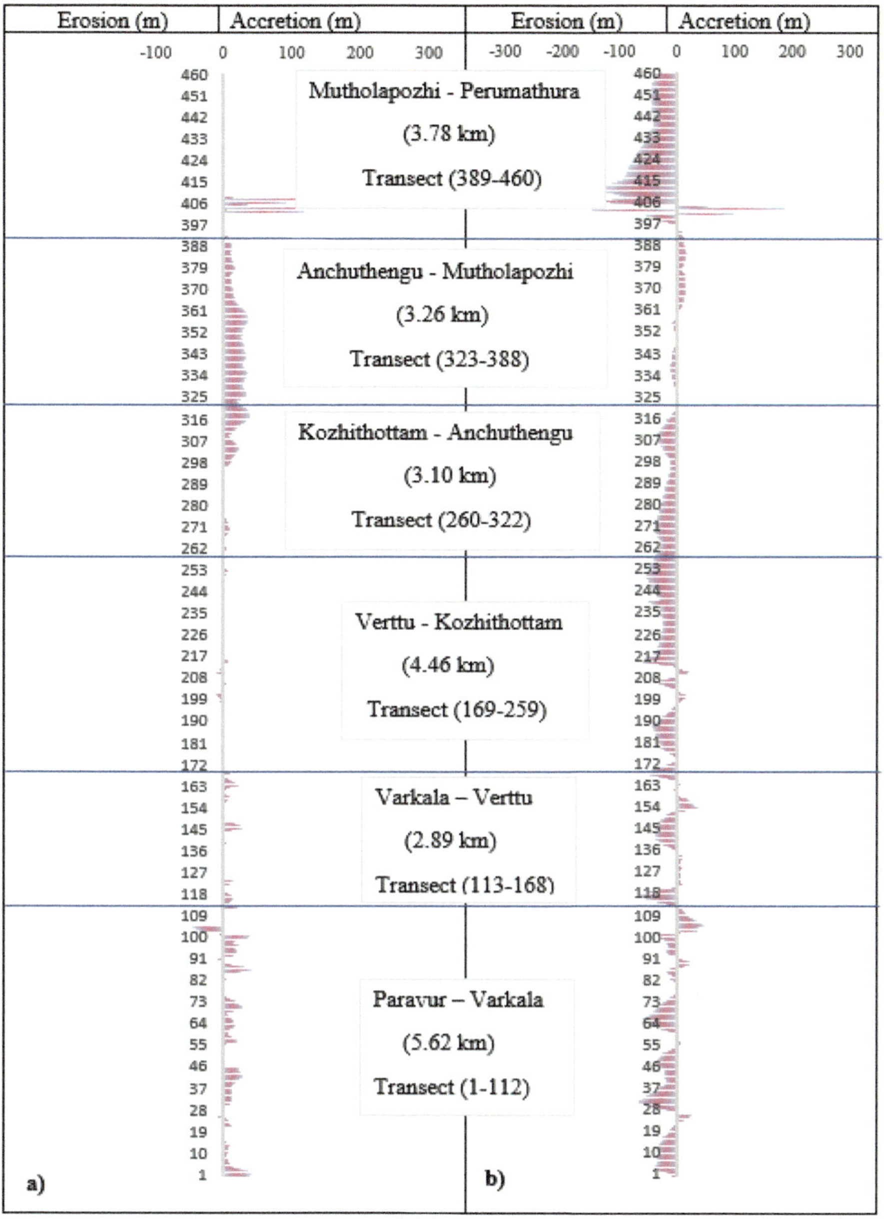

Fig. 5 Village wise estimation of NSM for Chirayinkeezhu Taluk **a** 2003–2014 and **b** 2014–2023

recent spell as well, with 87.91% erosion with a mean EPR erosion value of – 2.48 m/yr. However, there was a decrease in the mean erosion (– 0.39 m/yr) for the case of the recent spell, and the trend of high erosion persists for the past spell. Also, the overall NSM value of – 27.42 m reveals the coastal retreat.

A further coastal stretch proceeded by Kozhithottam-Anchuthengu with a shoreline of 3.10 km and was bounded by 63 transects (260–322). In the past spell, the coast was dominated by accretion (71.43%). The maximum erosion and accretion EPR values are – 1.01 m/yr and 4.51 m/yr, respectively, with a mean EPR accretion value of 1.75 m/yr articulating it as an accretional coast. The overall NSM result shows 11.31 m, revealing that the shoreline is advancing. However, during the recent spell, the coast was completely transformed, with 95.24% erosion. A maximum EPR value of – 3.58 m/yr and a mean value of – 2.14 m/yr demonstrate that the coast is experiencing a very high erosion rate presently. Further, the overall NSM value was (– 23.6 m), unveiling that the coast is retreating with shoreline displacement ranging from – 39.46 m to 3.12 m (Fig. 3).

The coast is preceded by Anchuthengu-Mutholapozhi, which has a coast length of 3.26 km covered by 66 transects (323–388). In the past spell, the entire shoreline experienced cent accretion. The maximum and minimum EPR values range from 4.24 m/yr to 1.24 m/yr. The mean accretional value of 2.8 m/yr suggests that it is a highly accretional coast. The overall NSM value (25.3 m) indicates an advancing shoreline with a maximum value of 38.23 m. Conversely, the recent spell showed erosion (51.51%) and accretion (48.49%) at close equilibrium. A maximum EPR value is low in this stretch (– 1.12 m/yr), but the mean EPR value shows 0.25 m/yr, indicating a stable coast. It is inferred from the overall NSM value of 2.85 m that the shoreline is advancing. A comparison between the past and the recent spell strongly shows a transformation from a stable coast to a highly accretional coast (Fig. 4).

The last stretch in the taluk is Mutholapozhi-Perumathura, with a length of 3.78 km bounded by 72 transects (389–460). The past spell (87.50%) was predominant erosion with a maximum EPR value of – 12.4 m/yr. Although the accretion accounted for just 11.27%, the maximum EPR value was 28.07 m/yr in a small pocket of 200 m stretch around the Mutholapozhi port. The overall NSM value of 10.04 m is suggestive of an advancing shoreline. A similar trend was observed with high erosion (86.11%) during the recent spell. Comparing the past and recent spells showed a very high erosion rate in the past two decades. Further, the overall NSM value reveals a retreating shoreline (– 51.24 m). The coast of Chirayinkeehu taluk (23.11 km) stretch had transformed into an erosional coast (– 1.97 m/yr) from a stable coast (– 0.37 m/yr) in the past two decades. Additionally, the overall NSM values for past and recent (9.85 to – 21.78 m) indicate the metamorphosis of the advancing shoreline to the retreating shoreline (Fig. 5).

4.2 *Thiruvananthapuram*

The studies and their values of the Thiruvananthapuram taluk are presented in Figs. 6, 7, 8 and 9. The entire taluk has a coast length of 32.55 km and is bounded by 648 transects (461–1108). Among ten stretches, the first one is between Pudukurichi-Pattanthura, which covers a length of 4.74 km and is bounded by 98 transects (461–558). The past and recent spell EPR data articulates that erosion is predominant with 94.89% (− 1.93 m/yr) and 97.95% (− 3.01 m/yr), respectively. It indicates that this coastal section is susceptible to very high erosion. Although the EPR values suggest that the coast is experiencing erosion at varying rates. The overall NSM values reveal that the past (0.08 m) and recent spell (− 33.01 m) were from an advancing shoreline to a retreating shoreline.

The Pattanthura-Pallithura section covers a length of 4.87 km and consists of 98 transects (559–656). In this, the past spell showed a predominant accretion of 76.53%. The EPR values indicate that erosion (− 177 m/yr) is much higher than accretion (1.95 m/yr). Finally, the NSM value reveals (5.42 m) that it is an advancing shoreline. However, the recent spell has a 100% erosion with a maximum (− 5.65 m/yr) and a mean (− 2.84 m/yr), articulating a very high erosion. The overall NSM value is negative (− 32.26 m), clearly indicative of a retreating shoreline. The next coastal stretch was Pallithura-Thumba, the smallest shoreline in Thiruvananthapuram taluk, covered by 40 transects (657–696) with a coast length of 2.88 km only. In this section, accretion (75%) was predominant for the past spell. The maximum EPR shows that

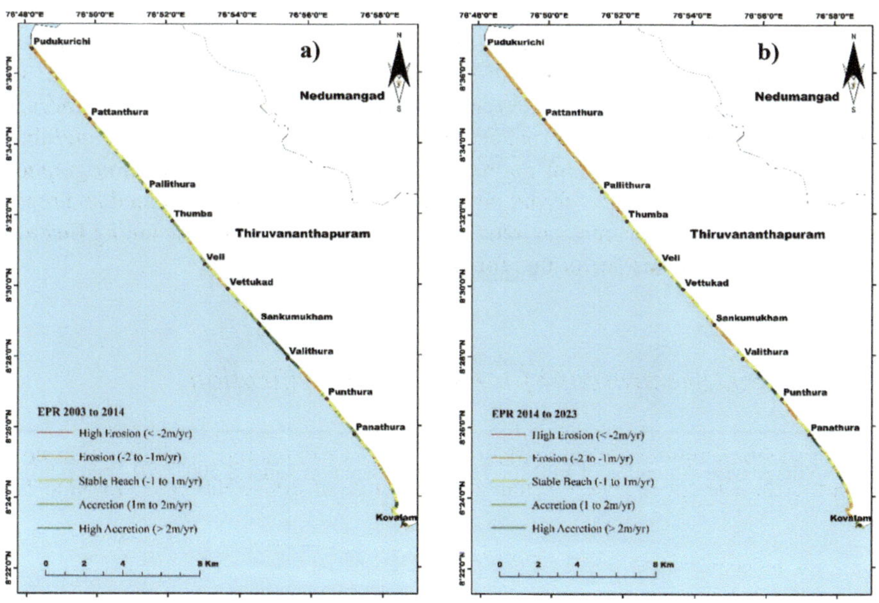

Fig. 6 Spatial distribution of EPR for Thiruvananthapuram Taluk **a** 2003–2014 and **b** 2014–2023

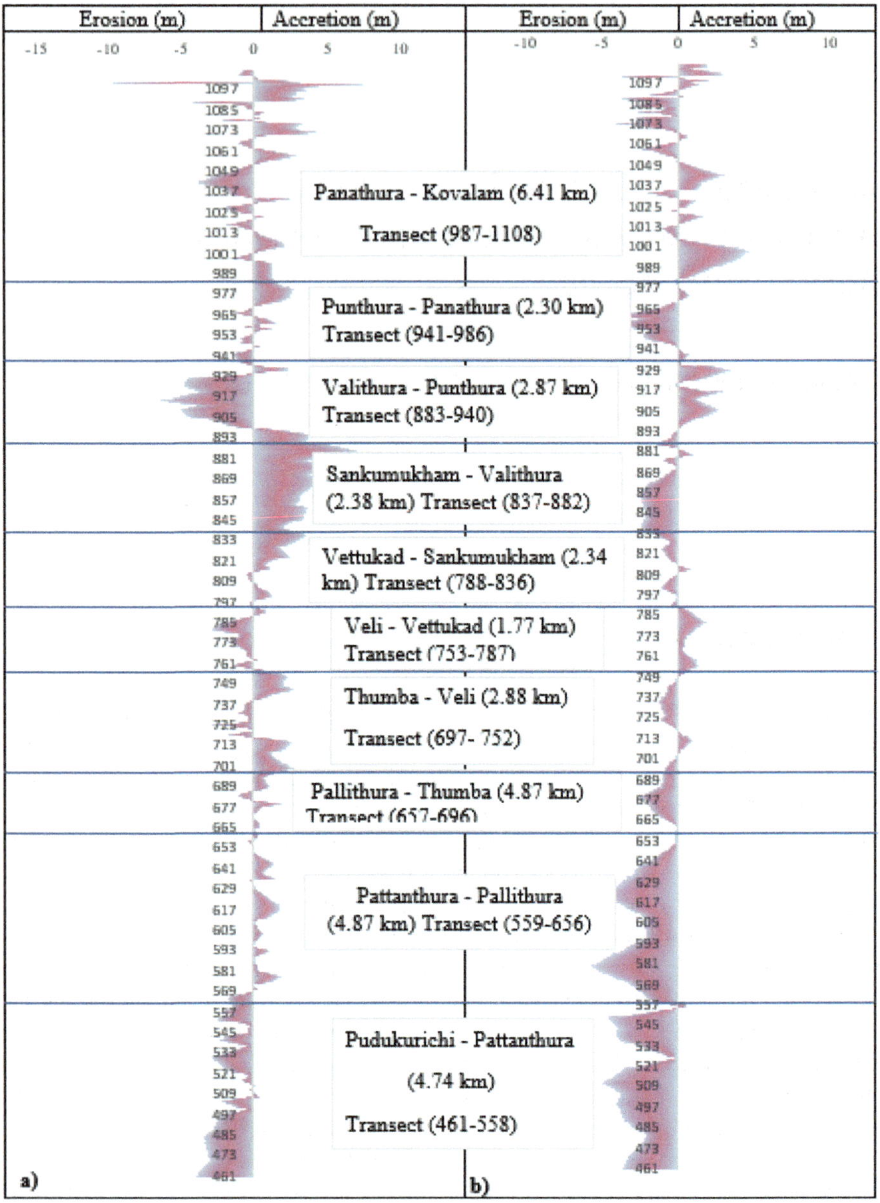

Fig. 7 Village wise estimation of EPR for Thiruvananthapuram Taluk **a** 2003–2014 and **b** 2014–2023

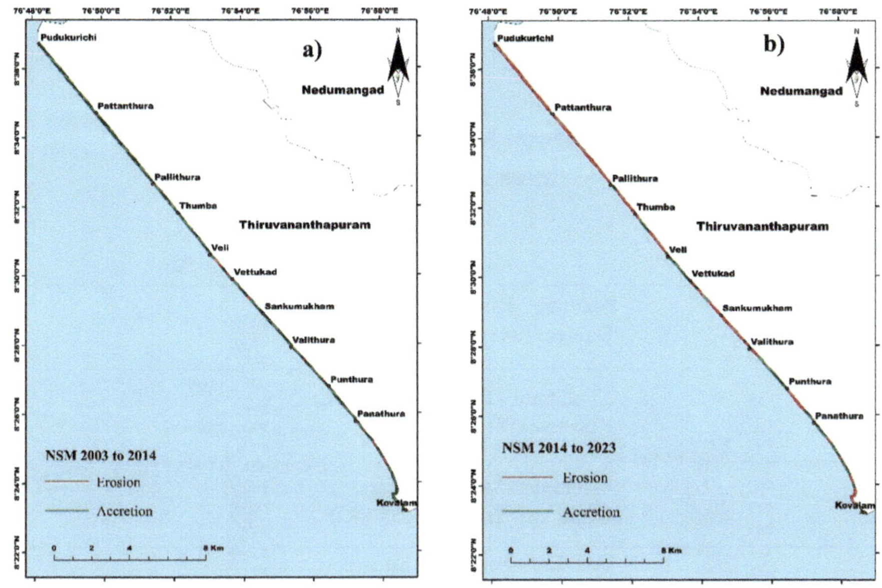

Fig. 8 Spatial distribution of NSM for Thiruvananthapuram Taluk **a** 2003–2014 and **b** 2014–2023

erosion (− 1.02 m/yr) and accretion (1.9 m/yr) were identified in this stretch (Fig. 6). The overall EPR of 0.32 m/yr indicates that the coastal stretch between these villages is stable. On the contrary, during the recent spell, a total reversal of the phenomenon was observed with 92.5% erosion. The overall EPR value (− 1.12 m/yr) suggests that the coast experienced erosion. A comparison of overall NSM values of the past spell (3.6 m) and recent spell (− 11.92 m) revealed the mutation of an advancing coast to a retreating coast, respectively. The Thumba-Veli stretch has a shoreline of 2.88 km bounded by 56 transects (697–752). A comparison between the past and recent spells indicates a pattern of reversal where erosion is dominant, like accretion (62.5%) and erosion (66.07%). A maximum accretion of 2.81 m/yr was recorded during the past spell, with an overall EPR value of 0.83 m/yr, articulating that the coast was stable between these villages. Although erosion dominated in the recent spell (− 0.7 m/yr), the overall EPR value of − 0.29 m/yr indicates the coast between this stretch is also stable, and further NSM indicates the shoreline shifted from advancement to retreatment, as noted from the past spell (10.46 m) to the recent spell (− 3.33 m), respectively.

The coastal stretch of Veli-Vettukad villages covers shorelines of 1.77 km with 35 transects (753–787). This stretch consists of dominance of erosion in the past spell, and recent times showed a 100% accretion. The overall EPR value (− 0.76 m/yr) suggests that the coast was stable with maximum erosion (− 2.82 m/yr) and accretion (2.32 m/yr). However, during the recent spell, the overall EPR value of 1.01 m/yr indicates that accretion was initiated with the maximum (1.9 m/yr) and minimum (0.33 m/yr). Further, the NSM values of this section show that shoreline changes

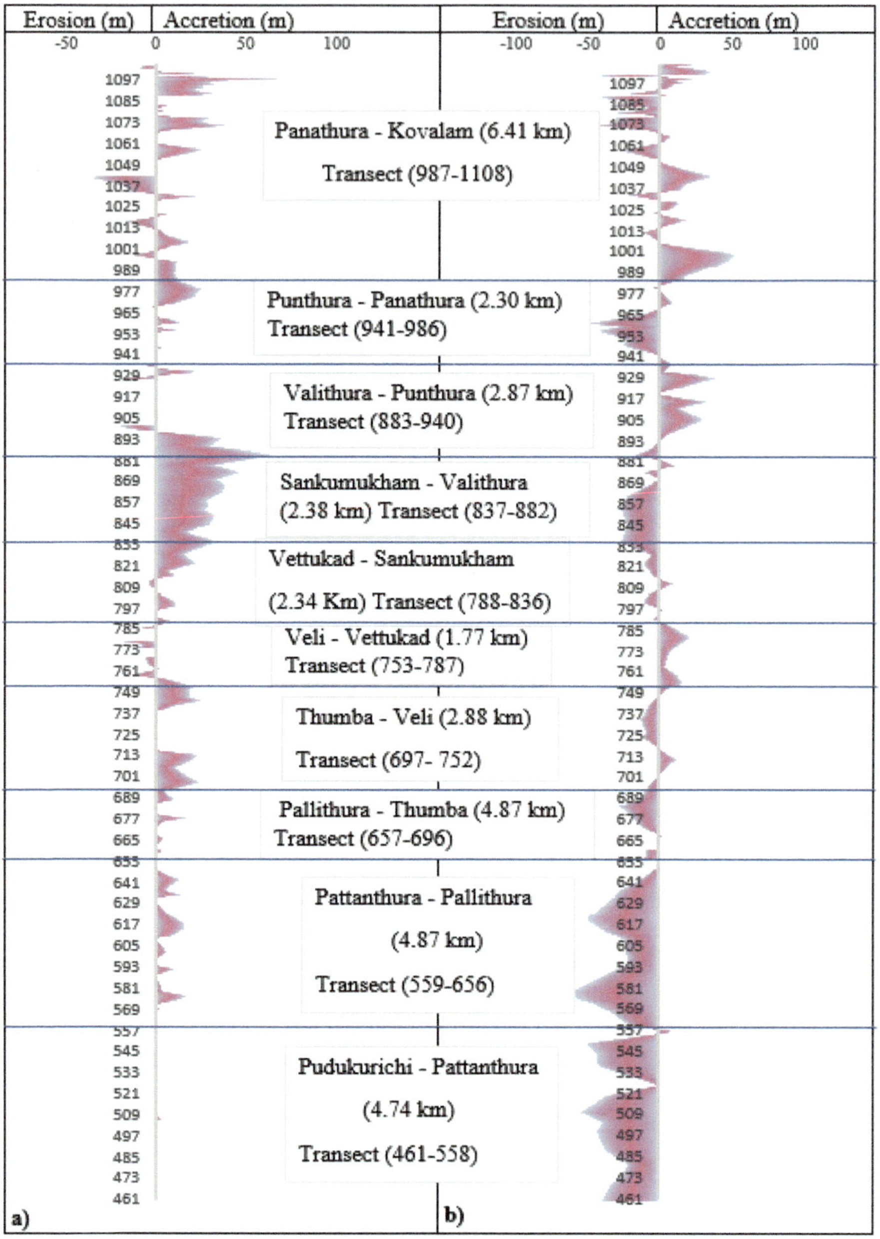

Fig. 9 Village wise estimation of NSM for Thiruvananthapuram Taluk **a** 2003–2014 and **b** 2014–2023

occur from retreating (– 0.5 m) in the past to advancement (11.13 m) in the recent one. The entire section of the coast shows the erosion face, and the recent one is opposite to the past one (Fig. 7). Like the previous coastal stretch, a similar kind of EPR reversal trend was observed in this section. The Vettukad-Sankumukham has a length of 2.34 km and is bounded by 49 transects (788–836). The past and recent sets of data reveal almost equal effects of accretion (75.51%) and erosion (77.55%). The maximum erosion, accretion, and average EPR values were – 0.88 m/yr, 3.58 m/yr, and 1.01 m/yr, respectively, and were observed during the past spell, enunciating a stable coast. Although the overall EPR value of – 0.51 m/yr during the recent spell articulates a stable coast, the prevalence of erosion was dominant. The maximum erosion EPR value of – 2.46 m/yr was recorded during the recent spell. Also, the overall NSM values 8.72 m (past spell) and – 4.88 m (recent spell) indicate that the coast was transformed from an advancing shoreline to a retreating shoreline, respectively (Fig. 8).

Further, in the Sankumukham-Valithura, a similar trend reversal of accretion and erosion was observed as in the previous section. This stretch has a shoreline of 2.38 km with 46 transects (837–882). In the past spell, accretion was cent percent observed with a maximum rate of EPR of 5.24 m/yr, and it has very high accretion. On the other hand, in the recent spell the coast is transformed into an erosional (89.13%) with a maximum EPR erosion value of – 2.52 m/yr. The average EPR value of – 1.47 m/yr indicates the onset of erosion. Further, the comparison of the average NSM value of the past spell (32.47 m) and recent spell (– 16.99 m) indicates the mutation from advancing to retreating coast, respectively. The next section of the coast is preceded by Valithura-Punthura with a shore length of 2.87 km and 58 transects (883–904). The past spell showed dominated erosion (63.79%) with maximum EPR value of erosion and accretion of – 6.36 m/yr and 7.05 m/yr, respectively, and it has a stable coast with an average value of – 0.93 m/yr. In the recent spell, the trend changed to predominant accretion with 77.59%. The maximum EPR value erosion and accretion were – 1.58 m/yr and 3.48 m/yr, respectively. The overall EPR value of 1.04 m/yr indicates the onset of accretion. Although the EPR value indicates the transformation of the erosional coast to the accretional coast, the overall NSM value discloses that the coast was advancing in both spells at a slow pace. The rate of NSM between the two spells was observed to be 0.38 m. The Punthura-Panathura is the second smallest stretch of coast in Thiruvananthapuram taluk, with a shoreline of 2.3 km covered by 46 transects (941–986). The accretion (56.52%) dominated during the past spell, with maximum accretion and erosion of 2.81 m/yr and – 1.67 m/yr. The overall EPR value of 0.49 m/yr pronounces that the coast was stable. In the recent spell, the onset of erosion (65.21%) was observed with maximum erosion and accretion of – 4.54 m/yr and 1.49 m/yr, respectively. The overall EPR value of – 1.11 m/yr indicates that the coast is susceptible to erosion. A comparison between the overall NSM values of the past spell (7.37 m) and the recent spell (– 12.53) indicates the transformation of the advancing shoreline to the retreating shoreline, respectively. The last stretch in Thiruvananthapuram taluk is Panathura-Kovalam, with a coast length of 6.41 km covered by 121 transects (987–1108), and the stretch is the largest in the taluk. The accretion (54.55%) dominated in the past spell with maximum erosion (– 9.71 m/

yr) and accretion (7.47 m/yr). The average EPR value trends (0.26 m/yr) to the coast were stable. Further, during the recent spell, the dominance of accretion (60.33%) continued with the maximum erosion (− 4.11 m/yr) and accretion (4.74 m/yr). The stable state of the coast continued from the past spell with the EPR value of 0.42 m/yr. The average positive NSM values of both spells indicate shoreline advancement. In general, the average EPR values of the past (0.15 m/yr) and the recent spell (− 0.97 m/yr) suggest that the coastal area of Thiruvananthapuram taluk is stable. However, the overall NSM value of the past spell (7.67 m) indicates that the shoreline was advancing. The recent spell (− 10.78 m) suggests that the coast was retreating (Fig. 9).

4.3 Neyyatinkara

The Neyyatinkara taluk has a coast length of 17.88 km with transects of 332 (1109–1440), and the outcome of the studies is presented in Figs. 10, 11, 12 and 13. The first stretch covers the villages of Vizhinjam-Chowara, and the longest coast length of 7.43 km is bounded by 124 transects (1109–1231). In this, the past spell showed that the predominant accretion (71.78%) andthe maximum accretion of 59.72 m/yr and erosion of − 5.49 m/yr. However, the average EPR value of 9.8 m/yr indicates a high level of accretion. A trend reversal, dominated by erosion (55.65%), was observed in the recent spell. The maximum erosion was − 6.68 m/yr and accretion was 7.19 m/yr. The average EPR value (− 0.14 m/yr) concludes that this stretch of coast was stable. But the average NSM values suggest that the coast was in a retreating state (− 1.41 m). The Chowara-Karumkulam with a shoreline of 2.68 km covered by 53 transects (1232–1284). The past spell has 67.92% erosion with a range of EPR values from − 3.77 m/yr to − 0.02 m/yr. The average EPR value of − 0.81 m/yr indicates the coast was stable. A substantial decrease in the erosional (9.43%) pattern was observed in the recent spell. The maximum erosion and accretion values were − 3.9 m/yr and 0.98 m/yr, respectively. The EPR (− 0.35 m/yr) and NSM results have indicated that, as in the previous stretch of the Chowara-Karumkulam villages, the coast shifts from advancing (1.36 m) to retreating (− 4.19 m). The next stretch is the Karumkulam-Poovar, with a shore length of 3.61 km within 74 transects (1285–1358). The past spell reveals 57.75% of erosion, and the rest has accretion (43.25%). Although erosion was dominant in this stretch, the average EPR value of − 0.09 m/yr suggests a stable coast (Fig. 10).

The erosional trend also continued in the recent spell, with 81.08% of the coast experiencing erosion. The maximum erosion and accretion were − 4.65 m/yr and 1.86 m/yr, respectively. The EPR value of − 1.2 m/yr is indicative of the onset of erosion. The average NSM values of the past and the recent spell indicate that the coast shifted from advancing (5.02 m) to a retreating coast (− 13.56 m). The last section of this taluk is Karumkulam-Poovar, with a coast length of 4.16 km within 82 transects (1359–1440). In the past spell, 92.68% of the coast was experiencing accretion with an EPR value of 7.58 m/yr and the average value of EPR was 3.19 m/

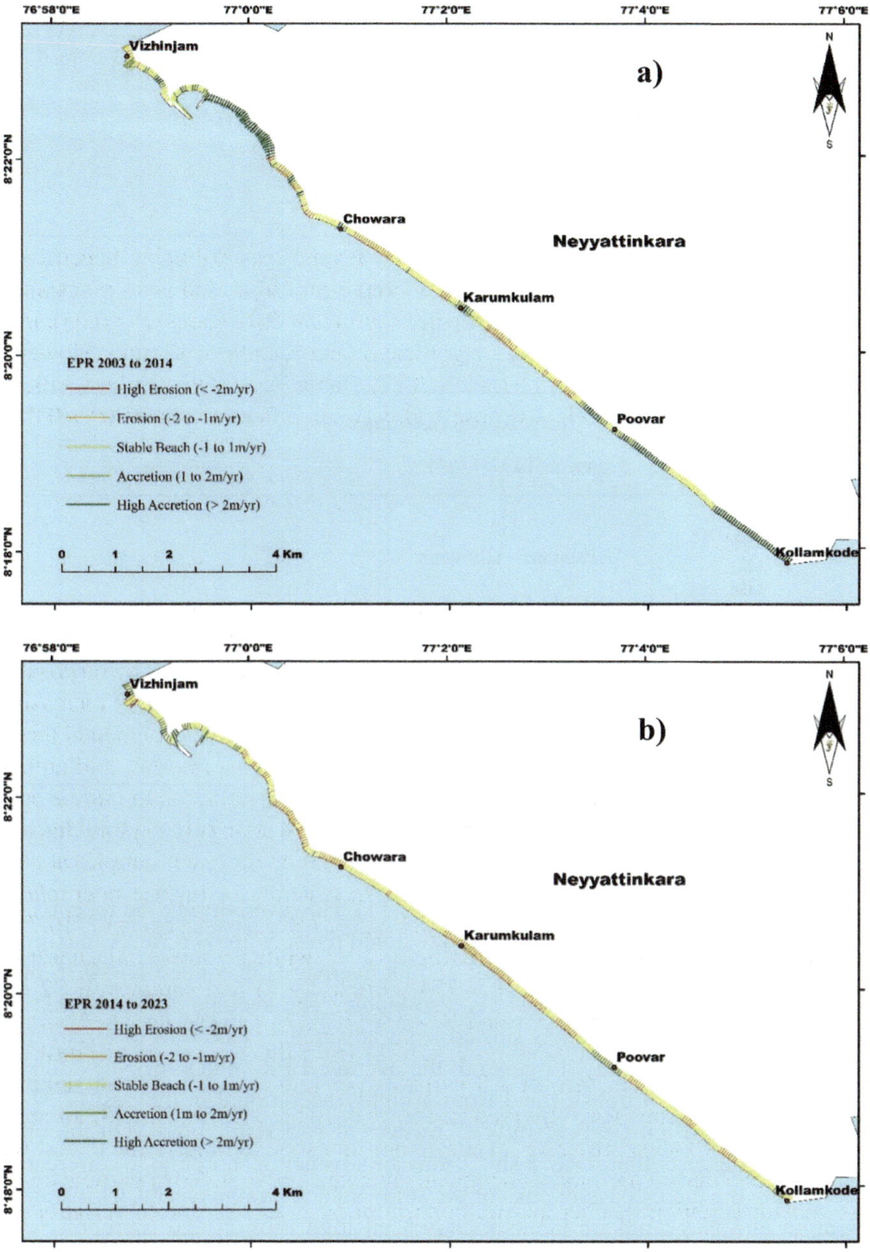

Fig. 10 Spatial distribution of EPR for Neyyatinkara Taluk **a** 2003–2014 and **b** 2014–2023

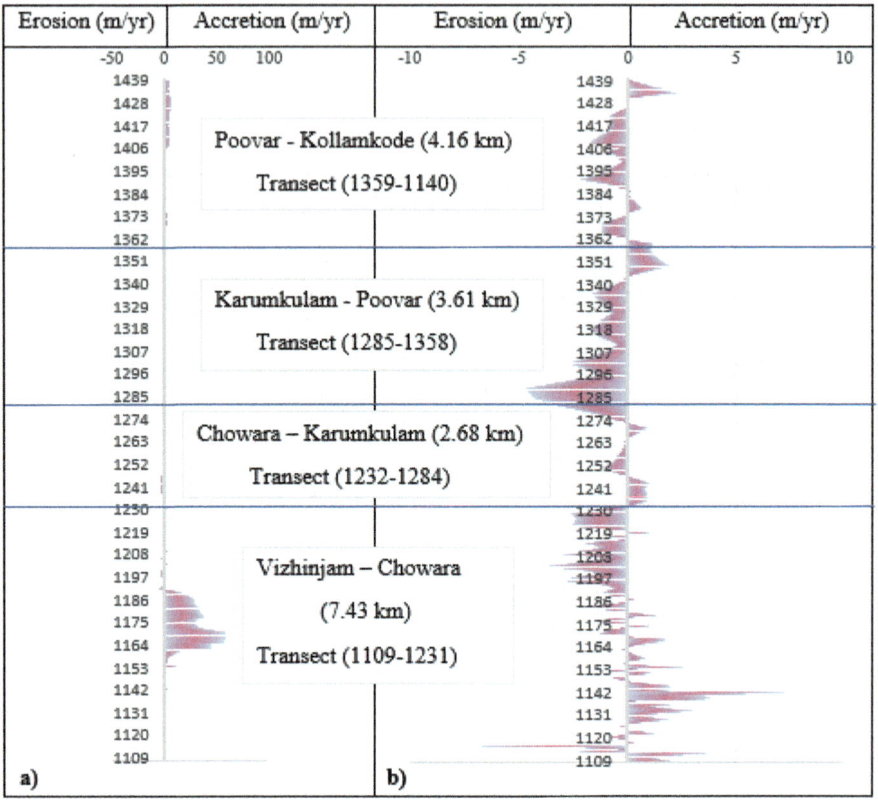

Fig. 11 Village wise estimation of EPR for Neyyatinkara Taluk **a** 2003–2014 and **b** 2014–2023

yr, indicating a very high accretion rate (Fig. 11). However, during the recent spell the accretional drastically dropped to an erosional phase with 64.63%. A maximum erosion of − 2.15 m/yr was recorded, and an average EPR value of − 0.36 m/yr was observed, suggesting the coast is also stable. The average NSM showed for the past and the recent spell has shifted from advancing (28.69 m) to a retreating coast (− 3.99), respectively. In general, the average EPR values for Neyyatinkara taluk suggested that it transformed from a highly accretional in the past (7.37 m/yr) and recent erosion (− 0.47 m/yr); however, it is a stable coast. On the other hand, NSM suggests there was a shift from an advancing shoreline to a retreating one. Among the three coastal taluks of the study area, Chirayinkeezhu, Thiruvananthapuram and Neyyatinkara taluks experienced high erosion and accretion rates, respectively, during the past spell (Fig. 12). It was inferred from the overall EPR that the coasts of Chirayinkeezhu (− 0.37 m/yr) and Thiruvananthapuram (0.15 m/yr) taluks are stable. On the other hand, there are very high accretion rates (4.28 m/yr) in Neyyatinkara. However, a dramatic change was observed during the recent spell

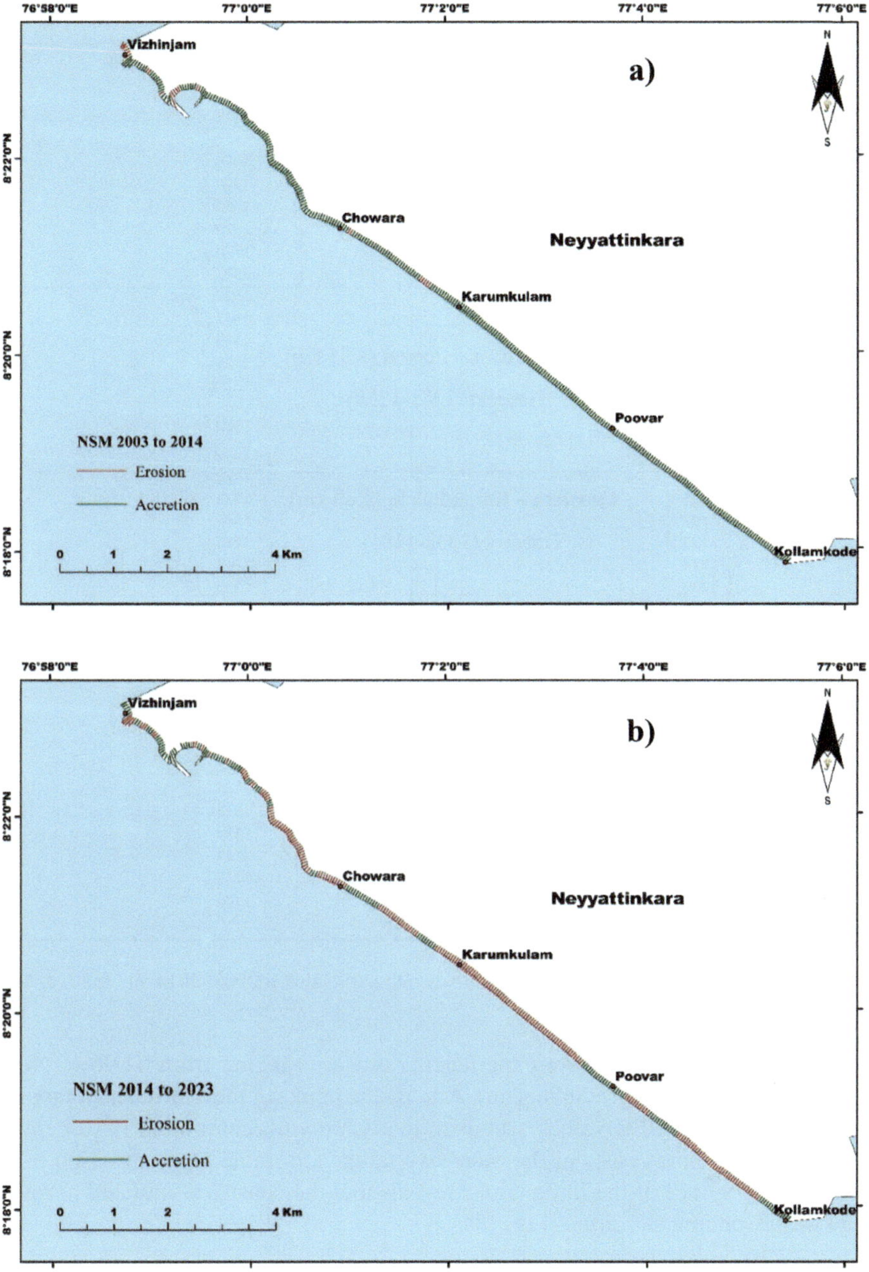

Fig. 12 Spatial distribution of NSM for Neyyatinkara Taluk **a** 2003–2014 and **b** 2014–2023

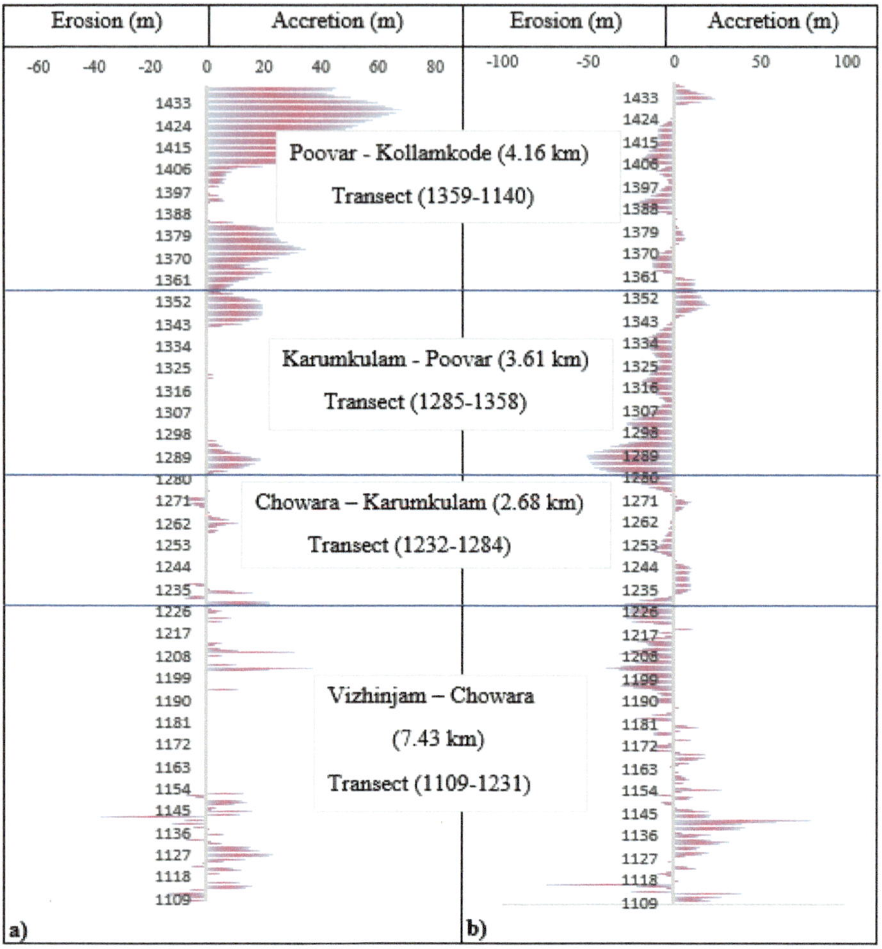

Fig. 13 Village-wise estimation of NSM for Neyyatinkara Taluk **a** 2003–2014 and **b** 2014–2023

in all three taluks, which were experiencing erosion ranging from 64.96% (Neyyatinkara) to 75.37% (Chirayinkeezhu). Among the taluks of the Thiruvananthapuram district, the overall EPR values articulate that Chirayinkeezhu taluk is susceptible to erosion, while the coasts of the other two taluks are stable. A comparison of the overall NSM values of the three taluks reveals that the coast was mutated to retreat from the advancing shoreline (Fig. 13).

5 Discussion

A wide range of oceanographic parameters influence the SLC in Thiruvananthapuram. The parameters include wave energy, tidal sedimentation, longshore and cross-shore processes, storm parameters, sea-level rise, and offshore wave energy fluxes (Baldock et al., 2017; Splinter et al., 2017; Vitousek et al., 2017; Zhong et al., 2017; Collins et al., 2018; Cooper et al., 2018; Ding et al., 2018; Robinet et al., 2018; D'Anna et al., 2020; Ibaceta et al., 2020; Vitousek et al., 2021; Chen et al., 2021; Robinet et al., 2020; Abdurrahman et al., 2021; Ibaceta et al., 2022; Lim et al., 2022; Koroy et al., 2023). These factors all contribute to the equilibrium of shoreline response and long-term shoreline behaviour. The parameters also have an impact on SLC at a variety of chronological and spatial scales, ranging from seconds to decades and from the surf zone to the entire ocean basin (Vitousek et al., 2021). The SLC is also influenced by variations in incident wave energy, gradients in longshore sediment transport, and cross-shore transport triggered by wave energy disequilibrium with the shoreline location (Robinet et al., 2018). Furthermore, the SLC is influenced by morphological hysteresis in the evolution of beach profiles under wave climate sequences, as well. Additionally, the SLC is sensitive to palaeogeographic changes influencing tidal inflow versus outflow (Baldock et al., 2017; Collins et al., 2018). Shoreline variability is influenced not only by significant sea-level change but also by changing wave-forcing characteristics, which are associated through physical processes with time-varying model-free parameters (Ibaceta et al., 2020, 2022). Moreover, shoreline behaviour can quickly adjust to changing storm seasonality, indicating that small changes in wave forcing can cause significant changes in shoreline response (Splinter et al., 2017). Kerala ranks third (23%) in erosion among the coastal states and union territories of India (Ratheesh et al., 2023). Thiruvananthapuram is one of the three districts experiencing high rates of erosion in the state (Dwivedi et al., 2023). The amplified erosion is attributed to factors such as the development of coastal infrastructure along the coast such as breakwaters, groins, harbours, etc., (Noujas & Thomas, 2015; Ratheesh et al., 2023), sand mining (Ratheesh et al., 2023) and monsoonal waves (Kurian et al., 2009). The monsoonal waves are the vital causative factor that affects the beach stability. The entire coast of Kerala, including Thiruvananthapuram district was preponderated by tenacious WSW trending waves with a periodicity of 10 s during the southwest monsoon (Mallik et al., 1987). This oblique WSW wave interaction with the coast is the determinantal factor for longshore sediment transportation (Noujas et al., 2019). Cross-shore wind stress generated by Ekman transport is the reason for predominant longshore drift (Jineesh et al., 2015; Gurumoorthi & Venkatachalapathy, 2017).

Among other districts on the west coast of India, Thiruvananthapuram exhibits high erosion (Ratheesh et al., 2023). The wave refraction values appear to be higher, articulating immense wave energy triggering littoral transport trending north to south during the southwest monsoon from Goa to Kanyakumari (Reddy & Varadachari, 1973; Mallik et al., 1987; Samsuddin & Suchindan, 1987). While during the northeast monsoon a reversal in the littoral drift trending south to north was observed throughout the entire Kerala coast (Samsuddin & Suchindan, 1987). This littoral drift causes a high proportion of erosion rather than accretion in the study area, and the findings of the present investigation are fortified by those in Thiruvananthapuram district (Shaji, 2014). Similar results articulated by several researchers in the Kanyakumari district further south of the study area (Mujabar & Chandrasekar, 2011; Saravanan et al., 2013; Kaliraj et al., 2014, 2015; Natesan et al., 2015; Kadavan & Joseph, 2017; Manjulavani et al., 2018; Sam & Gurugnanam, 2022; Sundar et al., 2022). The coast appears varied around the mouth of four rivers via., Adimalathura, Punthura, Poovar, and Veil wherein loads of sediments from the upstream are escorted and accreted around the coastal frontier of the study area. A similar scenario, unlike the study area, was reported by Sam and Gurugnanam (2022) in Kanyakumari district. Further, river-derived accreted sediments are acted upon by prevailing waves. The stretch between Varakala and Veli forms a part of a high-energy sandy shoreline; it also comprises the Mutholapozhi tidal inlet. The aforementioned coastal stretch was dynamically stable, with the exception of morphological changes by tidal inlet oscillation until the establishment of a fishing harbour at Muthalapozhi inlet. Two breakwaters were constructed on either side between 2002 and 2008 for easy access throughout the year. However, the harbour continued to be unusable due to the spit formation around the mouth during the fair season and waves breaking during the monsoonal season. Thus, the establishment of breakwaters has consequently led to serious erosion immediately north of the inlet, as well as accretion on the southern side (Noujas et al., 2019). The findings of the present study (2003–2014) contradict the results of Noujas et al. (2019), wherein accretion is encountered on the coastal stretches of villages such as Kozhithottam- Anchuthengu and Anchuthengu-Mutholapozhi towards the north of Mutholapozhi tidal inlet. On the other hand, erosion was observed further south of the Mutholapozhi tidal inlet between the coastal stretches of Mutholapozhi-Perumathura and Pudukurichi-Pattanthura. The field evidence of intense erosion has been presented in Fig. 14. Mallik et al. (1987) carried out a survey throughout the entire Kerala coast, and the study identified 20 major convergent and 19 major divergent cells, respectively.

Further, their study indicated two of each cell in the Thiruvananthapuram district. The divergent cell signifies erosion, while the convergent cell indicates accretion. However, in the current study during the past spell, only three divergent (erosion) cells were found along the coastal stretches of Verttu-Kozhithottam, Kozhithottam- Anchuthengu, and Anchuthengu-Mutholapozhi while the rest of the coast in Thiruvananthapuram district exhibited convergent cells (Accretion). On the contrary, during the recent spell, eleven divergent cells were observed along the coastal stretches of Paravur to Anchuthengu, Mutholapozhi-Thumba, Sankumukham-Valithura, Punthura-Panathura, and Karumkulam-Poovar (Fig. 14). Similar concordant findings were reported by Shaji (2014) from the study area. This high rate of erosion is attributed to various sensitive coastal geomorphic features like channel bars, spits, etc., (Shaji, 2014). Also, 80% of the coastal villages of the Thiruvananthapuram district are comprised of sand beaches (Kadavan & Joseph, 2017) that are vulnerable to high erosion rates (Kumar et al., 2006). In addition, dense human population manipulating the coastal land cover coupled with cyclones

Fig. 14 Field photos depicting coastal erosion at **a** Anchuthengu, **b** Sankumukham **c** Collapsed seawall at Valithura, **d** Perumathura, **e** Undercutting erosion exposing laterite cliff at Varakala, and **f** Veil

during the south-west monsoon and north-east monsoon accelerate the rate of erosion (Shaji, 2014; Kadavan & Joseph, 2017; Sundar et al., 2022; Sekar et al., 2023; Princy et al., 2023). Further, a comparison of the past spell and the recent spell is suggestive of either anthropogenic influence or the impact of climate change. Henceforth, to confirm the impact of climate change, a similar kind of future endeavour after a couple of decades is highly recommended.

6 Conclusion

Erosion and accretion rates for two decades (2003–2023) were determined over a 73.54 km coastal stretch of the Thiruvananthapuram district. The DSAS software was employed to analyse the spatio-temporal variations in the shoreline of the study area. The coastal villages such as Paravur-Varkala-Verttu, Kozhithottam-Anchuthengu, Pudukurichi-Pattanthura-Pallithura-Thumba, Sankumukham-Valithura, Punthura-Panathura, Vizhinjam-Chowara, and Karumkulam-Poovar experienced high rates of erosion (>− 2 m/yr). Generally, the annual mean EPR of the Thiruvananthapuram district was estimated to be − 1.17 m/yr with a maximum erosion and accretion of − 17.45 m/yr and 16.97 m/yr, respectively. Further, the decadal mean NSM of the study area was estimated to be − 13 m, with maximum retreating and advancing shorelines of 192.2 m and 186.98 m of the coast, respectively. The significant increase in the erosional patterns of the past spell (2003–2014) and the recent spell (2003–2023) demands constant monitoring, and an urgent management solution is the need of the hour. A similar kind of future study, after two decades or more in conjugation with climate variables would prove candid to assess the impact of global warming and climate change at the regional level. Further, owing to the social relevance, the findings of the present study would assist stakeholders, policymakers, and government bodies to take immediate preventive measures to minimise coastal erosion.

References

Abdurrahman, U., Pratyaksa, I., Suprijo, T., & Park, H. (2021). Beach slope distribution mapping using UAV in the Cirebon coastal area. *E3s Web of Conferences, 324*, 01006. https://doi.org/10.1051/e3sconf/202132401006

Al-Attar, S. M. I., & Basheer, A. M. (2023). Multi-temporal shoreline analysis and future regional perspective for Kuwait coast using remote sensing and GIS techniques. *Heliyon, 9*, Article e20001. https://doi.org/10.1016/j.heliyon.2023.e20001

Baldock, T., Birrien, F., Atkinson, A., Shimamoto, T., Wu, S., Callaghan, D., & Nielsen, P. (2017). Morphological hysteresis in the evolution of beach profiles under sequences of wave climates—Part 1; observations. *Coastal Engineering, 128*, 92–105. https://doi.org/10.1016/j.coastaleng.2017.08.005

Bojang, A., Oyedotun, T. D. T., Sawa, B. A., & Ismail, M. (2023). Spatio-temporal coastline dynamics of the Gambia littoral zone from 1989 to 2019. *Geosystems and Geoenvironment, 2*, Article 100194. https://doi.org/10.1016/j.geogeo.2023.100194

Chen, L., Weng, J., Naimullah, M., Hsiao, P., Tseng, C., Lan, K., & Chuang, C. (2021). Distribution and catch rate characteristics of narrow-barred Spanish Mackerel (Scomberomorus commerson) in relation to oceanographic factors in the waters around Taiwan. *Frontiers in Marine Science, 8*, Article 770722. https://doi.org/10.3389/fmars.2021.770722

Collins, D., Avdis, A., Allison, P., Johnson, H., & Piggott, M. (2018). Controls on tidal sedimentation and preservation: Insights from numerical tidal modelling in the late oligocene–miocene south China Sea, southeast Asia. *Sedimentology, 65*(7), 2468–2505. https://doi.org/10.1111/sed.12474

Cooper, J., Green, A., & Loureiro, C. (2018). Geological constraints on mesoscale coastal barrier behaviour. *Global and Planetary Change, 168*, 15–34. https://doi.org/10.1016/j.gloplacha.2018.06.006

D'Anna, M., Idier, D., Castelle, B., Cozannet, G., Rohmer, J., & Robinet, A. (2020). Impact of model free parameters and sea-level rise uncertainties on 20-years shoreline hindcast: The case of Trucvert beach (sw France). *Earth Surface Processes and Landforms, 45*(8), 1895–1907. https://doi.org/10.1002/esp.4854

Dewi, R. S., & Bijker, W. (2020). Dynamics of shoreline changes in the coastal region of Sayung, Indonesia. *Egyptian Journal of Remote Sensing and Space Science, 23*, 181–193. https://doi.org/10.1016/j.ejrs.2019.09.001

Ding, Y., Frey, A., Kim, S., & Permenter, R. (2018). Probabilistic shoreline change modeling and risk estimation of erosion. *Coastal Engineering Proceedings, 36*, 1. https://doi.org/10.9753/icce.v36.papers.1

Dwivedi, S. C., Pampattiwar, T. S., Pandey, C. A., Parida, R. B., Mitra, D., & Kumar, N. (2023). Characterization of the coastal vulnerability in different geological settings: A comparative study on Kerala and Tamil Nadu coasts using fuzzy AHP. *Sustainability, 15*, 9543. https://doi.org/10.3390/su15129543

Elkotby, R. M., Sarhan, A. T., & El-Gamal, M. (2023). Assessment of human interventions presence and their impact on shoreline changes along Nile delta. *Egypt. Oceanologia, 65*, 595–611. https://doi.org/10.1016/j.oceano.2023.06.008

Esmail, M., Mahmod, W. E., & Fath, H. (2019). Assessment and prediction of shoreline change using multi-temporal satellite images and statistics: Case study of Damietta coast, Egypt. *Applied Ocean Research, 82*, 274–282. https://doi.org/10.1016/j.apor.2018.11.009

Ferreira, T. A. B., da Silva, André Giskard Aquino, Perez, Y. A. R., Stattegger, K., & Vital, H. (2021). Evaluation of decadal shoreline changes along the parnaíba delta (NE Brazil) using satellite images and statistical methods. *Ocean and Coastal Management, 202*. https://doi.org/10.1016/j.ocecoaman.2020.105513

Gens, R. (2010). Remote sensing of coastlines: Detection, extraction and monitoring. *International Journal of Remote Sensing, 31*, 1819–1836. https://doi.org/10.1080/01431160902926673

Gurumoorthi, K., & Venkatachalapathy, R. (2017). Hydrodynamic modeling along the southern tip of India: A special emphasis on Kanyakumari coast. *Journal of Ocean Engineering and Science, 2*, 229–244. https://doi.org/10.1016/j.joes.2017.08.004

Ibaceta, R., Splinter, K., Harley, M., & Turner, I. (2020). Enhanced coastal shoreline modeling using an ensemble Kalman filter to include nonstationarity in future wave climates. *Geophysical Research Letters, 47*(22). https://doi.org/10.1029/2020gl090724

Ibaceta, R., Splinter, K., Harley, M., & Turner, I. (2022). Improving multi-decadal coastal shoreline change predictions by including model parameter non-stationarity. *Frontiers in Marine Science, 9*. https://doi.org/10.3389/fmars.2022.1012041

Jineesh, V. K., Muraleedharan, K. R., Lix John, K., Revichandran, C., Hareesh Kumar, P. V., & Naveen Kumar, K. R. (2015). Mesoscale process-induced variation of the West India Coastal Current during the winter monsoon. *Environmental Monitoring and Assessment, 187*, 8. https://doi.org/10.1007/s10661-015-4701-5

Kadavan, P. A., & Joseph, E. J. (2017). Shoreline changes at selected stretches on the Thiruvananthapuram-Kochi coast, southwest India. *ISH Journal of Hydraulic Engineering., 24*(3), 298–310. https://doi.org/10.1080/09715010.2017.1376292

Kaliraj, S., Chandrasekar, N., & Magesh, N. S. (2014). Impacts of wave energy and littoral currents on shoreline erosion/accretion along the south-west coast of Kanyakumari, Tamil Nadu Using DSAS and Geospatial Technology. *Environmental Earth Science, 71*, 4523–4542. https://doi.org/10.1007/s12665-013-2845-6

Kaliraj, S., Chandrasekar, N., & Magesh, N. S. (2015). Evaluation of coastal erosion and accretion processes along the southwest coast of Kanyakumari, Tamil Nadu using geospatial techniques. *Arabian Journal of Geoscience., 8*, 239–253. https://doi.org/10.1007/s12517-013-1216-7

Kankara, R. S., Murthy, R. V. M., & Rajeevan, M. (2018). *National assessment of shoreline changes along Indian coast: Status report for 26 years (1990–2016).* NCCR Publication. Available at NCCR Web site http://www.nccr.gov.in

Kilar, H. (2023). Shoreline change assessment using DSAS technique: A case study on the coast of Meriç Delta (NW Türkiye). *Regional Studies in Marine Science, 57*, Article 102737. https://doi.org/10.1016/j.rsma.2022.102737

Koroy, K., Wattimena, M., & Wahab, I. (2023). Analysis of changes in the coastaline of tabailenge island north Morotai district, Morotai island regency. *East Asian Journal of Multidisciplinary Research, 2*(3), 1251–1260. https://doi.org/10.55927/eajmr.v2i3.3500

Kumar, A., & Thambidurai, P. (2023). Impact of urbanization and river morphology on groundwater system in Patna urban area, Bihar, India. In P. Thambidurai, A. K. Dikshit (Eds.), *Impacts of urbanization on hydrological systems in India* (pp. 167–185). Springer. https://doi.org/10.1007/978-3-031-21618-3_9

Kumar, A., & Thambidurai, P. (2022). Relationship of topography, weathering, and water table in Usri watershed of Chota Nagpur Plateau, India. *Arabian Journal of Geosciences, 5–504*, 1–16. https://doi.org/10.1007/s12517-022-09624-z

Kumar, V. S., Pathak, K. C., Pednekar, P., Raju, N. S. N., & Gowthaman, R. (2006). Coastal processes along the Indian coastline. *Current Science, 91*, 530–536.

Kurian, N., Rajith, K., Shahul, H., Nair, S., Murthy, R., Arjun, S. & Shamji, V. (2009). Wind waves and sediment transport regime off the south-central Kerala coast, India. *Natural Hazards, 49*, 325–345.

Lim, C., Kim, T., Kim, J., & Lee, J. (2022). A study on the influence of sand median grain size on the short-term recovery process of shorelines. *Frontiers in Marine Science, 9*. https://doi.org/10.3389/fmars.2022.906209

Louati, M., Saïdi, H., & Zargouni, F. (2015). Shoreline change assessment using remote sensing and GIS techniques: A case study of the Medjerda delta coast, Tunisia. *Arabian Journal of Geosciences, 8*, 4239–4255. https://doi.org/10.1007/s12517-014-1472-1

Luijendijk, A., Hagenaars, G., Ranasinghe, R., Baart, F., Donchyts, G., & Aarninkhof, S. (2018). The state of the world's beaches. *Scientific Reports, 8*, 1–11. https://doi.org/10.1038/s41598-018-24630-6

Mageswaran, T., Sachithanandam, V., Sridhar, R., Mahapatra, M., Purvaja, R., & Ramesh, R. (2021). Impact of sea level rise and shoreline changes in the tropical island ecosystem of Andaman and Nicobar region, India. *Natural Hazards, 109*, 1717–1741. https://doi.org/10.1007/s11069-021-04895-3

Mallik, T. K., Samsuddin, M., Prakash, T. N., Vasudevan, V., & Machado, T. (1987). Beach erosion and accretion- An example from Kerala, Southwest, coast of India. *Environmental Geology & Water Science, 10*(2), 105–110.

Manjulavani, K., Supriya, V. M., Suhrullekha, M., & Harish, B. (2018). Detection of shoreline change using geo-spatial techniques along the coast between Kanyakumari and Tuticorin. In *IEEE international conference power control. Signals instrumentation engineering* (pp. 2822–2825). ICPCSI. https://doi.org/10.1109/ICPCSI.2017.8392235

Mirdan, M. M., Tolba, R. E., Abdellah, S., & Gala, M. E. (2023). Digital shoreline analysis system techniques for stability detection: An applied case study on Port Said, Egypt. *The Egyptian Journal of Aquatic Research, 49*(4), 460–470. https://doi.org/10.1016/j.ejar.2023.08.001

Mujabar, P. S., & Chandrasekar, N. (2011). A shoreline change analysis along the coast between Kanyakumari and Tuticorin, India, using digital shoreline analysis system. *Geo-Spatial Information Science., 14*, 282–293.

Natesan, U., Parthasarathy, A., Vishnunath, R., Kumar, G. E. J., & Ferrer, V. A. (2015). Monitoring long-term shoreline changes along Tamil Nadu, India using geospatial techniques. *Aquatic Procedia, 4*, 325–332. https://doi.org/10.1016/j.aqpro.2015.02.044

Noujas, V., Kurian, N. P., & Thomas, K. V. (2019). Comparison of coastal hydrodynamics in different energy regime coasts along southwest coast of India. *Journal of Earth System Science, 128*(93), 1–24.

Noujas, V., & Thomas, K. (2015). Erosion hotspots along southwest coast of India. *Aquatic Procedia, 4*, 548–555.

Princy, J. R., Ramesh, M., Jyothi, J., Krishna, P. S. S., & Nair, S. (2023). Non-monsoonal coastal erosion due to the tropical cyclone (OCKHI) and its impacts along Thiruvananthapuram coast, Southwest Coast of India—A geospatial approach, Chapter-29. In Jayaraju, N., et al. (Eds.), *Coasts, estuaries and lakes* (pp. 471–486). https://doi.org/10.1007/978-3-031-21644-2_29

Raj, B. N., Gurugnanam, V., Sudhakar, P., & Francis, G. (2019). Estuarine shoreline change analysis along the Ennore river mouth, south east coast of India, using digital shoreline analysis system. *Geodesy and Geodynamics, 10*, 205–212. https://doi.org/10.1016/j.geog.2019.04.002

Ratheesh, R., Rajput, P., Bhatti, H., Rajawat, A. S., & Rajak, R. D. (2023). Quantification of shoreline changes along the entire Indian coast using Indian Remote Sensing satellite images of 2004–06 and 2014–16. *Current Science, 124*(5), 578–584.

Reddy, M. P. M., & Varadachari, V. V. R. (1973). Sediment movement in relation to wave refraction near Marmogao harbor: Supplement to. *Mahasagar, 3*(1), 27–44.

Rezaee, S. M., Golshani, A., & Abedi, S. (2022). Shoreline changes at Fereydounkenar Port in light of Caspian Sea's water level fluctuations. *Regional Studies in Marine Sciences, 53*, Article 102393. https://doi.org/10.1016/j.rsma.2022.102393

Robinet, A., Castelle, B., Idier, D., D'Anna, M., & Cozannet, G. (2020). Simulating the impact of sea-level rise and offshore bathymetry on embayment shoreline changes. *Journal of Coastal Research, 95*(sp1), 1263. https://doi.org/10.2112/si95-244.1

Robinet, A., Idier, D., Castelle, B., & Marieu, V. (2018). A reduced-complexity shoreline change model combining longshore and cross-shore processes: The LX-shore model. *Environmental Modelling and Software, 109*, 1–16. https://doi.org/10.1016/j.envsoft.2018.08.010

Sam, C. S., & Gurugnanam, B. (2022). Coastal transgression and regression from 1980 to 2020 and shoreline forecasting for 2030 and 2040, using DSAS along the southern coastal tip of Peninsular India. *Geodesy and Geodynamics, 13*, 585–594. https://doi.org/10.1016/j.geog.2022.04.004

Samra, M. A. R., & Ali, R. R. (2021). Applying DSAS tool to detect coastal changes along Nile Delta, Egypt. *The Egyptian Journal of Remote Sensing and Space Sciences, 24*, 463–470. https://doi.org/10.1016/j.ejrs.2020.11.002

Samsuddin, M., & Suchindan, K. G. (1987). Beach erosion and accretion in relation to seasonal longshore current variation in the northern Kerala coast, India. *Journal of Coastal Research, 3*(1), 55–62.

Sanhory, A., El-Tahan, M., Moghazy, M. H., & Reda, W. (2022). Natural and manmade impact on Rosetta eastern shoreline using satellite Image processing technique. *Alexandria Engineering Journal, 61*, 6247–6260. https://doi.org/10.1016/j.aej.2021.11.053

Santos, C. A. G., Nascimento, T. V. M., do Mishra, M., & da Silva, R. M. (2021). Analysis of long- and short-term shoreline change dynamics: A study case of João Pessoa city in Brazil. *Science of Total Environment, 769*, 144889 https://doi.org/10.1016/j.scitotenv.2020.144889

Saravanan, S., Chandrasekar, N., & Joevivek, V. (2013). Temporal and spatial variation in the sediment volume along the beaches between Ovari and Kanyakumari (SE India). *International Journal of Sediment Research., 28*, 384–395. https://doi.org/10.1016/S1001-6279(13)60048-7

Sekar, G. L., Androws, X., Annaidasan, K., Kumar, A., Kannan, R., Muthusankar, G., & Balasubramani, K. (2023). Assessment of shoreline changes and associated erosion and accretion pattern in coastal watersheds of Tamil Nadu, India. *Natural Hazard Research, 4*(2), 231–238. https://doi.org/10.1016/j.nhres.2023.09.008

Selvan, C. S., Kankara, S. R., Markose, J. V., Rajan, B., & Prabhu, K. (2016). Shoreline change and impacts of coastal protection structures on Puducherry, SE coast of India. *Natural Hazards, 83*, 293–308. https://doi.org/10.1007/s11069-016-2332-y

Shaji, J. (2014). Coastal sensitivity assessment for Thiruvananthapuram, west coast of India. *Natural Hazards, 73*, 1369–1392. https://doi.org/10.1007/s11069-014-1139-y

Splinter, K., Turner, I., Reinhardt, M., & Ruessink, G. (2017). Rapid adjustment of shoreline behavior to changing seasonality of storms: Observations and modelling at an open-coast beach. *Earth Surface Processes and Landforms, 42*(8), 1186–1194. https://doi.org/10.1002/esp.4088

Sundar, V., Sannasiraj, S. A., Murali, K., & Singaravelu, V. (2022). Impact of coastal structure on shorelines along the southeast and southwest coasts of India. *ISH Journal of Hydraulic Engineering, 29*, 592–610. https://doi.org/10.1080/09715010.2022.2115319

Thieler, E. R., Himmelstoss, E. A., Zichichi, J. L., & Ayhan, E. (2009). Digital shoreline analysis system (DSAS) version 4.0. An ArcGIS extension for calculating shoreline change. U.S. Geological Survey open-file report 2008–1278. https://doi.org/10.3133/ofr20081278

Vitousek, S., Barnard, P., Limber, P., Erikson, L., & Cole, B. (2017). A model integrating longshore and cross-shore processes for predicting long-term shoreline response to climate change. *Journal of Geophysical Research Earth Surface, 122*(4), 782–806. https://doi.org/10.1002/2016jf004065

Vitousek, S., Cagigal, L., Montaño, J., Rueda, A., Méndez, F., Coco, G., & Barnard, P. (2021). The application of ensemble wave forcing to quantify uncertainty of shoreline change predictions. *Journal of Geophysical Research Earth Surface, 126*(7), 1–43. https://doi.org/10.1029/2019jf005506

Zhong, X., Yu, P., & Chen, S. (2017). Fractal properties of shoreline changes on a storm-exposed island. *Scientific Reports, 7*(1), 8274. https://doi.org/10.1038/s41598-017-08924-9

Analysing Shoreline Changes: A Case Study of Ernakulam Coast in Kerala, India

Anupama Chandran, Mahalingam Bose, and Ali Raza Moosvi

1 Introduction

India is one of several countries experiencing severe coastal erosion in the past few years, and the erosion is critically significant due to the presence of coastal villages and towns along the coastal lowlands (Dar & Dar, 2009). Around 45% of Kerala's coast is eroding even though the west coast is considered generally stable compared to India's east coast (Kankara et al., 2018). The studies revealed that the coast of Kerala is one of the most active coastal stretches of India after the West Bengal coast (Mohanty et al., 2017; Sheela Nair et al., 2018). Kerala faces severe problems from coastal erosion due to the dense population along the coast, including the loss of life, land, and property, emphasizing the need for scientific inquiry into this phenomenon within the state (Sreelakshmi et al., 2024). Coastal erosion is investigated by monitoring and measuring the spatio-temporal changes in shorelines. The term 'shoreline' defines the boundary that separates land and the sea (Al-Mansoori & Al-Marzouqi, 2016). Because of the dynamic nature and uncertainty associated with the shoreline, several indicators, including high tide line, low tide line, wet/dry boundary and instantaneous water line, are widely used as a 'proxy' to indicate the position of the shoreline (Boak & Turner, 2005; Gens, 2010). The term 'shoreline change' refers to the displacement of shoreline position over time (Alemayehu et al., 2014; Shanghude et al., 2015). Shorelines continuously migrate in both landward and seaward directions, resulting in accretion and erosion, respectively. These changes respond to natural causes such as waves, storms, tides, rising sea levels, and human activities (Kankara et al., 2015). Constant monitoring of shorelines provides insights into various coastal processes. Remote sensing and GIS are employed to map shoreline positions at different spatiotemporal scales. Traditionally, it involves the field surveys or interpretation of aerial photographs (Guariglia et al., 2009). Jimenez et al.

A. Chandran · M. Bose (✉) · A. R. Moosvi
Department of Geography, Central University of Karnataka, Kalaburagi, India
e-mail: mahabose@gmail.com

P. Thambidurai and A. K. Dikshit (eds.), *Coastal Environments of India*,
https://doi.org/10.1007/978-3-031-97341-3_11

(1997) reported the suitability of aerial photographs in computing shoreline changes for a short-term period with beach profile analysis. Similarly, the shoreline changes were studied by Matin and Hasan (2021) through a field survey along the coast of Bangladesh to obtain precise information for regional studies. These methods are expensive, time-consuming, and labour-intensive.

Many more technologies have been adopted through satellite data (Nayak, 2002), such as IRS, SPOT, and Landsat data, for mapping various coastal themes. Satellites provide synoptic coverage that greatly helps shoreline monitoring (Natesan et al., 2012). Further, researchers have been widely using satellite data to monitor shoreline evolution due to its availability of frequent medium-resolution images free of cost. Several studies have utilised multitemporal Landsat images to examine shoreline changes in different periods across the world (Li & Gong, 2016; Sarwar & Woodroffe, 2013; Sowmya et al., 2019; Xu, 2018). There are numerous techniques for estimating the rate of shoreline changes using satellite data in a GIS environment. Marfai et al. (2007) used visual interpretation to delineate the shorelines of the Semarang coastal area from 1908 to 2003 and applied superimposed techniques by overlay analysis for change detection. Monalisha and Panda (2018) calculated rates of both erosion and accretion on the Ganjam coast in India by overlying shorelines of different periods and digitizing polygon features between them in ArcGIS software. Matin and Hasan (2021) calculated the net area change along the Bangladesh coast by identifying the erosion and accretion between 1989 and 2019 by creating a polygon between the two-time periods' shorelines. Similarly, (Mujabar & Chandrasekar, 2011) adopted the linear change detection method to explore changes that occurred on the southeast coast of India by analysing the length of overlaid shorelines over different years. Xu (2018) applied transect-based and area-based methods to monitor shoreline change in Texas, USA; the transect-based method calculates the changes using the transects, and the area-based method compares the changes in land/water areas using thematic maps. In addition, the advancements in the GIS field developed a range of statistical tools to quantify shoreline change using satellite data, such as Beach Tools (Hoeke et al., 2001), SCARPS (Jackson, 2004), AMBUR (Jackson et al., 2012) and DSAS (Thieler & Danforth, 1994; Himmelstoss et al., 2018). The USGS develops DSAS to estimate historical shoreline changes statistically within a GIS environment. It is an extension tool available freely for ESRI ArcGIS software. In DSAS, transects are generated perpendicular to a reference baseline by intersecting multi-temporal shorelines. The change is estimated by considering the distance from the baseline to every point of intersection on the shorelines (Parthasarathy & Deka, 2019). The DSAS has numerous applications in coastal environment research. Mishra et al. (2021) analysed the consequences of cyclone in Balukhand-Konark Wildlife Sanctuary of India by employing DSAS to monitor the shoreline and LULC change. Kale et al. (2019) calculated the erosion rates using DSAS to understand the effect of large-scale dams on shoreline changes along the Yeşilırmak Delta of Turkey. Jebaraj et al. (2021) analysed the shoreline change on the coast of southern Odisha coast and northern Andhra Pradesh using DSAS to identify the erosional hotspots. In another study, decadal shoreline changes were estimated concerning the Indian Ocean Tsunami in

2004 and its impacts on the Aceh coast, Sumatra Island of Indonesia (Meilianda et al., 2023).

The DSAS employs various statistics to calculate the Indian shoreline change rate, such as Shoreline Change Envelope (SCE), Net Shoreline Movement (NSM), End Point Rate (EPR), Linear Regression Rate (LRR), and Weighted Linear Regression Rate (WLR) (Himmelstoss et al., 2018). These methods within DSAS have been applied in numerous studies along the Indian coast to analyse shoreline changes. Kar et al. (2021) assessed the changes along the Odisha coast from 1978 to 2017 with EPR, LRR, SCE, and NSM. Another study that employed EPR and NSM elaborates on the shoreline dynamics on the Odisha coast that may be attributed to natural cyclones and floods as well as anthropogenic activities (Murali et al., 2015). Similarly, Kannan et al. (2016) employed EPR and LRR approaches to monitor shoreline change in the Nellore Coast of Andhra Pradesh and observed significant changes mainly due to the morphological processes in the river mouth. Further, Mahapatra et al. (2014) analysed shoreline change from 1972 to 2011 using LRR, which revealed that about 69.31% of the coast of South Gujarat is under erosion. LRR and EPR are the two most used methods in previous works. The EPR method requires only two shorelines, thus making it more suitable for shoreline change analysis in short-term periods, while LRR is suitable for long-term analysis using multiple shorelines (Quang et al., 2021; Zambrano-Medina et al., 2023). LRR incorporates every intersection point between the different shorelines and the transects (Parthasarathy & Deka, 2019). Recently, Sreelakshmi et al. (2024) used LRR to interpret shoreline changes between 2002 and 2021, aiming to assess the risks associated with coastal hazards in central Kerala. The study observed erosion on many parts of the Ernakulam coast, such as Aniyal, Manaserry, Edavanakkad, and Chellanam, while accretion is noticed on the Puduveypu to Elamkunnappuzha stretch. A shoreline change study conducted by Fasil et al. (2023) around the islands in the Cochin Estuary states that human activities, tidal factors, and tsunamis are responsible for shoreline changes in these islands. The 2004 Indian Ocean tsunami triggered destruction and many changes in morphology in the coastal districts of Kerala, including Ernakulam. It was identified as a potential contributing factor to morphological changes in these Island regions (Kumar et al., 2014). There are a few more studies conducted in Ernakulam and the surrounding regions looking into various coastal processes that uncovered the vulnerability of the district for the last few decades (Kumar et al., 1990; Mani & Dinesh, 2015; Thiruvenkatasamy & Baby Girija, 2014).

In light of the challenging scenario in coastal areas of Ernakulam, this study attempts to comprehensively analyse the spatio-temporal changes in shoreline positions on Ernakulam coast during 1988–2023 using remote sensing images and GIS. Landsat images were used to delineate historical shorelines, which were later used as inputs in DSAS, and appropriate statistical methods were employed to estimate the long-term changes for 35 years and short-term changes in Ernakulam during the study period. This study explores the significance of regional shoreline studies in understanding coastal environments.

2 Study Area

Ernakulam is one of the nine coastal districts of Kerala, India, and lies between latitude 09°47′13″ and 10°10′44″ N and longitude 76°10′05″ and 77°05′24″ E. The region experiences a tropical monsoon climate, with annual temperatures typically ranging from 23 to 31 °C. This district, which occupies the central part of the state, has the shortest shoreline after the Kollam district. The Ernakulam coast, stretching from Munambam to Chellanam, has long, narrow beaches and estuaries. The Cochin Estuary, also known as Kochi Kayal, is one of the largest wetland ecosystems in India. Coastal alluvial soil and laterites cover the coastal lowlands due to the presence of Kayal. Being the state's industrial-technological centre, the coastal region of Ernakulam is always under immense development pressure. In this study, the entire coastal area of Ernakulam district, which spans approximately 46 km, is investigated for shoreline changes during the last 35-year period (Fig. 1).

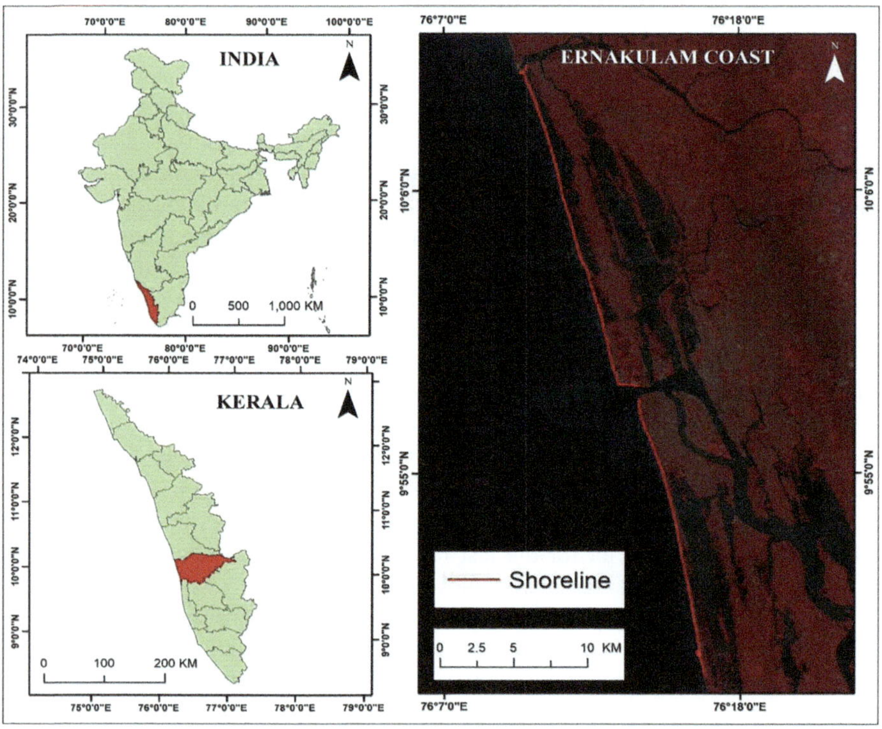

Fig. 1 Study area map

3 Materials and Methods

The present study was carried out to estimate the shoreline dynamics on the Ernakulam coast of Kerala; Multi-temporal Landsat data for 1988, 1992, 2000, 2008, 2015, and 2023 were downloaded from the USGS open source. The images obtained are of uneven time intervals due to the unavailability of images with less cloud cover. However, all the images are obtained for the dry season (December and January) to minimize any seasonal error. The detailed methodology employed in this study is presented in Fig. 2.

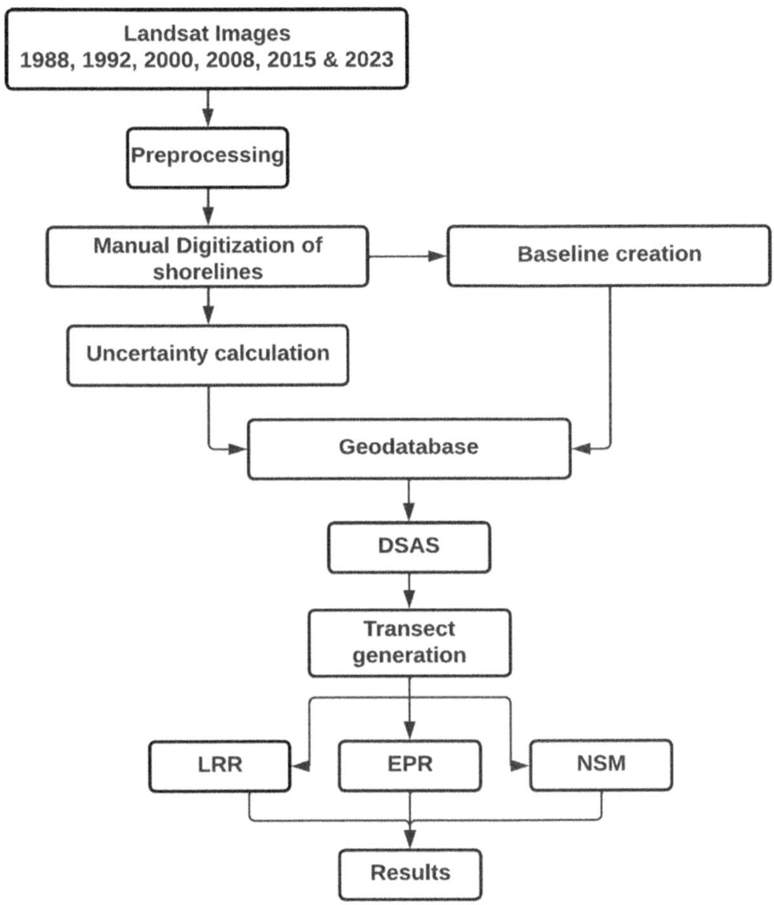

Fig. 2 Research methodology flow chart

3.1 Shoreline Extraction

The historic shorelines were obtained from satellite images of the years 1988, 1992, 2000, 2008, 2015, and 2023 by manual on-screen digitization method using ArcGIS v10.7 software. Manual Digitization through visual image interpretation is the most simple and common method for shoreline extraction (Daoudi & Niang, 2021). The near-infra-red bands of the images, which are capable of separating the land–water boundary distinctly (Matin & Hasan, 2021), were used to identify the instantaneous shoreline. According to Gens (2010), "the instantaneous shoreline is the position of the land–water interface at one instant in time." In this study, the land-sea boundary line visible at the time of the image taken was considered the shoreline for the corresponding years.

3.2 Uncertainty Calculation

The accuracy of shoreline positions and shoreline change rate is influenced by several uncertainties (Basheer Ahammed & Pandey 2022). The different types of uncertainties, namely, Georeferencing error (Eg), Pixel error (Ep), Digitization error (Ed), Tidal error (Et), and Seasonal error (Es), are sourced from previous studies. The Eg refers to the geometric error that is given in the metadata file of Landsat images (Darwish & Smith, 2023). The Ep is the uncertainty associated with image resolution (Kar et al., 2021). All the images used in the present study have a uniform spatial resolution. The Ed represents the uncertainty in the manual digitization of shorelines by visual image interpretation (Islam & Crawford, 2022). In this study, the Ed has kept at least half a pixel size (Dutta et al., 2022; Murray et al., 2023). The influence of tides on shorelines depends on the range of tidal fluctuations (Elnabwy et al., 2020) and also the resolution of the satellite images (Guariglia et al., 2009). Considering the minimal tidal fluctuations (< 2 m) in the study area and the 30 m resolution of the images, the Et was neglected. Also, the images were collected from the same season, nullifying the errors due to seasonal changes. Hence, the Es has also been neglected (Dutta et al., 2022). The total uncertainty value is calculated using three uncertainty types that are relevant to this study, as follows (Daoudi & Niang, 2021),

$$Ut = \sqrt{Eg^2 + Ep^2 + Ed^2}$$

where Ut is total uncertainties, E.g. is georeferencing error, Ep is pixel error, and Ed is a digitization error. Table 1 shows the total uncertainty of shorelines in different years individually.

According to Hapke et al. (2011), the annualized uncertainty with respect to the end-point rate at every individual transect is determined using the following formula:

Table 1 Uncertainties associated with shorelines

Uncertainty (in meter)	1988	1992	2000	2008	2015	2023
Eg	4.608	4.569	4.132	4.119	3.793	5.25
Ep	30	30	30	30	30	30
Ed	15	15	15	15	15	15
Total Uncertainty (Ut)	33.86	33.85	33.79	33.79	33.75	33.95

$$Ua = \frac{\sqrt{Ut_1^2 + Ut_2^2}}{T}$$

where Ua is the annualised uncertainty, Ut^1 and Ut^2 are the total uncertainties of two different periods, and T is the time period between them. In this study, the annualized error obtained for the study period was 2.41 m/y.

3.3 Shoreline Change Analysis

The present study analysed shoreline changes in the Ernakulam coast for the long-term period of 35 years and also short-term periods in the DSAS using ArcGIS v 10.7. The methodology for the change analysis is adapted from the 'DSAS Version 5.0 User Guide: USGS Open-File Report 2018' (Himmelstoss et al., 2018). There are four main steps in DSAS analysis: shoreline extraction, creation of baseline, transect generation, and computing the rate of change ((Natesan et al., 2012). The DSAS cast equally spaced lines called transects perpendicularly to the baseline. This study established a baseline in the offshore region using the buffer method for a distance of 2000 m seaward from the existing shorelines. The baseline and digitized shoreline positions in the vector format were then stored in a geodatabase for further processes. After employing the attribute automator function within the DSAS toolbar to add the fields to the attribute tables of both the baseline and shorelines, transects were generated to 3000 m length from baseline with 100 m spacing and a smoothing distance of 500 m. Three statistical methods such as, NSM, EPR, and LRR, were then applied to calculate the change in rates of shoreline. The NSM is the distance between the two shorelines of different periods. The EPR is NSM divided by the time interval, while the LRR is calculated by fitting a linear regression model (Burningham & Fernandez-Nunez, 2020).

4 Results and Discussion

4.1 Analysing Long-Term Shoreline Changes

The shoreline dynamics along the Ernakulam coast from 1988 to 2023 were calculated by DSAS using multi-temporal shorelines. Perpendicular to the baseline, 542 transects were created, intersecting the shorelines. LRR and EPR statistics provided the average change in Meters/year, while NSM provided the total distance in meters gained/lost during the entire study period. Whereas the values less than 0 show coastal erosion, the values greater than 0 show accretion. Based on that, the results of LRR and EPR were classified into 7 categories, as shown in Fig. 3. According to LRR, the average erosion rate and accretion rate estimated on the coast of Ernakulam between 1988 and 2023 were 0.63 m/y and 3.48 m/y, respectively. Out of 542 transects, 131 transects were erosional, which accounts for 24.17%. The Munambam, the northern end of the Ernakulam coast, had the highest erosion rate of − 11.73 m/y. Another zone of very high erosion was identified in the Fort Kochi region. The coastal stretch from Munambam to Cherai also experienced high erosion throughout the study period. The rest of the coast is characterized by moderate erosion to moderate accretion. A few pockets of very high accretion were identified along the breakwater at Munambam and Puthuvype, Fort Kochi–Beach Road stretch, and Chellanam fishing harbour. An overall 411 transects (75.83%) showed positive values indicating accretion in the study area. The maximum accretion was estimated at Puthuvype, with a rate of 25.09 m/y. The EPR estimated an average of 3.86 m/y for accretion and 0.5 m/y for erosion during the study period. The EPR indicates that 35.24% (191 transects) of the Ernakulam shoreline was under erosion, and 64.76% (351 transects) was under accretion condition. Like the LRR result, the maximum erosion (7.97 m/y) and maximum accretion (28.3 m/y) were identified on the Munambam coast along the breakwater. Apart from Munambam, very high erosional areas were identified on the coast of Fort Kochi. A zone of high erosion was identified on the Munambam–Cherai stretch, Aniyal beach, and Nayarambalam–Njarackal stretch. In contrast, moderate erosion was identified in Cherai, Nayarambalam, Njarackal, Malipuram, Manassery, Cheriyakadavu and Chellanam. The coastal stretches between Cherai–Kuzhupilly, Fort Kochi–Beach Road, Saudi–Manassery, and Puthenthodu have experienced moderate accretion to moderate erosion. The very high accretional zones were mainly observed along three coasts: Munambam, Puthuvype, and Chellanam Fishing Harbour.

The NSM provided both landward and seaward gain/loss along the coast of Ernakulam over the time in meters. According to NSM, the average erosion rate for 1988–2023 was 17.49 m. As observed earlier, the highest erosion rate was identified in the Munambam, with a distance of 279.14 m. Out of 542 transects, 191 (35.24%) had landward movement showing erosion, while 351 (64.76%) had seaward movement showing accretion. The average accretion rate and maximum accretion were 134.99 m and 991.25 m, respectively. Additionally, the shoreline was classified into three classes to show the total gain/loss of coastal land based on the NSM

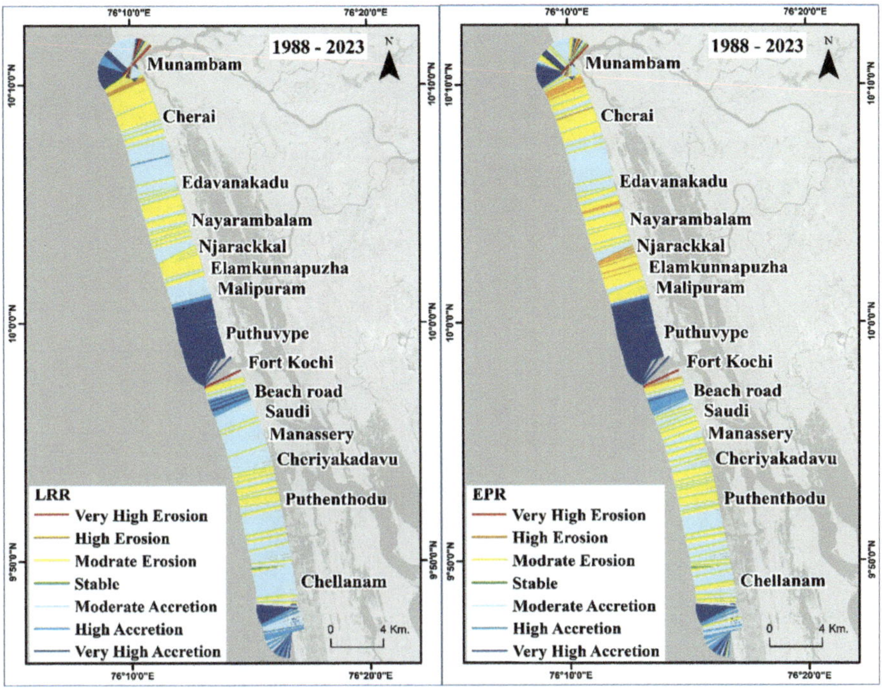

Fig. 3 Shoreline change estimated using LRR and EPR from 1988 to 2023

values (Fig. 4). It is identified that the entire coast of Ernakulam has witnessed both erosion and accretion with no stable coast. Coastal stretches such as Cherai, Aniyal, Nayarambalam, Elakunnapuzha, Malipuram, Manassery, and a few areas in the Kannamali–Chellanam stretch had lost land at an average rate of − 17.49 m. In contrast, stretches such as Munambam, Kuzhupilly, Puthuvype, and Chellanam fishing harbour regions have gained an average rate of 134.99 m during the last 35 years. The findings revealed that the shoreline of Ernakulam witnessed significant erosion and accretion, and the absence of a stable coast indicates the dynamic nature of this coast in the long term.

4.2 Analysing Short-Term Shoreline Changes

The total study period of 35 years was segmented into sub-periods to understand the changes in each period. Subsequently, DSAS was employed to compute short-term changes across five intervals: 1988–1992, 1992–2000, 2000–2008, 2008–2015, and 2015–2023. The EPR statistic that estimates the progress between two successive shorelines (Niang, 2020) was adopted for computing short-term changes. The results of the EPR for five different subperiods are shown in Fig. 5a–e. During the sub-period

Fig. 4 Map showing net shoreline movement from 1988 to 2023

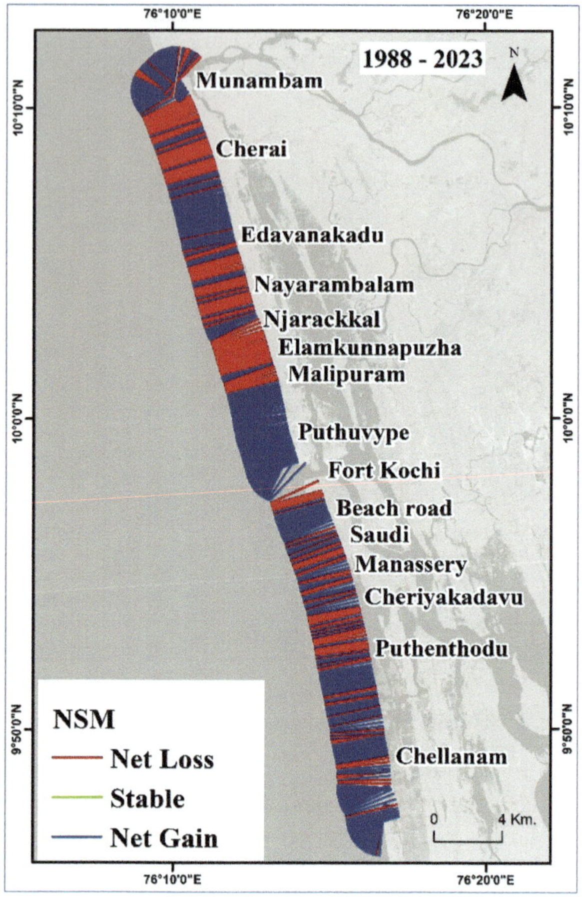

1988–1992 (Fig. 5a), considerable erosion occurred at a rate of 67.65% along the Ernakulam coast, with only around 32.35% under accretion. Apart from the zone of very high accretion on the Puthuvype and Munambam stretches, the coast was dominated by moderate to very high erosion. One of the major erosional zones was identified over a coastal stretch between Aniyal and Elamkunnapuzha, and on the Fort Kochi and Chellanam coasts. The average erosion and accretion rates during this short period were 3.28 m/y and 15.64 m/y, respectively. In contrast to this period, 81.2% of the Ernakulam coast accreted significantly over the next few years, from 1992 to 2000 (Fig. 5b). Aside from a few areas along Chellanam, Fort Kochi, Munambam, and Malipuram coastal stretches that experienced moderate to very high erosion, the rest of the coast witnessed accretion at a rate of 9.75 m/y. During this sub-period, 18.8% of the coast eroded at a rate of 8.02 m/year. The average erosion rate in 2000–2008 (Fig. 5c) was only 1.35 m/y, which is nominal compared to the previous period; nonetheless, in eight years, the area under erosion expanded from 18.8 to 29.49%. Interestingly, a trend of erosion has been observed at the southernmost portion of the Puthuvype coast. However, a vast area of the coast experienced accretion at an average of 3.61 m/year with approximately 70.51%. In the next sub-period, 2008–2015 (Fig. 5d), the erosion rate boosted to 4.97 m/y, with 66.16% of the area under erosion. The accretion rate has also increased drastically to 12.95 m/y, but only 33.84% of the area was accretional. Over the next subperiod, 2015–2023 (Fig. 5e), there has been another significant change in the accretional and erosional trends. The average erosion rate declined to 2.07 m/y, with 49.53% of the area under erosion, while the accretion rate also dropped to 1.8 m/y, with 50.47% of the area accreting. One of the very high erosional hotspots during this period was identified along the Puthuvype coastal stretch.

The present study discovered considerable shoreline changes on the Ernakulam coast over short and long periods. A field investigation with the help of a GPS mobile application was carried out to locate and validate the results with ground truth. As stated earlier, one of the notable changes was the very high accretion on the Puthuvype coast (Fig. 6a), located at the immediate north of the Cochin inlet entrance, which is highly influenced by morphological processes and human interference. As reported by Sheela Nair et al. (2018), the accumulation of dredged soil in this region from the Cochin ship channels due to the northerly sediment transport phenomenon is the reason for this unique formation. However, Kurian et al. (2007) assessed shoreline change from 1985 to 2000 and stated that accretion on the coastal stretch of Njarackal–Puthuvype, except for the Elamkunnapuzha coast and low erosion during those years. Even though the Puthuvype coastal stretch exhibited a tendency of accretion over the study period from 1988 to 2023, another interesting result is the erosional trend observed in the southern end of this stretch during the sub-periods 2000–2008 and 2015–2023. The Indian Ocean Tsunami of 2004 and human activities are the potential reasons for such changes in this area (Fasil et al., 2023). Human interference, such as the construction of breakwaters, groynes, fishing harbours, and other coastal infrastructures, can disrupt the various coastal processes (Selvan et al., 2019). The construction of breakwaters, seawalls, and groynes obstruct the natural flow of water and sediments, causing accretion on the updraft side and erosion on the downdrift side.

Once the sediment flow is disrupted, the erosion becomes irreversible. There was accretion along the breakwater in Munambam (Fig. 7a) and erosion in the region immediately south of it (Fig. 7b). Thiruvenkatasamy and Baby Girija (2014) observed considerable changes in the bathymetry of the Munambam area after the construction of the breakwater in 2000. The sand bar formation in the mouth due to the littoral drift from the north direction has disappeared, and the inlet depth increased from a range of 0–5 m to a maximum of about 11 m, improving the navigability of the inlet. Narayana and Priju (2006) observed an erosional trend in Fort Kochi and Chellanam, while other regions had an accretion trend from 1966 to 1995. This study also found that Fort Kochi Beach (Fig. 6b) was constantly eroding during the study period at different rates over time; it is located in the south of Cochin inlet mouth thus, highly exposed to erosion due to the presence of the Cochin ship channel (Sajinkumar et al., 2020). The coast immediately north of the groyne in Chellanam experiences erosion (Fig. 8a), whereas the area around Chellanam Fishing Harbour (Fig. 8b) has shown accretion in

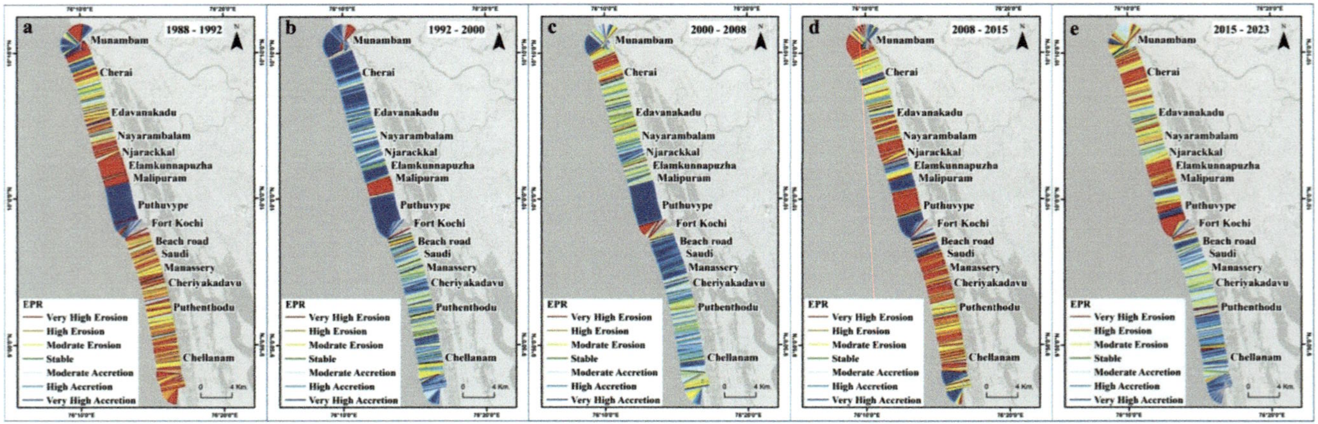

Fig. 5 **a** Shoreline change estimated using EPR for sub-period 1988–1992 **b** Shoreline change estimated using EPR for sub-period 1992–2000 **c** Shoreline change estimated using EPR for sub-period 2000–2008 **d** Shoreline change estimated using EPR for sub-period 2008–2015 **e** Shoreline change estimated using EPR for sub-period 2015–2023

Fig. 6 Google Earth image and field photos showing **a** accretion and **b** erosion on the two sides of the mouth of the Cochin inlet

recent times, mainly due to the breakwater and constructed to aid the fishing harbour. Mudbanks, a unique feature found on the southwest coast of India, may also have an impact on shoreline changes in the study area. These mudbanks act like groins, causing sand accumulation on the updrift side and erosion on the downdrift side. During the mudbank period, the southerly current dominates, leading to accretion on the south side and erosion on the north side of the mudbanks (Noujas et al., 2016). As Kunte et al. (2000) stated, the Kerala coast experiences strong southward littoral currents early in the southwest monsoon, but there are fewer littoral sediments due to mudbanks and barrier islands, causing severe erosion in the monsoon. Due to the dense population and rapidly advancing industrial and economic landscape of the study area, such changes along the coastal area are inevitable. Overall, this study has assessed the shoreline dynamics on the Ernakulam coast, evidencing the significant changes observed by the previous literature. It is noted that the dynamic shoreline has been exposed to vigorous changes in the study area during the last three decades due to natural factors and human activities.

Fig. 7 Google Earth image and field photos showing **a** accretion along the Munambam breakwater and **b** erosion to the south

Fig. 8 Google Earth image and field photos showing **a** seawall, groyne and **b** breakwater in Chellanam

5 Conclusion

Shoreline changes are one of the most vulnerable aspects of coastal areas. Monitoring and assessing spatiotemporal changes in shorelines have become much more convenient with the emergence of remote sensing and GIS. The present study investigated shoreline dynamics along the coast of Ernakulam for 35 years, from 1988 to 2023, using the shoreline data delineated from multitemporal Landsat images in a GIS environment. The results revealed that 24.17% of the shoreline was eroding, while 75.83% was accreting. The maximum erosion was observed at the coast of Munambam and Fort Kochi, and maximum accretion was identified at Puthuvype on Ernakulam Coast. The analysis of short-term changes revealed a dynamic shoreline in Cherai, Elamkunnapuzha, and Manassery, with instances of erosion followed by accretion and vice versa during selected years. The earlier studies conducted on the Kerala coast also revealed results similar to those of the present study, where the changes happened due to natural factors and human interference, such as the tsunami of 2004 and the construction of harbours, groynes, and breakwaters. It is observed that the Ernakulam coast was highly influenced by human activities rather than natural activity. Erosion and accretion are continuous processes along the coast, and thus, the shoreline will never be stationary. However, monitoring changes and proper management would help to protect the coastal environment.

Acknowledgements The authors would like to acknowledge the Department of Geography, Central University of Karnataka, for the invaluable support throughout the study.

Conflict of Interest. There is no conflict of interest.

References

Ahammed Basheer, K. K., & Pandey, A. C. (2022). Assessment and prediction of shoreline change using multi-temporal satellite data and geostatistics: A case study on the eastern coast of India. *Journal of Water and Climate Change, 13*(3), 1477–1493. https://doi.org/10.2166/wcc.2022.270

Al-Mansoori, S., & Al-Marzouqi, F. (2016). Coastline extraction using satellite imagery and image processing techniques. *International Journal of Current Engineering and Technology, 6*(4), 1245–1251.

Alemayehu, F., Richard, O., Kinyanjui, J., & Oliverv, W. (2014). Assessment of shoreline changes in the period 1969–2010 in Watamuarea, Kenya. *Global Journal of Science Frontier Research: Environment & Earth Science, 14*(H6), 19–31.

Boak, E. H., & Turner, I. L. (2005). Shoreline definition and detection: A review. *Journal of Coastal Research, 214*, 688–703. https://doi.org/10.2112/03-0071.1

Burningham, H., & Fernandez-Nunez, M. (2020). Shoreline change analysis. In *Sandy beach morphodynamics* (pp. 439–460). https://doi.org/10.1016/b978-0-08-102927-5.00019-9

Daoudi, M., & Niang, A. J. (2021). Detection of shoreline changes along the coast of Jeddah and its impact on the geomorphological system using GIS techniques and remote sensing data (1951–2018). *Arabian Journal of Geosciences, 14*(13). https://doi.org/10.1007/s12517-021-07605-2

Dar, I. A., & Dar, M. A. (2009). Prediction of shoreline recession using geospatial technology: A case study of Chennai Coast, Tamil Nadu, India. *Journal of Coastal Research, 256*, 1276–1286.

Darwish, K., & Smith, S. (2023). Landsat-based assessment of morphological changes along the Sinai Mediterranean Coast between 1990 and 2020. *Remote Sensing, 15*(5), 1392. https://doi.org/10.3390/rs15051392

Dutta, D., Kumar, T., Jayaram, C., & Akram, W. (2022). Shoreline change analysis of Hooghly estuary using multi-temporal landsat data and digital shoreline analysis system. *IntechOpen.* https://doi.org/10.5772/intechopen.103030

Elnabwy, M. T., Elbeltagi, E., El Banna, M. M., Elshikh, M. M. Y., Motawa, I., & Kaloop, M. R. (2020). An approach based on Landsat images for shoreline monitoring to support integrated coastal management—A case study, Ezbet Elborg, Nile Delta, Egypt. *ISPRS International Journal of Geo-Information, 9*(4), 199. https://doi.org/10.3390/ijgi9040199

Fasil, M., Gopinath, G., Surendran, U., Prem, P., Al-Ansari, N., & Mattar, M. A. (2023). Using RS and GIS techniques to assess and monitor coastal changes of coastal islands in the marine environment of a humid tropical region. *Water, 15*(21), 3819–3819. https://doi.org/10.3390/w15213819

Gens, R. (2010). Remote sensing of coastlines: Detection, extraction and monitoring. *International Journal of Remote Sensing, 31*(7), 1819–1836. https://doi.org/10.1080/01431160902926673

Guariglia, A., Buonamassa, A., Losurdo, A., Saladino, R., Trivigno, M. L., Zaccagnino, A., & Colangelo, A. (2009). A multisource approach for coastline mapping and identification of shoreline changes. *Annals of Geophysics, 49*(1), 295–304.

Hapke, C. J., Himmelstoss, E. A., Kratzmann, M. G., List, J. H., & Thieler, E. R. (2011). National assessment of shoreline change: Historical shoreline change along the New England and Mid-Atlantic coasts. U.S. Geological Survey OpenFile Report 2010–1118, 57.

Himmelstoss, E. A., Henderson, R. E., Kratzmann, M. G., & Farris, A. S. (2018). *Digital Shoreline Analysis System (DSAS) version 5.0 user guide*. U.S. Geological Survey Open-File Report, 1179, 110. https://doi.org/10.3133/ofr20181179

Hoeke, R., Zarillo, G., & Synder, M. (2001). *A GIS based tool for extracting shoreline positions from aerial imagery (BeachTools)*.

Islam, M. S., & Crawford, T. W. (2022). Assessment of spatio-temporal empirical forecasting performance of future shoreline positions. *Remote Sensing, 14*(24), 6364–6364. https://doi.org/10.3390/rs14246364

Jackson, C. W., Alexander, C. R., & Bush, D. M. (2012). Application of the AMBUR R package for spatio-temporal analysis of shoreline change: Jekyll Island, Georgia, USA. *Computers & Geosciences, 41*, 199–207. https://doi.org/10.1016/j.cageo.2011.08.009

Jackson, C. W. (2004). *Quantitative shoreline change analysis of an inlet-influenced transgressive barrier system; figure eight island, North Carolina*. Master's Thesis, University of North Carolina at Wilmington, Wilmington, NC, USA, 86.

Jebaraj, S., Kr Mishra, S., Ghetiya, S., Nayak, R., & Shaik, A. (2021). Estimation of shoreline change along the Odisha and Andhra Pradesh coast using Landsat data (2002–2017). *Journal of Remote Sensing & GIS, 10*(7), 1–6.

Jimenez, J. A., Sanchez-Arcilla, A., Bou, J., & Ortiz, M. A. (1997). Analysing short-term shoreline changes along the Ebro delta (Spain) using aerial photographs. *Journal of Coastal Research, 13*(4), 1256–1266.

Kale, M. M., Ataol, M., & Tekkanat, I. S. (2019). Assessment of shoreline alterations using a digital shoreline analysis system: A case study of changes in the Yeşilırmak Delta in northern Turkey from 1953 to 2017. *Environmental Monitoring and Assessment, 191*(398), 1–13. 191(6). https://doi.org/10.1007/s10661-019-7535-8

Kankara, R. S., Ramana Murthy, M. V., & Rajeevan, M. (2018). *National assessment of shoreline changes along Indian coast—A status report for 26 years 1990–2016* (p. 76). NCCR Publication.

Kankara, R. S., Selvan, S. C., Markose, V. J., Rajan, B., & Arockiaraj, S. (2015). Estimation of long and short term shoreline changes along Andhra Pradesh coast using remote sensing and GIS techniques. *Procedia Engineering, 116*, 855–862. https://doi.org/10.1016/j.proeng.2015.08.374

Kannan, R., Ramanamurthy, M., & Kanungo, A. (2016). Shoreline change monitoring in Nellore Coast at East Coast Andhra Pradesh District using remote sensing and GIS. *Journal of Fisheries & Livestock Production, 04*(01), 1–6. https://doi.org/10.4172/2332-2608.1000161

Kar, P. K., Mohanty, P. K., Pradhan, S., Balaji Behera, Padhi, S. K., & Mishra, P. (2021). Shoreline change along Odisha coast using statistical and geo-spatial techniques. *Proceedings of the Indian Academy of Sciences. Earth and Planetary Sciences/Journal of Earth System Science, 130*(4), 209. https://doi.org/10.1007/s12040-021-01703-1

Kumar, P. K. D., Gopinath, G., Murali, R. M., & Muraleedharan, K. R. (2014). Geospatial analysis of long-term morphological changes in cochin estuary, SW Coast of India. *Journal of Coastal Research, 298*, 1315–1320. https://doi.org/10.2112/jcoastres-d-12-00244.1

Kumar, S., Vethamony, P., & Murty, C. (1990). Wave-induced nearshore flow patterns in the vicinity of Cochin Harbour, India. *Ocean & Shoreline Management, 13*, 111–125.

Kunte, P. D., Wagle, B. G., & Sugimori, Y. (2000). Remote assessment of net shore drift along the west coast of India. *PORSEC Proceedings, II*, 685–689.

Kurian, N. P., Ramachandran, K. K., Sheela Nair, L., Thomas, K. V., & Hameed, T. S. S. (2007). Long-term shoreline change along the south central Kerala Coast. In *Fourth Indian National conference on harbour and ocean engineering*, NITK, Surathkal.

Li, W., & Gong, P. (2016). Continuous monitoring of coastline dynamics in western Florida with a 30-year time series of Landsat imagery. *Remote Sensing of Environment, 179*, 196–209. https://doi.org/10.1016/j.rse.2016.03.031

Mahapatra, M., Ratheesh, R., & Rajawat, A. S. (2014). Shoreline change analysis along the coast of South Gujarat, India, using digital shoreline analysis system. *Journal of the Indian Society of Remote Sensing, 42*(4), 869–876. https://doi.org/10.1007/s12524-013-0334-8

Mani, M. R., & Dinesh, K. P. K. (2015). Implications of sea level rise scenarios on land use /land cover classes of the coastal zones of Cochin, India. *Journal of Environmental Management, 148*, 124–133. https://doi.org/10.1016/j.jenvman.2014.06.010

Marfai, M. A., Almohammad, H., Dey, S., Susanto, B., & King, L. (2007). Coastal dynamic and shoreline mapping: Multi-sources spatial data analysis in Semarang Indonesia. *Environmental Monitoring and Assessment, 142*(1–3), 297–308. https://doi.org/10.1007/s10661-007-9929-2

Matin, N., & Hasan, G. M. J. (2021). A quantitative analysis of shoreline changes along the coast of Bangladesh using remote sensing and GIS techniques. *CATENA, 201*, Article 105185. https://doi.org/10.1016/j.catena.2021.105185

Meilianda, E., Mauluddin, S., Pradhan, B., & Sugianto, S. (2023). Decadal shoreline changes and effectiveness of coastal protection measures post-tsunami on 26 December 2004. *Applied Geomatics, 15*(3), 743–758. https://doi.org/10.1007/s12518-023-00514-x

Mishra, M., Santos, S., Rana, N. K., Kar, D., & Parida, N. R. (2021). Monitoring vegetation loss and shoreline change due to tropical cyclone Fani using Landsat imageries in Balukhand-Konark wildlife sanctuary, India. *Journal of Coastal Conservation, 25*, 53. https://doi.org/10.1007/s11852-021-00840-5

Mohanty, P. C., Mahendra, R. S., Nayak, R. K., & Kumar, T. S. (2017). Impact of sea level rise and coastal slope on shoreline change along the Indian coast. *Natural Hazards, 89*(3), 1227–1238. https://doi.org/10.1007/s11069-017-3018-9

Monalisha, M., & Panda, G. (2018). Coastal erosion and shoreline change in Ganjam coast along east coast of India. *Journal of Earth Science & Climatic Change, 09*(04), 1–6. https://doi.org/10.4172/2157-7617.1000467

Mujabar, P. S., & Chandrasekar, N. (2011). Shoreline change analysis along the coast between Kanyakumari and Tuticorin of India using remote sensing and GIS. *Arabian Journal of Geosciences, 6*(3), 647–664. https://doi.org/10.1007/s12517-011-0394-4

Murali, R. M., Dhiman, R., Choudhary, R., Seelam, J. K., Ilangovan, D., & Vethamony, P. (2015). Decadal shoreline assessment using remote sensing along the central Odisha coast, India. *Environmental Earth Sciences, 74*(10), 7201–7213. https://doi.org/10.1007/s12665-015-4698-7

Murray, J., Adam, E., Woodborne, S., Miller, D., Xulu, S., & Evans, M. (2023). Monitoring shoreline changes along the southwestern coast of South Africa from 1937 to 2020 using varied remote sensing data and approaches. *Remote Sensing, 15*(2), 317. https://doi.org/10.3390/rs15020317

Narayana, A. C., & Priju, C. P. (2006). Landform and shoreline changes inferred from satellite images along the central Kerala coast. *Journal Geological Society of India, 68*, 35–49.

Natesan, U., Thulasiraman, N., Deepthi, K., & Kathiravan, K. (2012). Shoreline change analysis of Vedaranyam coast, Tamil Nadu, India. *Environmental Monitoring and Assessment, 185*(6), 5099–5109. https://doi.org/10.1007/s10661-012-2928-y

Nayak, S. (2002). Use of satellite data in coastal mapping. Indian Cartographer, 147–157.

Noujas, V., Thomas, K. V., & Badarees, K. O. (2016). Shoreline management plan for a mudbank dominated coast. *Ocean Engineering, 112*, 47–65. https://doi.org/10.1016/j.oceaneng.2015.12.006

Parthasarathy, K. S. S., & Deka, P. C. (2019). Remote sensing and GIS application in assessment of coastal vulnerability and shoreline changes: A review. *ISH Journal of Hydraulic Engineering*, 1–13. https://doi.org/10.1080/09715010.2019.1603086

Quang, D. N., Ngan, V. H., Tam, H. S., Viet, N. T., Tinh, N. X., & Tanaka, H. (2021). Long-term shoreline evolution using DSAS technique: A case study of Quang Nam Province, Vietnam. *Journal of Marine Science and Engineering, 9*(10), 1124. https://doi.org/10.3390/jmse9101124

Sajinkumar, K. S., Bincy, H. S., Bouali, E. H., Oommen, T., Vishnu, C. L., Anilkumar, Y., Thrivikramji, K. P., & Keerthy, S. (2020). Picturing beach erosion and deposition trends using PSInSAR: An example from the non-barred southern west coast of India. *Wetlands Ecology and Management, 29*(5), 775–788. https://doi.org/10.1007/s11273-020-09706-3

Sarwar, Md. G. M., & Woodroffe, C. D. (2013). Rates of shoreline change along the coast of Bangladesh. *Journal of Coastal Conservation, 17*(3), 515–526. https://doi.org/10.1007/s11852-013-0251-6

Selvan, S. C., Kankara, R. S., Prabhu, K., & Rajan, B. (2019). Shoreline change along Kerala, south-west coast of India, using geo-spatial techniques and field measurement. *Natural Hazards, 100*(1), 17–38. https://doi.org/10.1007/s11069-019-03790-2

Shanghude, Y., Mburu, W., Aurthurton, R., Dubi, A., Gachuiri, S., Kangwe, J., Magori, C., Msuya, F., Mwaipop, R., Nyandwi, N., Ochiewo, J., Onganda, H., Mtui, R. S., Sanga, I., Shalli, M., & Uku, J. (2015). Shoreline change in Tanzania and Kenya: Manual assessment and design of mitigation strategies. *WIOMSA, 16*, 85.

Sheela Nair, L., Prasad, R., Rafeeque, M. K., & Prakash, T. N. (2018). Coastal morphology and long-term shoreline changes along the southwest coast of India. *Journal of the Geological Society of India, 92*, 588–595. https://doi.org/10.1007/s12594-018-1072-x

Sowmya, K., Sri, M. D., Bhaskar, A. S., & Jayappa, K. S. (2019). Long-term coastal erosion assessment along the coast of Karnataka, west coast of India. *International Journal of Sediment Research, 34*(4), 335–344. https://doi.org/10.1016/j.ijsrc.2018.12.007

Sreelakshmi, M., Balachandran, S., Abdurazak, F., Gopinath, G., & Joseph, S. (2024). From perils to preparedness: Decoding coastal hazard risks on Kerala's central coast through the IPCC framework. *Natural Hazards, 120*, 8639–8666. https://doi.org/10.1007/s11069-024-06543-y

Thieler, E. R., & Danforth, W. W. (1994). Historical shoreline mapping (II): Application of the digital shoreline mapping and analysis systems (DSMS/DSAS) to shoreline change mapping in Puerto Rico. *Journal of Coastal Research, 10*(3), 600–620.

Thiruvenkatasamy, K., & Baby Girija, D. K. (2014). Shoreline evolution due to construction of rubble mound jetties at Munambam inlet in Ernakulam-Trichur district of the state of Kerala in the Indian peninsula. *Ocean & Coastal Management, 102*, 234–247. https://doi.org/10.1016/j.ocecoaman.2014.09.026

Xu, N. (2018). Detecting coastline change with all available Landsat data over 1986–2015: A case study for the state of Texas, USA. *Atmosphere, 9*(3), 107. https://doi.org/10.3390/atmos9030107

Zambrano-Medina, Y. G., Plata-Rocha, W., Monjardin-Armenta, S. A., & Franco-Ochoa, C. (2023). Assessment and forecast of shoreline change using geo-spatial techniques in the Gulf of California. *Land, 12*(4), 782. https://doi.org/10.3390/land12040782

Assessment and Prediction of Shoreline Change Using Geospatial Techniques: A Case Study of North Goa Coast of India

Siddhi Madhusudan Gaonkar, F. M. Nadaf, and Vikas Balajirao Kapale

1 Introduction

In simpler terms, the shoreline is where the land meets the water (Ahmed, 2011; Ekon, 2017; Ericson et al., 2006; Gens, 2010) a plethora of ocean resources encircles this interface (Wang et al., 2023). Less than 15% of the earth's total land area consists of coastlines, which are densely populated worldwide, as 60% of the global population lives there (Cenci et al., 2018; McGranahan et al., 2007). Many coastal cities and towns depend on the ocean for their livelihoods (Kummu et al., 2016). Coasts are dynamic in nature, and hence they are time and space dependent. Natural factors like waves, currents, and coastal landforms, along with human activities such as sand mining, urban development, tourism, and building coastal structures, contribute to alterations in the coastline (Alberico et al., 2022; Di Stefano et al., 2013; Fang et al., 2018; Kourosh Niya et al., n.d.; Liao et al., 2022). Around the world, coastal regions are at risk from various natural disasters such as floods, erosion, hurricanes, cyclones, tsunamis, and saltwater intrusion due to their close location to the ocean. These dangers result in significant economic, social, ecological, and human impacts (Sahoo & Bhaskaran, 2018). The coastal shoreline undergoes a continuous natural process, influenced by human activities, leading to positive or negative impacts on adjacent coastal regions. Short-term changes in shoreline positions are mainly seasonal and periodic, while long-term changes are typically a result of fluctuations in sea level or alterations in coastal sediment supply (Cai et al., 2009;

S. M. Gaonkar
Cluster Research Centre of Geography (affiliated to Goa University), Government College of Arts, Science & Commerce, Khandola, Goa, India

F. M. Nadaf (✉)
Government College of Commerce and Economics, Borda, Margao, Goa, India
e-mail: fmnadaf@gmail.com

V. B. Kapale
Department of Civil Engineering, Indian Institute of Technology, Kanpur, India

P. Thambidurai and A. K. Dikshit (eds.), *Coastal Environments of India*,
https://doi.org/10.1007/978-3-031-97341-3_12

Klein & Lichter, 2006). Approximately 80% of coastal regions worldwide are facing erosion, with erosion rates ranging from 1 cm/year to 10 m/year (Pilkey & Hume, 2001). The analysis of shoreline changes has been the focus of numerous studies around the globe (Addo, 2013; Maiti & Bhattacharya, 2009; Zuzek et al., 2003). In recent years, coastal areas globally have experienced fast-paced urban development, and its aftermath increase in population altering the coasts (Kilibarda et al., 2014). Shoreline change is a prevalent natural hazard that is heavily researched and presents a considerable risk to the coastal environment, human life, and property (Bagli & Soille, 2003; Yadav et al., 2018). Geospatial technology is experiencing a technological revolution that has greatly improved the accuracy, precision, and affordability of tools like GIS, GPS, and RS, making them applicable to coastal zones (Ahammed & Pandey, 2022; Craglia et al., 2012). The monitoring of shoreline change requires a long-term commitment and is based on temporal change modelling using a GIS database. Traditional field mapping techniques remain valuable but are often restricted by logistical challenges. As a result, remote sensing and geospatial technologies are becoming more popular due to their cost-effectiveness and reliability, and are being utilized more frequently to create valuable information for decision-making purposes across various industries (Klemas, 2010; Wright, 2009). Nevertheless, advancements in data, technologies, and theories within the broader field of remote sensing and geospatial technologies have given scientists valuable chances to progress their research on the coastal environment (Chang & Lai, 2014). Shoreline change monitoring requires a sustained effort and relies on temporal change modelling; hence GIS database is proven to be a reliable dataset to carry out shoreline change.

Goa is famous for its 105 km extensive shoreline characterized by small beaches enclosed by cliffs, estuaries, bays, and mangroves in some regions. The beaches in southern Goa are known for their long, straight shape and the existence of sand dunes. The largest estuarine system on this coast section is found in Goa, specifically in Mandovi and Zuari. Mudflats, marshy swamps, and wetlands are mostly found in close proximity to estuaries and creeks. (Nayak & Dsouza, 2010). In the 1970s, when tourism was identified as a lucrative industry for Goa, a region promoted as a top vacation spot. Between 1971 and 1991, around 80% of urban expansion occurred in the coastal talukas and continues to be concentrated there (Mascarenhas, 1998; Sawkar et al., 2005). With the rapid development along the open seafront of Goa, the impact of heavy developmental activities on the sandy areas, including the dunes, remains uncertain. Truly, the ever-changing characteristics of Goa's shoreline, combined with unregulated expansion and progress, require a holistic and lasting strategy for coastal management. The shoreline of Goa is undergoing changes caused by natural factors like erosion, sediment accumulation, and sea level increase. Moreover, human actions such as tourism, urbanization, and industrialization place considerable stress on coastal ecosystems. A comprehensive approach involving environmental, economic, and social aspects is necessary for the enduring viability of Goa's coastal regions.

2 Objectives

This study aims to analyse changes along the coastline, concentrating on accretion and erosion patterns and identifying the most sensitive areas requiring urgent intervention. It also projects future shoreline modifications by application of historical shoreline change data with the aid of mathematical modelling. This projection targets two key timelines: a forecast for the short term, to 2030, and for the long term, until 2050.

3 Study Area

The state of Goa is situated along the central west coast, with a 105 km coastal stretch, and is characterized by the presence of alternately positioned headlands, estuaries, bays, and globally renowned beaches (Kunte & Wagle, 1994). The region includes Arambol, Anjuna, Baga, Vagator, Calangute, Miramar, Colva, Agonda, Palolem, Patnem, and Galgibag, as well as an exceptional estuarine rivers and mangrove ecosystem (Nadaf, 2019). From an administrative standpoint, the state is separated into two districts: North Goa and South Goa. There is a total of 12 talukas, with 7 of them being located along the coast, namely Pernem, Bardez, Tiswadi, Mormugao, Salcete, Quepem, and Canacona. For the research, the North Goa district has been taken into consideration. The study area lies between 15°24′30″ N and 15°45′30″ N latitude to 73°41′30″ E – 73°55′30″ E longitude. It includes 3 coastal talukas, i.e. Pernem, Bardez, and Tiswadi, comprising 17 coastal villages (Fig. 1). Compared to other Indian states, Goa has a small coastline but it plays a crucial role in coastal ecology and the Indian economy. Goa's serene coastal region attracts a lot of domestic and international tourists. In addition, the coastal stretch provides an excellent location for fishing and a natural port for transportation (Kunte & Wagle, 1994). Around 80% of the people in Goa reside in coastal regions at low altitudes, putting them at a high risk of flooding due to the gradual increase in sea levels of 1.5 mm annually in several of its shorelines. (Banerjee, 2024). Up until the 1970s, the coast of Goa was mainly pure and untouched since most activities along the coast were situated behind sand dunes. There was a unique balance between humans and nature. In the past twenty years, an increase in tourism promotion, population growth, and construction projects has led to a significant rise in demand for resources, leading to crowded coasts, higher concentration of buildings, and strain on infrastructure (Niya, 2000). Thus, resulting in erosion and accretion at the coast. Coastal dynamics progressively shift the location of the shoreline, which causes flooding in low-lying regions (Ahammed & Pandey, 2022). Since most selected regions have low elevation and steep slopes, they are extremely vulnerable to changes in shoreline position. Thus, the study, in turn going to identify hotspots of erosion and eventually its position soon.

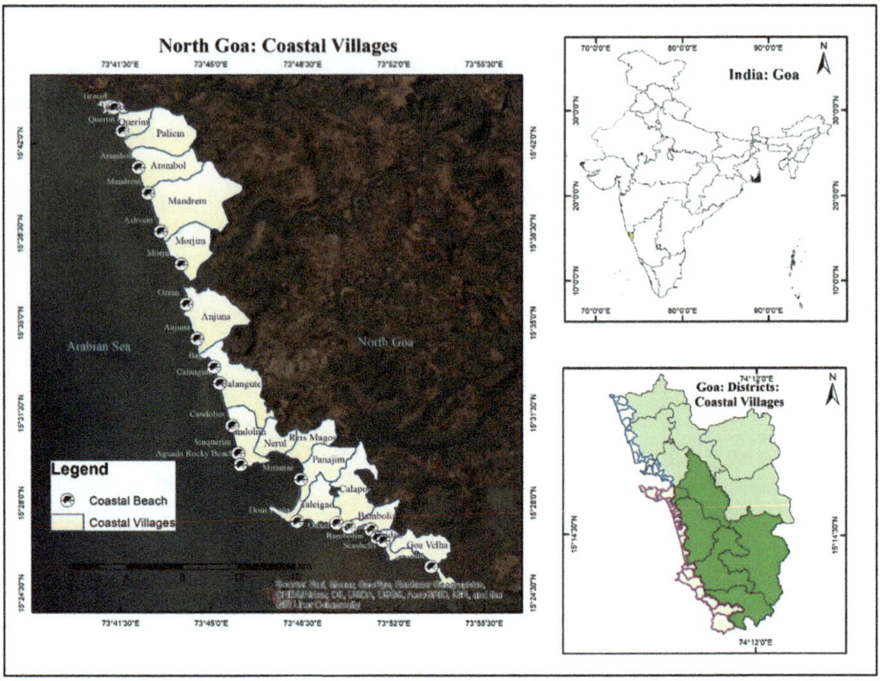

Fig. 1 Study area map of North Goa coastal villages

4 Material and Methods

4.1 Data Collection and Shoreline Extraction

The North Goa shoreline position was analysed from satellite images (Landsat TM, ETM, and OLI/TIRS) for a span of 30 years at a spatial resolution of 30 m from USGS Earth Explorer (Table 1) (Earth Explorer, 2022). The region was enveloped by one scene with path 147 and row 049. The shoreline for each specific time was created using a mosaic procedure. Using ArcGIS 10.3, Operations of NDWI, NDVI, and land–water boundary were performed to distinctly separate the land–water boundary.

A positive NDVI and NDWI value suggests that the location is land. Conversely, their adverse values represent bodies of water. NDVI and NDWI, with a value of 0, represent the boundaries between land and water. Similarly, land water bound was separated with a deep observation in the Blue, Green, and NIR bands of the selected dataset (Ahammed & Mahendra, 2016; Rajawat et al., 2014).

$$\text{NDVI} = (\text{NIR} - \text{R})/(\text{NIR} + \text{R}) \tag{1}$$

$$\text{NDWI} = (\text{NIR}-\text{SWIR})/(\text{NIR} + \text{SWIR}) \tag{2}$$

Table 1 Details of satellite dataset

Satellite	Sensor	Resolution	Date of acquisition
Landsat 5	TM	30	19/01/1992, 20/02/1992, 08/04/1992, 04/12/1992 and 29/12/1992 20/02/1998 and 05/12/1998
Landsat 7	ETM	30	15/01/2014, 31/01/2014, 16/02/2014, 04/03/2014, 21/04/2014 and 28/09/2014
Landsat 8	OLI/TIRS	30	03/01/2022, 21/01/2022, 29/01/2022, 06/02/2022, 10/03/2022, 27/04/2022, 29/05/2022, 28/10/2022 and 05/11/2022

where

NDVI Normalized Difference Vegetation Index
NIR Near Infrared
Red Visible light with a wavelength of 0.630–0.680 _m
NDWI Normalized Difference Water Index
Green Visible light with a wavelength of 1.560–1.660 _m
SWIR1 Short-Wave Infrared 1.

Several types of research utilizing Landsat satellite images, like those by Ahammed and Mahendra (2016); Ahammed and Pandey (2022), have examined coastal transformations through different interpretation methods, such as visual interpretation and object-oriented classification techniques. Reclassified images were used to convert raster into polyline and on-screen digitization of the shoreline. The rough edges (blocky shape) of each selected shoreline were smoothed (Fig. 2) and merged as a single polyline in one length.

4.2 Shoreline Changes

The Digital Shoreline Analysis System (DSAS), an extension of ArcGIS, was adopted to calculate and analyse the shoreline change. According to earlier research (Abd-Elhamid et al., 2023; Niya et al., 2013), DSAS has been known as a precision method for identifying changes in the shoreline (Thieler et al., 2009). Together, 1064 transects with 100-m spacing were used to cover 58.63 km of North Goa shoreline. The Net Shoreline Movement (NSM) model is utilised to determine the total distance of shoreline displacement over the entire study period. EPR and LRR are methods used to estimate the rates of shoreline change. Figure 3 presents a flowchart that provides an overview of the methodology employed in this study.

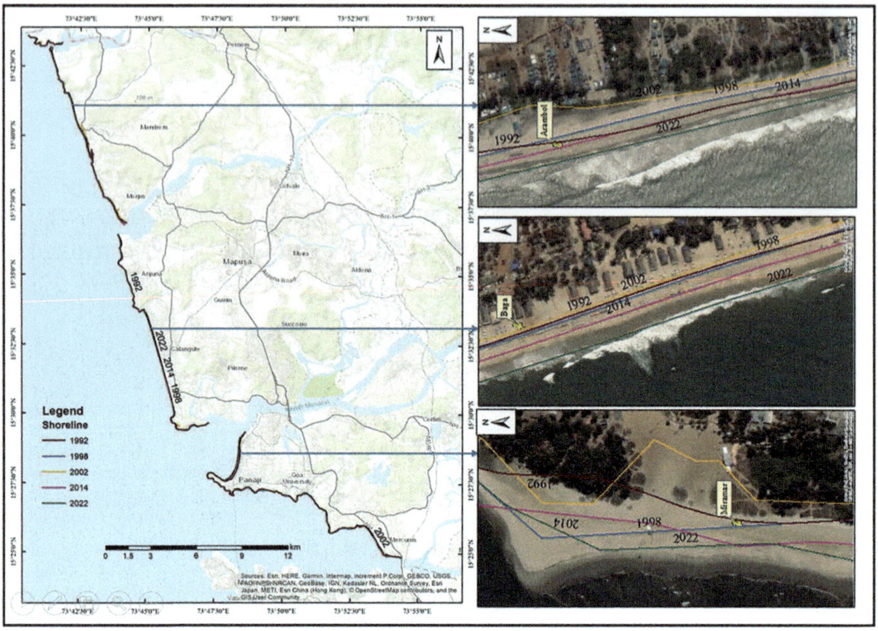

Fig. 2 Shoreline extraction from 1992 to 2022

4.3 Shoreline Prediction Using LRR

The Least-square regression model has been utilised to forecast the shoreline. The LRR model uses this formula to estimate the shoreline: The slope-intercept form represents a straight-line equation. The slope-intercept form can represent a straight-line equation as

$$y = mx + c \tag{3}$$

where m is the slope of the line and c is the intercept.

An equation can be set for the distance of the shoreline from the baseline with respect to time (ElKotby et al., 2023).

$$d_f = LRR \times \Delta t + d_r \tag{4}$$

where d_f: distance between the baseline and future shoreline, Δt: time difference between the future shoreline and the recent shoreline in years, d_r: recent shoreline position by using the least-square approach of the linear regression model, we estimate the parameters LRR and intercept. Here, LRR is the linear regression rate, which is the slope of the best-fit line.

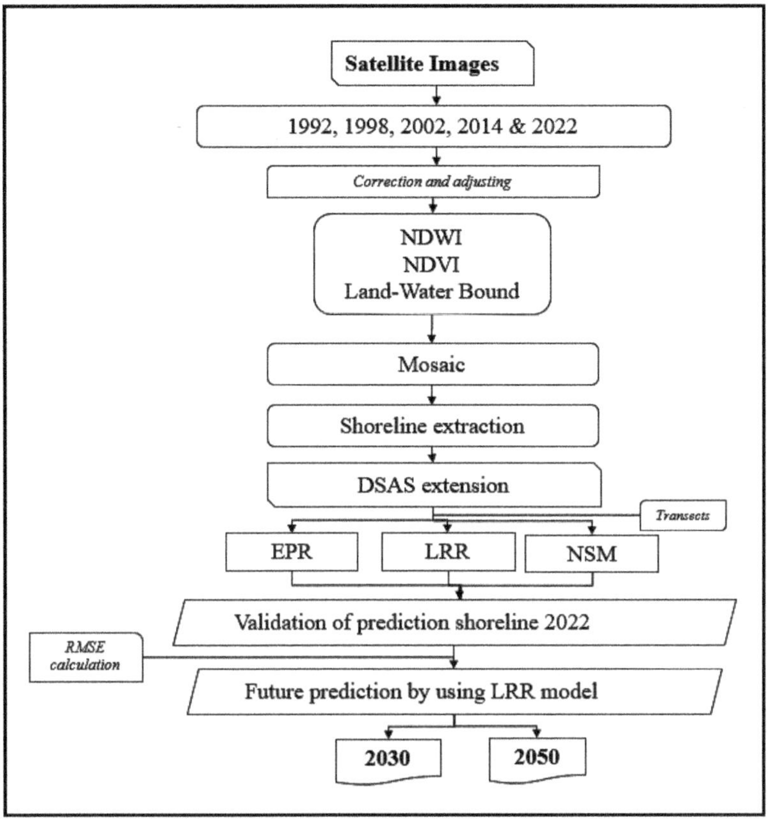

Fig. 3 Methodology flowchart

The least-square regression model provides the best-fit line; it is not necessary for the best-fit line to pass through all or any points; however, it is the most suitable method for shoreline change detection and prediction as the least-squares approach minimizes the residuals. The least-square linear regression method is advantageous over the endpoint rate, as EPR only accounts for the two extreme data points. In contrast, using the available dataset, the Least-square regression method gives the best-fit line. Example: If we want to predict the shoreline for 2030 and the recent data that we have is for 2020, then, $\Delta t = 10$ years, and d_r shall be the distance between the baseline and shoreline of the year 2020. Figure 4 shows a single transect line to explain the regression model. In Fig. 4, the slope of the line is negative, i.e. the distance between the baseline and the shoreline increases with time for this particular transect line. The LRR value of – 3.1604 m/yr indicates the shoreline's landward movement, and this method could justify the coast's erosion. The regression model is constructed for each transect line using MATLAB programming language. The inbuilt *fitlm* function provides RMSE, R^2, slope, intercept, F-statistics, and p-value.

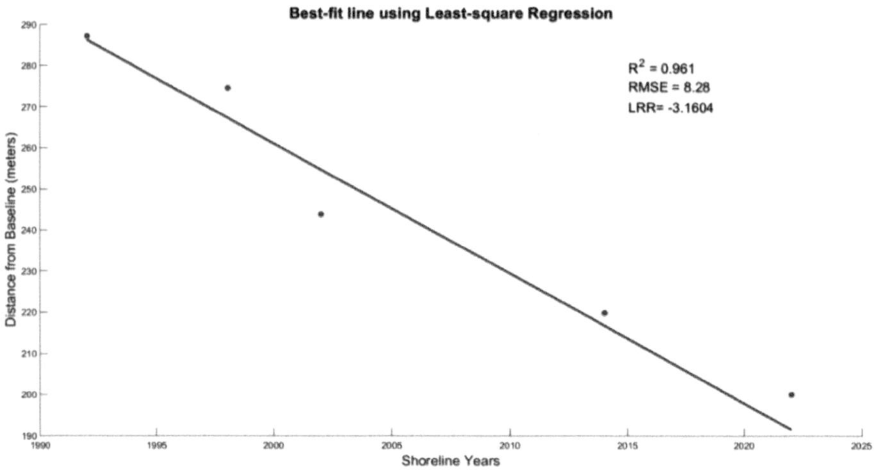

Fig. 4 Least-square regression

5 Result and Discussion

5.1 Shoreline Change

The shorelines undergo constant changes over time due to erosion or accretion, representing a natural process that occurs in the long run (Kannan & Ramanamurthy, 2016; Ranasinghe, 2016). However, human activities in vulnerable coastal areas have also had a substantial impact on the coastlines worldwide. In recent years, coastal areas globally have seen rapid urbanization and the resulting increase in population (Zoran et al., 2014).

Linear Regression Rate (LRR) Model Between 1992–2022

By analysing historical data, the rate of shoreline change was determined from 1992 to 2022 to forecast future shoreline alterations, using a total of 1064 transects. The picturesque coast of North Goa draws visitors from across the globe. The beaches are major tourist destinations, contributing significantly to the local economy. It has been noted that major parts of North Goa are highly eroding. Study estimated shoreline change rate by using EPR, LRR, and NSM for the period of 30 years. The EPR and LRR model (Fig. 5c) represents the change over a period of time. The significant erosion and accretion rates are denoted by the highlighted blue boxes on the map (Fig. 5a and b) and graph (Fig. 5c). To analyse the results, the shoreline was classified as Very High Erosion, High Erosion, Moderate Erosion, Stable, Moderate Accretion, High Accretion, and Very High Accretion. The study calculated the rate of shoreline change from 1992 to 2022 to predict future shoreline changes using historical data analysis. The research noted increased sediment accumulation in the coastal area of Mandovi and identified areas like Querim, Mandrem, and Sirdao as experiencing more erosion (Fig. 5c). While the northern and southern coasts of

North Goa displayed a more subdued coastline. The research determined the rate of shoreline change through the use of EPR, LRR, and NSM statistical methods, showing significant erosion along the North Goa coast with a positional shift of −136.13 m. In the same way, the greatest increase in land area occurred on the North Goa coast, with a movement of 201.36 m over a span of 30 years. According to the LRR, the West Coast had an average shoreline change rate of 0.89 m/y from 1992 to 2022, while the EPR calculated a rate of 0.73 m/y during the same time period (Table 2).

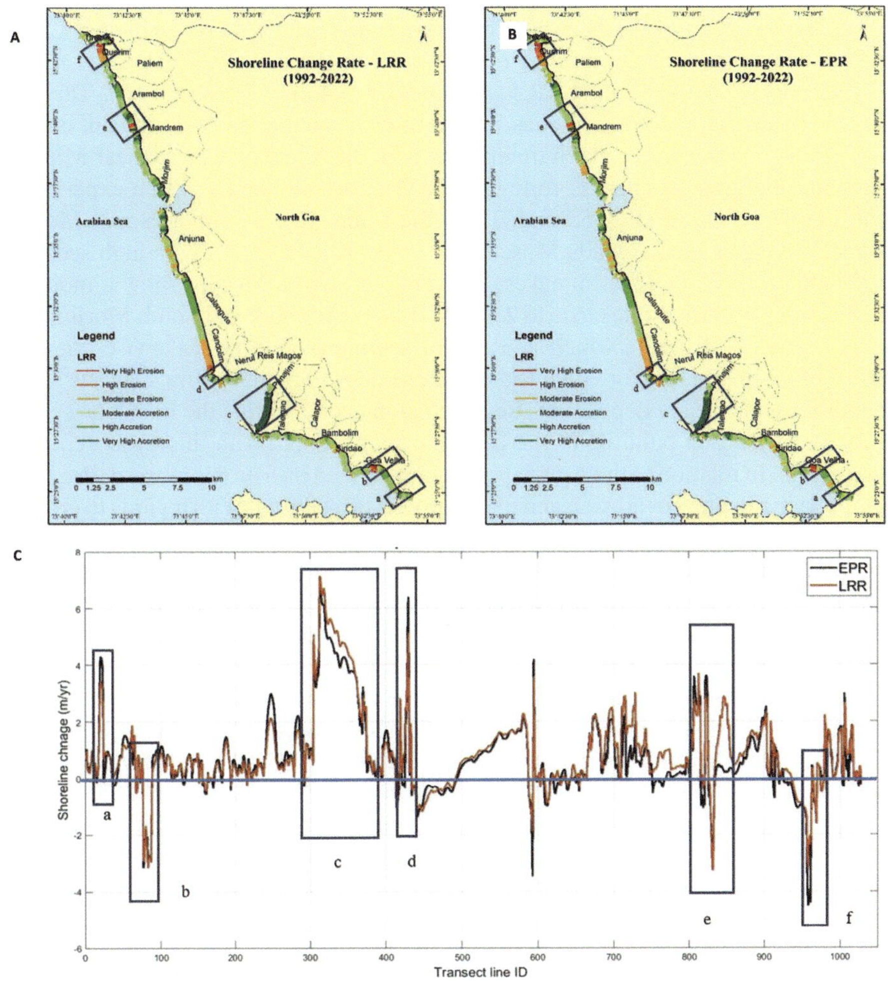

Fig. 5 Change in shoreline from 1992 to 2022: **a** Rate of historical change computed with LRR. **b** Rate of historical change computed with EPR, and **c** Comparison chart displaying EPR and LRR Change Rate

Table 2 Statistical data of shoreline change between 1992 and 2022

Shoreline Classification	LRR (m/yr)	EPR (m/yr)	NSM (m)
Highest Erosion	– 3.27	– 4.54	– 136.13
Highest Accretion	7.14	6.71	201.36
Average Erosion	– 4.27	– 4.65	– 141.38
Average Accretion	5.49	5.43	144.18
Mean	0.89	0.73	21.81
SD	1.42	1.40	41.89

Likewise, the mean erosion and mean accretion values were calculated for the whole coastline, with estimations of – 141.38 m (LRR – 4.27 m per year, EPR – 4.65 m/y) and 144.18 m (LRR 5.49 m/y, EPR 5.43 m/y), respectively. The findings indicate that the western coast of North Goa saw the most significant erosion and accretion rates in the given timeframe, with the Mandovi estuary near Miramar showing the greatest change, followed by Vagator and Ozran, which experienced substantial accretion (Fig. 5a and B). In the same way, Arambol, South Morjim, Baga, Calangute, and Aguada Museum were identified as areas with high sediment accumulation. While Querim caves, Ashvem, and North Morjim show a moderate increase with an LRR rate close to 2 m/y (Table 3). Arambol and South Morjim have a wide stretch of beach, which is ideal for sediment build-up. Likewise, the existence of nesting places for Olive Ridley Sea turtles in South Morjim suggests steady conditions and steady deposition. Beach erosion occurs during the monsoon season, followed by beach building in the post-monsoon period, with this cycle repeating every year. In relation to the Aguda Fort area is located on a promontory, with a rocky cliff serving to prevent erosion and maintain coastal stability. With regards to erosion, coastal areas of Querim, Mandrem, Sinquerim, Anjuna, and Candolim were found to be highly eroded (Table 3). The news has highlighted that enhanced wave activities and frequent tropical cyclones, coupled with climate change, play a role in shoreline changes. River sand mining along the Chapora, and Tiracol rivers and at Queri, Morjim, and Coco beaches alters the littoral drift of sands in certain areas (Iyer & Viegas, 2023). In terms of Morjim, man-made disasters such as increasing tourism activities and extensive sand mining extraction add points for erosion at a higher rate (Naik, 2014). The study carried out by NCCR and NCSM highlights that natural and human encroachment is a major reason for the same (Malkarnekar, 2023; Times, 2023; Venkateswaran, 2006). Illegal commercial establishments and structures like toilets, soak pits, and borewells in Anjuna within the No Development zone are also one of the reasons for shoreline change (Malkanekar, 2023; TheGoanEveryday, 2024). The coastal landscape of Miramar has experienced notable transformations in recent decades. The significant extraction of sand dunes and resulting urban growth have drastically changed the area, impacting the local environment significantly.

In May 1992, a large sand dune close to Miramar was nearly completely excavated for building needs. The sand from this dune was taken for local construction, causing the dune to be completely depleted. This former natural structure has been

Table 3 Major affected areas data of North Goa shoreline classification LRR 1992–2022

Sr. No.	Shoreline classification	LRR (m/yr)	Areas
1	Very high erosion	> − 2	North Querim, extreme North Mandrem
2	High erosion	− 2 to − 1	Central Querim, Mandrem, North Baga, North Morjim
3	Moderate erosion	− 1 to 0	Anjuna, Candolim, Aguada fort area
4	Moderate accretion	0–1	Near Querim caves, Ashvem and South Morjim
5	High accretion	1–2	Arambol, South Morjim, Baga, Calangute, Aguada Museum
6	Very high accretion	< 2	Vagator, Ozran, Miramar

replaced by a building. This event showcases the beginnings of human impact on the coastal scenery of Miramar (Mascarenhas, 1998). In January 1998, more changes to the dune landscape were noted when a new wide road was built across a dune area (Mascarenhas, 1998). This progress caused damage to the dunes, leading to sand being blown onto the road in various locations. The road construction has caused dunes to be disrupted, leading to moving sand that is now accumulating on the new road, posing maintenance difficulties and safety risks. Between 1992 and 2022, Miramar, specifically in the vicinity of Caranzalem, underwent notable alterations in its coastal morphology. The greatest accumulation measured was 201.36 m. This significant increase could be linked to different reasons, such as alterations in sediment input, wave behaviour, and human activities, such as building coastal structures and infrastructure. Accretion occurs naturally when sediments are deposited along the coast, causing the land area to increase. In the Miramar context, this accumulation may not be entirely natural. Human actions, such as extracting dunes and building, can change the way sediment is spread. Removing dunes and buildings in their place can result in more sediment being available in specific locations, which helps with the growth of land. Accretion may be viewed as advantageous in relation to land (Fig. 6).

Fig. 6 Google earth images of Miramar coastal area. *Source* Google Earth

Fig. 7 Google earth images of Terekhol coastal area. *Source* Google Earth

Over the last thirty years, South Terekhol, situated on the northern shoreline of Goa, has witnessed considerable erosion of its coast. Between 1992 and 2022, the most significant erosive activity measured in this region was a negative 136.13 m. Examination of Google Earth images and on-site data collected from 1992 to 2022 shows significant degradation of the mudflats in South Terekhol (Fig. 7). The images reveal a noticeable shift of the shoreline towards land, signifying a substantial amount of land being lost to the sea. The mudflats, vital to the local ecosystem, have been significantly impacted, leading to greater water intrusion onto the land. The situation at South Terekhol demonstrates the intricate interaction between natural and human factors that contribute to coastal erosion. The shore has moved significantly inland in the last thirty years, highlighting the pressing requirement for successful management and mitigation plans.

Querim effectively illustrates the delicate balance between natural and human actions. In August 1996, the undamaged vegetation present in Querim's stable dunes indicated an ecosystem that remained untouched by human interference. The plant life on these sand dunes plays a vital role in maintaining their stability and providing protection for the land from the direct force of the sea, showcasing a crucial defensive mechanism necessary for coastal areas. Currently, sand mining is causing erosion in this pocket. The degradation of the dunes in Candolim, Goa, is a major problem that involves ecological, legal and socio-economic aspects. The occurrences between August 1996 and January 1998 demonstrate a gradual decline in the dune ecosystem caused by human activities.

The creation of an unlawful road by destroying part of a dune in Candolim initiates the documented decline. Dunes act as natural defenses against coastal erosion and offer a living environment for different species. When a part of these dunes was destroyed, the original defense was weakened, making it more susceptible to erosion. The construction of countless beach shacks, some of which were built outside the designated beach areas, worsened the erosion of the dunes. This method frequently includes removing vegetation on the dunes, which is important for keeping the dunes stable. Using vast dune areas for parking various vehicles led to more flattening and vegetation depletion. These actions, when combined, have transformed the previously vibrant, green dunes into a barren, lifeless expanse. This adjustment leads to several enduring effects: The significance of maintaining a balance between progress and ecological protection is emphasized by the circumstances in Candolim. Having effective plans for managing and conserving these vital coastal ecosystems is crucial to safeguard them for future generations' benefit. The situations in Baga and

Calangute in Goa showcase important problems related to environmental damage and legal actions linked to coastal development projects. The Baga resort project included building a resort on a dune, with a large portion of it being ruined during development. Dunes function as natural defenses against coastal erosion and are crucial environments for a variety of plant and animal species. The destruction of these dunes not only disturbs the nearby ecosystem but also weakens the natural defense against rising sea levels and storm surges. In August of 1996, legal authorities stepped in to stop the project due to acknowledging the environmental harm it had caused (Mascarenhas, 1998). This legal proceeding highlights the importance of legal systems in protecting the environment. The intervention probably resulted from breaches of the Coastal Regulation Zone (CRZ) rules, which are designed to control building projects close to shorelines in order to avoid harming the environment. The CRZ notification, initially released in 1991 and later modified, classifies coastal regions into various zones, each having specific limitations on development actions. In December 1997, another instance of environmental damage took place in Calangute as dunes were flattened and taken away to make way for tourist entry to the beach (Mascarenhas, 1998). This activity showcases the clash between the advancement of tourism and the preservation of the environment. Although tourism contributes significantly to the economy of Goa by creating employment opportunities and generating income, it frequently results in environmental degradation. The occurrences in Baga and Calangute are not random events but instead are part of a larger trend of coastal development in Goa. These examples highlight the importance of more rigorous implementation of environmental rules and an increased understanding of the ecological importance of coastal characteristics.

5.2 Model Validation and Shoreline Prediction (2022)

We can divide the data into training and test data in order to evaluate the correctness of the linear regression model. Shoreline changes for the years 1992, 1998, 2002, and 2014 are used for the model training. Further, estimated parameters are used to predict the shoreline for the year 2022. The RMSE metric gives us quantitative information about the deviation of the model from actual values. The regression model is constructed for each transect line, and the RMSE for all the shorelines ranges between 0.1986 m and 166.63 m, with mean and standard deviations of 21.50 m and 22.47 m, respectively. The overall RMSE is observed to be 34.50 m. The high RMSE is found in the areas of high accretion, i.e. Mandovi estuary, Sweet Lake, Arambol, and near Chapora estuary, Morjim. The lowest RMSE was found in the found in areas of high erosion i.e. Siridao Beach, Candolim, Anjuna, and Querim. The LRR model showed a strong positive correlation of 91.2% between actual and predicted shorelines, while the EPR model achieved an accuracy of 85.7% (Fig. 8). The EPR uses the earliest and most recent shorelines to estimate the shoreline shift slope, while LRR analyses the entire shoreline provided (Ahammed & Pandey, 2019). Due to the constantly changing nature of the shoreline, it is very difficult to predict the annual shift in its position. Nevertheless, LRR calculates the slope of the positional change by determining the mean positional shift using the shoreline data available in the model. The LRR model provides more precise information than the EPR model and has significantly lower uncertainty in the predicted and observed positional shift (Ahammed & Pandey, 2022).

5.3 Future Shoreline Prediction

The present research observed how the shoreline will move in the future in 2030 and 2050 based on actual changes in the shoreline. Hence, the effects of these extreme events are not measured in predicting future shorelines yet, the model will offer essential data for developing adaptation strategies and policy guidelines. As the model doesn't consider physical, geological, or extreme climatic factors, there will likely be variability in the projected shoreline location. Nonetheless, the model uses historical shoreline data to project the future shoreline, providing valuable insights into future trends. The speed at which the shoreline moves is a crucial element in forecasting the future direction of shoreline movement (Maiti & Bhattacharya, 2009).

The LRR forecasted an average shoreline shift of 7.10 m from 2022 to 2030 but predicts a rise to 24.84 m by 2050. The Mandovi estuary, Chapora estuary, the North part of Calangute, and Arambol will see the most sediment build-up in the study area, moving 57.12 m closer to the ocean. Between 2022 and 2050, the position will shift to 199.92 m along the coastline (Table 4). However, in the timeframe of 2022–2050, Querim, Mandrem, Anjuna, and Candolim are expected to experience higher erosion levels, resulting in a landward movement of − 91 m. Specifically, between 2022 and 2030, the coastline will shift by − 26.13 m (Figs. 9 and 10). However, the model provides insight into where the shoreline could be located. This research discovered that man-made embankments and natural features like mangroves and wetlands can protect the exposed coast.

6 Conclusion

The present study makes a detailed assessment of the shoreline changes in North Goa with reference to the patterns and rates of erosion and accretion over the past thirty years. The predicted loss of coastline for 2030 and 2050 suggests that North Goa faces a grave risk of severe erosion, resulting in a threat to tourism, the coastal ecosystem, and the stability of coastal infrastructure. This study uses analytical models such as the Linear Regression Rate (LRR) and the End Point Rate (EPR) to explain coastal behavior in which human activities (urban development, tourism, and sand mining) are major factors of coastal instability. The Digital Shoreline Analysis System (DSAS) was used to assess shoreline changes in North Goa using remote sensing and geospatial techniques. The period analysed was from 1992 to 2022, and

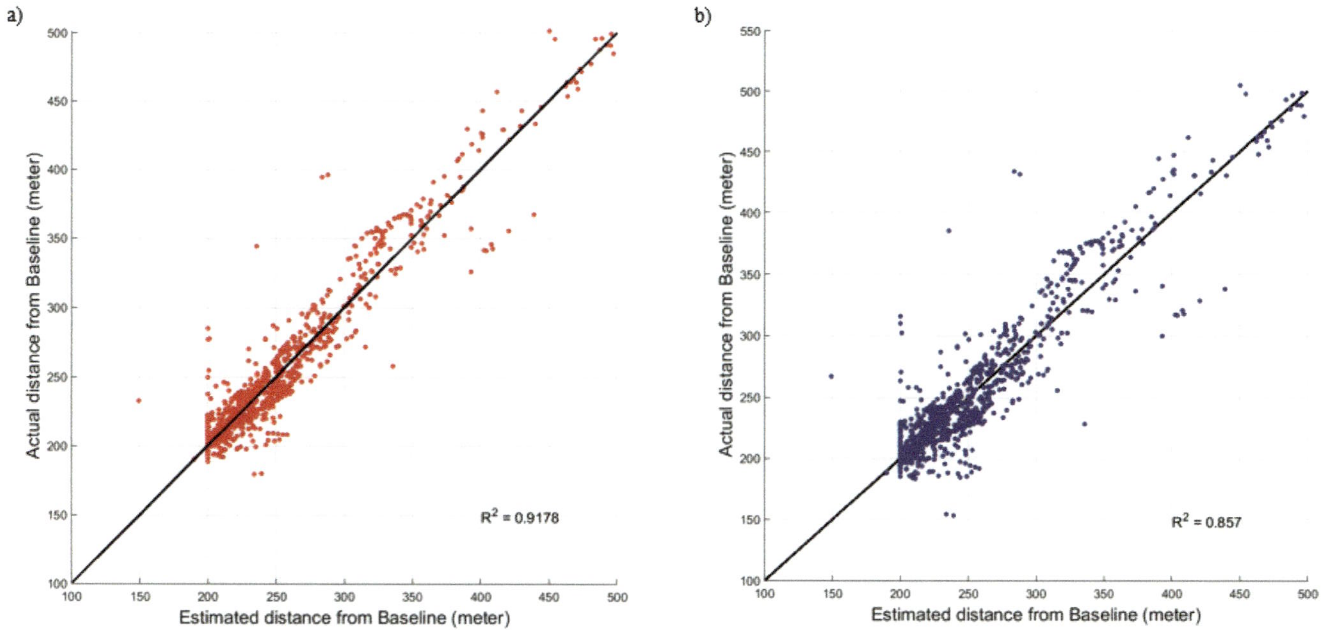

Fig. 8 Correlation between predicted and observed 2022 shoreline: **a** LRR model's and **b** EPR models

Table 4 Model predicted shoreline positional change between 2022–2030 and 2022–2050

Shoreline Positional Shift	2022–2030	2022–2050
Average Erosion	− 5.57	− 19.50
Average Accretion	9.93	34.77
Highest Erosion	− 26.13	− 91.44
Highest Accretion	57.12	199.92
Mean	7.10	24.84
SD	11.35	39.73

this analysis covered the rates of shoreline change, thus providing useful insight into coastal processes. The results suggest significant erosion along the North Goa coastline with an average positional shift of − 136.13 m per 30 years. In contrast, the greatest area of land increase within the 30 years was 201.36 m. At the popular beach places such as Querim, Mandrem, Anjuna, Candolim, and Siridao, massive erosion was observed, and at places like Miramar, Arambol, Morjim, and Calangute, significant sediment accumulation was seen. Their erosion and deposition patterns indicate the high risk of shoreline change, which could threaten coastal stability and the tourism-dependent local economy. Forecasting through the least-squares regression model suggests an average coastline shift of 7.10 m between 2022 and 2030, escalating to 24.84 m by 2050. The Mandovi estuary, Chapora estuary, the northern segment of Calangute, and Arambol are projected to experience the most sediment deposition, with movement of approximately 57.12 m toward the Sea. Combining remote sensing technology with geospatial techniques establishes a solid foundation for monitoring and managing coastal changes in North Goa. The insights gained from these studies are essential for safeguarding the region's natural and economic resources and ensuring the long-term viability of Goa's vibrant coastal areas. Together with geospatial techniques, remote sensing technology provides a good basis for monitoring and managing changes in North Goa. The lessons drawn from these studies are critical for ensuring the natural and economic resources of the region and ensuring the long-term viability of Goa's vibrant coastal areas.

Recommendations

Goa's coastline is both ecologically fragile and key to sustaining local communities and the tourism industry. Hence, it is important to protect it from harmful human activity. Following are some important recommendations for preserving the Goan coastline as follows, 1. *Coastal Zone Management:* In the areas where the rate of erosion is comparatively high, there is a need to successfully implement coastal zone management protocols by placing a prohibition on the construction, exploitation of coastal resources and misuse of dunes, coastal vegetation, mangroves and other eco-sensitive regions. 2. *Erosion Control Measures:* Government agencies need immediate steps to finance erosion control measures and habitat restoration. A priority must be accorded to protecting the vulnerable area through dune stabilization and the making of artificial reefs. We also need engineering actions in areas where natural shields are weakened. 3. *Adaptive Infrastructure Design:* To maintain a

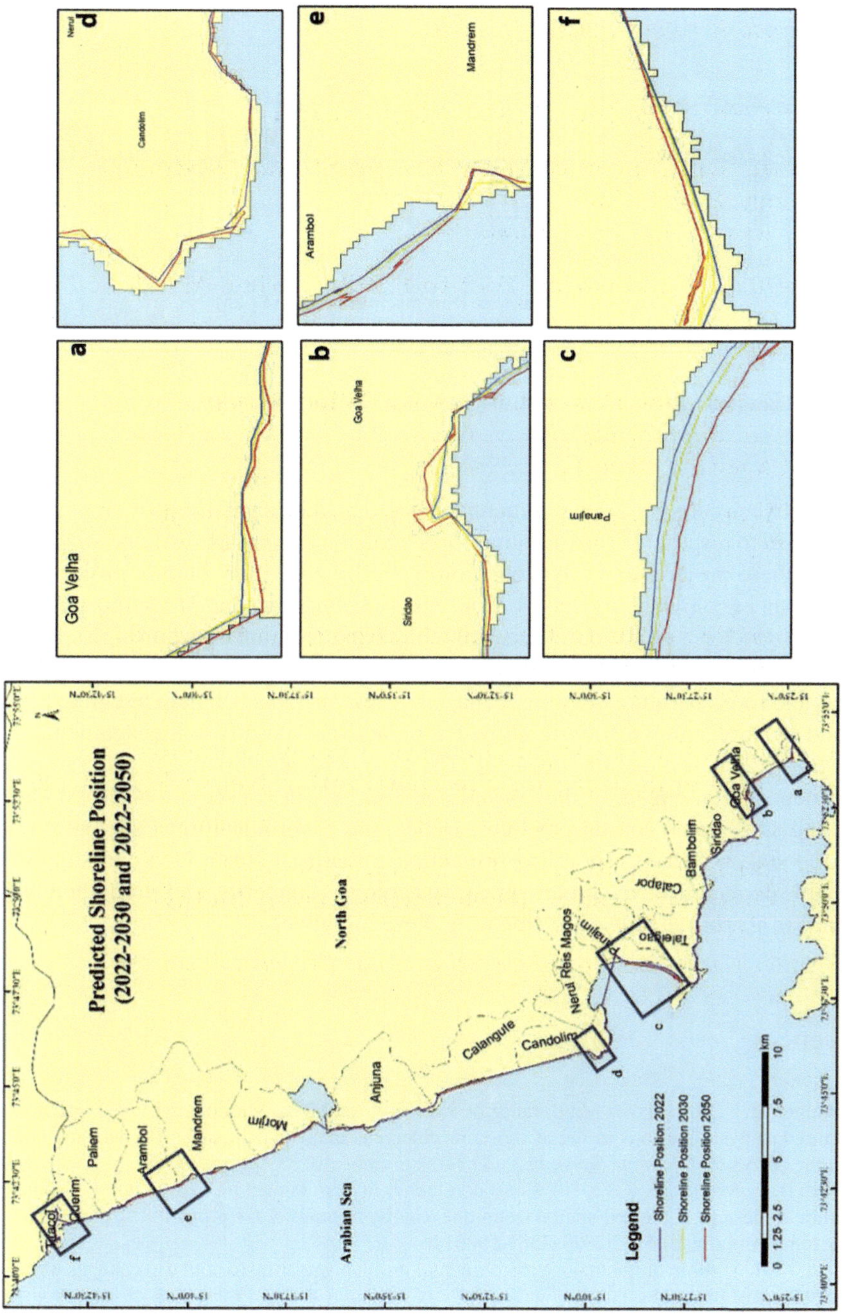

Fig. 9 The map displays the expected shoreline changes for years 2030 and 2050

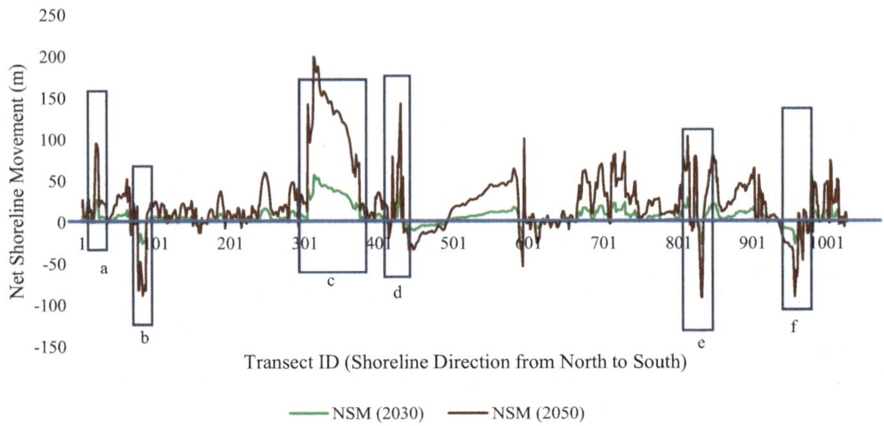

Fig. 10 The projected shorelines for 2030 and 2050 and coastline change from 2022 to 2030 and 2050

balance between economic development and ecological balance, infrastructure must be developed to respond to shifting shorelines while incorporating resilience into the design of these projects and the protection of buffer zones for tourism projects to reduce harsher ecological impacts. 4. *Continuous Monitoring and Modelling:* There is a need to develop a smart routine coastal monitoring program based on high spatial resolution satellite imagery and predictive models to monitor its current shoreline changes and forecast future possible scenarios. These insights will help integrate into state and local planning efforts to allow for proactive, data-driven management. 5. *Community Awareness and Engagement:* The success of any initiative or programme largely depends on engaging the coastal communities in conservation efforts and promoting sustainable coastal practices. Hence, the government must engage residents and stakeholders in shoreline protection initiatives. North Goa can support sustainable economic development, preserve ecological integrity, and strengthen the resilience of its shoreline by implementing these initiatives.

References

Abd-Elhamid, H. F., Zeleňáková, M., Barańczuk, J., Gergelova, M. B., & Mahdy, M. (2023). Historical trend analysis and forecasting of shoreline change at the Nile delta using RS data and GIS with the DSAS tool. *Remote Sensing, 15*(7). https://doi.org/10.3390/rs15071737

Ahammed, B., & Mahendra, R. (2016). Coastal vulnerability assessment for eastern coast of India, Andhra Pradesh by using geo-spatial technique. *Geoinformatics & Geostatistics: An Overview, 4*(3). https://doi.org/10.4172/2327-4581.1000146

Ahmed, A. (2011). Some of the major environmental problems relating to land use changes in the coastal areas of Bangladesh: A review. *Journal of Geography & Reg. Planning,* 1–8. http://www.academicjournals.org/JGRP

Ahammed, K., & Pandey, A. (2019). Geoinformatics-based assessment of coastal multi-hazard vulnerability along the East Coast of India. *Spatial Information Research, 27*(3), 295–307. https://doi.org/10.1007/s41324-018-00236-y

Ahammed, K., & Pandey, A. (2022). Assessment and prediction of shoreline change using multi-temporal satellite data and geostatistics: A case study on the eastern coast of India. *Journal of Water and Climate Change, 13*(3), 1477–1493. https://doi.org/10.2166/wcc.2022.270

Alberico, I., Casalbore, D., Pelosi, N., Tonielli, R., Calidonna, C., Dominici, R., & De Rosa, R. (2022). Remote sensing and field survey data integration to investigate on the evolution of the coastal area: The case study of Bagnara Calabra (Southern Italy). *Remote Sensing, 14*(10). https://doi.org/10.3390/rs14102459

Addo, K. (2013). Shoreline morphological changes and the human factor. Case study of Accra Ghana. *Journal of Coastal Conservation, 17*(1), 85–91. https://doi.org/10.1007/s11852-012-0220-5

Bagli, S., & Soille, P. (2003). Morphological automatic extraction of coastline from pan-{E}uropean {L}andsat {TM} images. In *Proceedings of the fifth international symposium on GIS and computer cartography for coastal zone management* (pp. 58–59). http://Www.Gisig.It/Coastgis/

Banerjee, A. (2024, February 26). Goa partners with World Bank to tackle sea level rise, coastal erosion and other climate change-related hazards. *Business Insider*. https://www.businessinsider.in/sustainability/news/goa-partners-with-world-bank-to-tackle-sea-level-rise-coastal-erosion-and-other-climate-change-related-hazards/articleshow/108015194.cms

Cai, F., Su, X., Liu, J., Li, B., & Lei, G. (2009). Coastal erosion in China under the condition of global climate change and measures for its prevention. *Progress in Natural Science, 19*(4), 415–426. https://doi.org/10.1016/j.pnsc.2008.05.034

Cenci, L., Disperati, L., Persichillo, M. G., Oliveira, E. R., Alves, F. L., & Phillips, M. (2018). Integrating remote sensing and GIS techniques for monitoring and modeling shoreline evolution to support coastal risk management. *Giscience and Remote Sensing, 55*(3), 355–375. https://doi.org/10.1080/15481603.2017.1376370

Chang, F. J., & Lai, H. C. (2014). Adaptive neuro-fuzzy inference system for the prediction of monthly shoreline changes in northeastern Taiwan. *Ocean Engineering, 84*, 145–156. https://doi.org/10.1016/j.oceaneng.2014.03.020

Craglia, M., de Bie, K., Jackson, D., Pesaresi, M., Remetey-Fülöpp, G., Wang, C., Annoni, A., Bian, L., Campbell, F., Ehlers, M., van Genderen, J., Goodchild, M., Guo, H., Lewis, A., Simpson, R., Skidmore, A., & Woodgate, P. (2012). Digital Earth 2020: Towards the vision for the next decade. *International Journal of Digital Earth, 5*(1), 4–21. https://doi.org/10.1080/17538947.2011.638500

Di Stefano, A., De Pietro, R., Monaco, C., & Zanini, A. (2013). Anthropogenic influence on coastal evolution: A case history from the Catania Gulf shoreline (eastern Sicily, Italy). *Ocean and Coastal Management, 80*, 133–148. https://doi.org/10.1016/j.ocecoaman.2013.02.013

EarthExplorer. (2022). https://earthexplorer.usgs.gov/

ElKotby, M. R., Sarhan, T. A., & El-Gamal, M. (2023). Assessment of human interventions presence and their impact on shoreline changes along Nile delta, Egypt. *Oceanologia, 65*(4), 595–611. https://doi.org/10.1016/j.oceano.2023.06.008

Ericson, J. P., Vörösmarty, C. J., Dingman, S. L., Ward, L. G., & Meybeck, M. (2006). Effective sea-level rise and deltas: Causes of change and human dimension implications. *Global and Planetary Change, 50*(1–2), 63–82. https://doi.org/10.1016/j.gloplacha.2005.07.004

Fang, J., Kong, F., Fang, J., & Zhao, L. (2018). Observed changes in hydrological extremes and flood disaster in Yangtze River Basin: Spatial-temporal variability and climate change impacts. *Natural Hazards, 93*(1), 89–107. https://doi.org/10.1007/s11069-018-3290-3

Gens, R. (2010). Remote sensing of coastlines: Detection, extraction and monitoring. *International Journal of Remote Sensing, 31*(7), 1819–1836. https://doi.org/10.1080/01431160902926673

Iyer, S. D., & Viegas, A. (2023). A novel way to protect Goa from SEA (Shoreline changes, erosion, and accretion). *O Herald*. https://www.heraldgoa.in/Edit/UPFRONT/A-novel-way-to-protect-Goa-from-SEA-Shoreline-changes-erosion-and-accretion/206274

Kannan, R., & Ramanamurthy, M. (2016). Shoreline change monitoring in Nellore coast at east coast Andhra Pradesh District using remote sensing and GIS. *Journal of Fisheries & Livestock Production, 04*(01). https://doi.org/10.4172/2332-2608.1000161

Kilibarda, Z., Graves, N., Dorton, M., & Dorton, R. (2014). Changes in beach gravel lithology caused by anthropogenic activities along the southern coast of Lake Michigan, USA. *Environmental Earth Sciences, 71*(3), 1249–1266. https://doi.org/10.1007/s12665-013-2529-2

Klein, M., & Lichter, M. (2006). Monitoring changes in shoreline position adjacent to the Hadera power station, Israel. *Applied Geography, 26*(3–4), 210–226. https://doi.org/10.1016/j.apgeog.2006.01.001

Klemas, V. V. (2010). Sensors and techniques for observing coastal ecosystems. In X. Yang (Ed.), *Remote sensing and geospatial technologies for coastal ecosystem 17 assessment and management* (p. 418). Springer. https://doi.org/10.1007/978-3-540-88183-4

Kourosh Niya, A., Asghar Alesheikh N, A. K., & Soltanpour Khaje Nasir, M. (n.d.). Shoreline change mapping using remote sensing and GIS case study: Bushehr province spatial analysis of Crimean Congo Hemorrhagic Fever (CCHF) view project forest fire spread modeling view project. www.ijrsa.org

Kummu, M., De Moel, H., Salvucci, G., Viviroli, D., Ward, P. J., & Varis, O. (2016). Over the hills and further away from coast: Global geospatial patterns of human and environment over the 20th-21st centuries. *Environmental Research Letters, 11*(3). https://doi.org/10.1088/1748-9326/11/3/034010

Kunte, P. D., & Wagle, B. G. (1994). Analysis of space-borne data for coastal zone information extraction of Goa coast, India. *Ocean and Coastal Management, 22*(3), 187–200. https://doi.org/10.1016/0964-5691(94)90031-0

Liao, Z., Su, K., Jiang, X., Zhou, X., Yu, Z., Chen, Z., Wei, C., Zhang, Y., & Wang, L. (2022). Ecosystem and driving force evaluation of northeast forest belt. *Land, 11*(8). https://doi.org/10.3390/land11081306

Maiti, S., & Bhattacharya, A. K. (2009). Shoreline change analysis and its application to prediction: A remote sensing and statistics-based approach. *Marine Geology, 257*(1–4), 11–23. https://doi.org/10.1016/j.margeo.2008.10.006

Malkarnekar, G. (December 5, 2023). 21 beaches in Goa face erosion, no accretion on 12 : Central data. *Times of India.* https://timesofindia.indiatimes.com/city/goa/21-beaches-in-goa-face-erosion-no-accretion-on-12-central-data/articleshow/105739853.cms

Mascarenhas, A. (1998). A report on coastal sand dune ecosystems of Goa: Significance, uses and anthropogenic impacts. In *NIO* (p. 5). https://www.researchgate.net/publication/273318244%0ACoastal

Mascarenhas, A. (2000). Human interreference along the coast of Goa. *Environmental Problems of Coastal Areas in India*, 145–171. https://www.researchgate.net/publication/27667643

McGranahan, G., Balk, D., & Anderson, B. (2007). The rising tide: Assessing the risks of climate change and human settlements in low elevation coastal zones. *Environment and Urbanization, 19*(1), 17–37. https://doi.org/10.1177/0956247807076960

Nadaf, F. M. (2019). Geographical analysis of the coastal landforms of Canacona, Goa. *Research Review International Journal of Multidisciplinary, 4*(2), 655–661.

Naik, K. (October 6, 2014). Now, its Arambol, Morjim beaches' turn to battle erosion. *Times of India.* https://timesofindia.indiatimes.com/City/Goa/Now-its-Arambol-Morjim-beaches-turn-to-battle-erosion/articleshow/44430774.cms

Nayak, G., & DSouza, J. (2010). Changes in erosional and depositional processes with time and management of Goa Coast, central west coast of India. In *EGU general assembly conference abstracts* (Vol. 12, p. 5759).

Niya, A. K., Asghar Alesheikh, A., Soltanpor, M., & Kheirkhahzarkesh, M. M. (2013). Shoreline change mapping using remote sensing and GIS. *International Journal of Remote Sensing Applications, 3*(3). www.Ijrsa.Org

Pilkey, O. H., & Hume, T. (2001). The shoreline erosion problem: Lessons from the past. *Water and Atmosphere, 9*(2), 22–23.

Rajawat, A. S., Chauhan, H. B., Ratheesh, R., Rhode, S., Bhanderi, R. J., Mahapatra, M., Kumar, M., Yadav, R., Abraham, S. P., Singh, S. S., Keshri, K. N., & Ajai. (2014). Assessment of coastal erosion along Indian coast on 1:25, 000 scale using satellite data. *International Archives of the Photogrammetry, Remote Sensing and Spatial Information Sciences—ISPRS Archives, 40*–8(1), 119–125. https://doi.org/10.5194/isprsarchives-XL-8-119-2014

Ranasinghe, R. (2016). Assessing climate change impacts on open sandy coasts: A review. *Earth-Science Reviews, 160*, 320–332. https://doi.org/10.1016/j.earscirev.2016.07.011

Sahoo, B., & Bhaskaran, P. K. (2018). Multi-hazard risk assessment of coastal vulnerability from tropical cyclones—A GIS based approach for the Odisha coast. *Journal of Environmental Management, 206*, 1166–1178. https://doi.org/10.1016/j.jenvman.2017.10.075

Sawkar, K., Noronha, L., Mascarenhas, A., & Chauhan, O. S. (2005). *Tourism and the environment issues of concern in the coastal zone of goa* (pp. 259–274). The Management of Tourism.

TheGoanEveryday. (2024). High Court to the rescue of Goa's coast yet again. TheGoanEveryday, February 13. https://www.thegoan.net/editorial/high-court-to-the-rescue-of-goas-coast-yet-again/109595.html

Thieler, E. R., Himmelstoss, E., Zichichi, J., & Ergul, A. (2009). DSAS 4.0 installation instructions and user guide. *U.S. Geological Survey Open-File Report 2008–1278, 3*, 291–294. http://pubs.usgs.gov/of/2008/1278/

Times, N. (February 13, 2023). 48-hour deadline to beach encroachers. *The Navhind Times*. https://www.navhindtimes.in/2023/02/13/goanews/48-hour-deadline-to-beach-encroachers/#google_vignette

Venkateswaran, S. (2006). Goa to pull down buildings violating coastal norms. TNN, August 17. https://economictimes.indiatimes.com/news/politics-and-nation/goa-to-pull-down-buildings-violating-coastal-norms/articleshow/1899169.cms

Wang, J., Wang, L., Feng, S., Peng, B., Huang, L., Fatholahi, S. N., Tang, L., & Li, J. (2023). An overview of shoreline mapping by using airborne LiDAR. *Remote Sensing, 15*(1), 1–27. https://doi.org/10.3390/rs15010253

Wright, D. J. (2009). Spatial data infrastructures for coastal environments. In X. Yang (Ed.), (pp. 92–112). Springer. https://doi.org/10.1007/978-3-540-88183-4

Yadav, A., Dodamani, B. M., & Dwarakish, G. S. (2018). Shoreline change threat to coastal zone: A case study of Karwar Coast, *2*(2), 18–30. https://doi.org/10.17501/2513258x.2018.2202

Kilibarda, Z., Graves, N., Dorton, M., & Dorton, R. (2014). Changes in beach gravel lithology caused by anthropogenic activities along the southern coast of Lake Michigan, USA. *Environmental Earth Science, 71*(3), 4–7. https://doi.org/10.1007/s12665-013-2529-2

Zuzek, P. J., Nairn, R. B., & Thieme, S. J. (2003). Spatial and temporal considerations for calculating shoreline change rates in the Great Lakes Basin. *Journal of Coastal Research SPEC. ISS., 38*, 125–146.

Abundance of Microplastic Pollution in Sediment of Puducherry Coast

B. Visnu, S. Bragadeeswaran, S. Kumaresan, and R. Parthasarathi

1 Introduction

Staggering range of animals, including crabs, corals, fish, sea turtles, marine mammals, as well as many other organisms, are in the marine environment. Over 80% of biodiversity on earth resides in the waters. There are various zones and marine environment peculiarities. These include open ocean, which is huge deep-water region farther offshore, as well as the continental shelf, which is the shallow section of the ocean extending from the shoreline, Seamounts, trenches, underwater canyons, and hydrothermal vents are further features. Seal reefs, estuaries, and mangroves support almost 75% of fish and other marine life. Due to food as well as nutrients buildup needed for nursing as well as feeding, reefs and estuaries serve as a shelter for finfish and shellfish (Marghany & Mansor, 2016). Rising environmental hazard has been plastic pollution in rivers along with oceans, moreover deposition on deltas, riverbanks, beaches, along with ocean has been developing rapidly. It has been estimated 60% of all plastics produced have been wasted in natural surroundings. Ecosystems, Aquatic life, along with human health have all been threatened by plastic waste (Meijer et al., 2021). Additionally, Plastic litter causes significant monetary losses due to vessels destruction along with fishing gear, Tourism's negative effects along with extra coastline cleaning operations, totalling US$1.26billion/year for Asian-Pacific Rim alone. Riverine plastic transport has still been poorly understood, especially in locations that have been predicted in contributing most to global plastic pollution in ocean. For creating effective prevention along with collection techniques, deeper paths as well as transport processes understanding of plastic trash to as well as within

B. Visnu (✉) · S. Bragadeeswaran · S. Kumaresan · R. Parthasarathi
Marine Biotoxinology Laboratory, CAS in Marine Biology, Annamalai University, Parangipettai, India
e-mail: visnubala1997@outlook.com

R. Parthasarathi
Department of Microbiology, Annamalai University, Chidambaram, India

P. Thambidurai and A. K. Dikshit (eds.), *Coastal Environments of India*,
https://doi.org/10.1007/978-3-031-97341-3_13

rivers, global riverine plastic pollution dispersion into ocean, has been needed(Meijer et al., 2021). Though plastic waste has not been biodegradable, durability of it has been uncertain, moreover it might persist for centuries (Andrady, 1994; Barnes et al., 2009; Dilkes-Hoffman et al., 2019 and Shen et al., 2020). As loss caused because of inappropriate waste management along with dumping onshore, offshore, plastic debris frequently attains marine environment where it might assemble. Plastic debris can drift for years moreover deteriorate because of weathering, physical abrasion, as well as UVradiation (Klemchuk, 1990; Derraik, 2002 Thompson et al., 2019; Andrady & Koongolla, 2022 and Dimassi et al., 2022).

Most of monomers, including ethylene, propylene, that have been utilized in making plastics come from fossil hydrocarbons. In nature, 0 regularly utilized plastics break down. As result, they gather in landfills or environment instead of breaking down. Plastic waste can only be permanently eliminated by destructive heat treatment, including pyrolysis or fire. Near-permanent contamination of natural environment with plastic debris has been becoming rising concern. Plastic debris has been discovered in all major ocean basins, with 8million Mt(metric tons) of plastic debris produced on land getting into ocean. Freshwater system, terrestrial ecosystem contamination has also been growing more common, as has been environment contamination having synthetic fibres. Plastic garbage has now been widespread in environment which has been recommended as geological indication of Anthropocene era (Geyer et al., 2017). Due to Due to poor management as well as disposal procedures, significant plastic trash amount pollutes environment by entering it via different channels. After entering environment, plastic waste could decompose gradually moreover produce large amount of smaller plastic debris called microplastics because of chemical, physical, as well as biological processes (Plastics Europe 2019). Microplastics could be categorized into 2primary groups: primary, secondary microplastics. Primary microplastics refer to micro-sized synthetic polymers utilised in processes like sandblasting media, chemical formulations, plastic products maintenance along with synthetic clothing production. Among primary microplastics are microbeads, which are plastic beads (size < 2 mm) made of PE(polyethylene), PP(polypropylene), PS(polystyrene) moreover have commonly been found in cosmetic along with healthcare products. Secondary microplastics have been fragmentation output of larger plastic items, occurring via different environmental procedures including photo-degradation, biodegradation, thermo-oxidative degradation, thermal degradation, along with hydrolysis. Nanoplastics, which have been plastic fragments smaller than 1 μm, fall within this category. Both nanoplastics along with microplastics pose potential risks for bioamplification along with bioaccumulation of different chemicals along with pollutants because of their significant surface-to-volume ratio. Microplastics formation involves several processes, primarily mechanical degradation, photodegradation, biodegradation, abrasion and the intentional use of microbeads. Mechanical degradation occurs when larger plastic items are exposed to environmental factors like sunlight and temperature fluctuations, causing them to become brittle and break into smaller fragments. Photodegradation, caused by UV radiation from sunlight, weakens plastics, making them more susceptible to mechanical forces. Some plastics can partially biodegrade through microbial action, but this process is slow and typically results in the creation of microplastics. Abrasion, seen in areas with high plastic pollution like oceans, produces tiny plastic particles when plastic waste rubs against surfaces due to wave and current movements. Additionally, microbeads intentionally added to personal care products can enter aquatic environments when washed down the drain, contributing to microplastic pollution.

Numerous studies have explored the presence and impact of microplastic pollution along the Indian coast (Table 1). Veerasingam et al. (2016) investigated microplastic pellets on Goa's beaches, while Kumar et al. (2016) focused on Marina Beach in Chennai for marine debris pollution. Karthik et al. (2018) expanded this research to 25 beaches between Chennai and Kanyakumari, considering different beach profiles. Specific island impacts were noted by Krishnakumar et al. (2018) on Nallathani Island's microplastic pollution. Dowarah and Devipriya (2019) studied microplastics in Puducherry, and Begum et al. (2020) assessed estuarine regions along Tamil Nadu's coast. Plastic debris accumulation was also documented in various locations: Ramesh and Nagalakshmi (2022) in Chennai, Sundar et al. (2021) in Kanyakumari and Singare (2012a, 2012b) in Mumbai's creeks and rivers. Vanapalli et al. (2021) provided a comprehensive view of microplastic sources and effects in India's aquatic ecosystems, while Pavithran (2021) focused on Kerala's northern coast. The studies Chilika Lake in Odisha, Ranjani et al. (2022) investigated seasonal variations in Chennai's beach sediments, and Velmurugan et al. (2023) examined microplastics along the east coast from Ennore to Kovalam. Additionally, Singh et al. (2023) reported on microplastic contamination in the soil in Bopal, central India. These studies collectively emphasize the widespread and diverse nature of microplastic pollution in both coastal and inland regions of India. Based on insights gained from a comprehensive review of existing literature, this work was undertaken with the primary objective of identifying and characterizing the sources and pathways of marine microplastics. Aiming to study the marine microplastics on the selected Continental Shelf off the Puducherry coast, the research focuses on isolating microplastics from bottom sediments, studying different types such as fibers, pellets, fragments, and foam, and understanding their abundance, size variation, and polymer composition. Developing accurate methods for quantifying and monitoring these microplastics is essential for assessing their distribution in various marine environments. This formulation, stemming from thorough research, provides a crucial foundation for advancing knowledge in marine pollution management and mitigation strategies.

2 Study Area

The study area for the microplastic investigation was meticulously chosen, considering several crucial factors. The primary criterion for site selection revolved around the strategic placement of reefs, as these ecosystems play a pivotal role in the accumulation and distribution of marine debris. Additionally, the impact of local fishing activities on microplastic prevalence was considered, recognising that areas with

Table 1 Previous works reported on the microplastic abundance in the sediment sample

Location	Microplastic in sediments	References
Belgium	91.9 particles/kg	Claessens et al. (2011)
Slovenia shoreline	Infralittoral zone: 155.6 particles/kg Shoreline: 133.3 particles/kg	Laglbauer et al. (2014)
French Belgian Dutch coastline	6 ± 5.7 particles/kg	Van Cauwenberghe et al. (2015)
Bostanu, Persian Gulf	1258 ± 291 particles/kg	Naji et al., (2017a, 2017b)
Gorsozan, Persian Gulf	122 ± 23 particles/kg	Naji et al., (2017a, 2017b)
Tahibu Lake	11–234.61 particles/kg	Su et al. (2016)
Bizerte Lagoon, Tunisia	7960 ± 6840 particles/kg	Abidli et al. (2017)
Scapa Flow, Scotland	730–2300 particles/kg	Blumenroder et al. (2017)
Thames river	349.50 ± 303.00 particles/kg	Horton et al. (2017)
Persian Gulf, Iran	61 ± 49 particles/kg	Naji et al., (2017a, 2017b)
Hong Kong Marine Environment	49–297 particles/kg	Tsang et al. (2017)
Three Gorges Reservoir	25 - 300 particles/kg	Di and Wang (2018)
Shanghai	802 ± 594 particles/kg	Peng et al. (2018)
Antua River, Portugal	225.25 ± 109.52 particles/kg	Rodrigues et al. (2018)
Bohai Sea, North Yellow Sea, South Yellow Sea	171.8 ± 55.4 particles/kg	Zhao et al. (2018)
	123.6 ± 71.6 particles/kg 72 ± 27.2 particles/kg	
Caranzalem Beach, Goa	50–300 items/m^2	Nigam (1982)
Alang-Sosiya ship-breaking yard, Gujarat	81.43 mg/kg	Reddy et al., (2004, 2006)
Goa (Keri, Vagator, Calangute, Colva, Mobor and Galgibaga beaches)	1345 items (January)	Veerasingam et al. (2016)
	1655 items (June)	
Tinnakara, Lakshadweep	603 items	Mugilarasan et al. (2017)
Tamil Nadu (from Kanyakumari to Ennore)	9–178 items/m^2 (HTL) 2–64 items/m^2 (LTL)	Karthik et al. (2018)
Rameswaram, Tamil Nadu	403 items	Vidyasakar et al. (2018)
Tamil Nadu (Marina Beach,	119–439 items/kg (HTL)	Sathish et al. (2019)
Kanyakumari, Thiruchendur, Manapad, Tuticorin)	33–179 items/kg (LTL)	
Puducherry Coast	720.30 ± 191.60 particles/kg	Dowarah & Devipriya, 2019
Nattika Beach, Kerala	70.15 items/kg (2017)	Ashwini and Varghese (2020)
	120.85 items/kg (2018)	
Andaman and Nicobar Islands (Port Blair Bay)	45.17 items/kg	Goswami et al. (2020)
Rameswaram, Tamil Nadu	55–259 items/kg	Jeyasanta et al., (2020a)

(continued)

Table 1 (continued)

Location	Microplastic in sediments	References
Tuticorin, Tamil Nadu	25–83 items/m^2	Jeyasanta et al. (2020b)
Maharashtra (Aksa, Juhu, Dadar, Girgaon)	43.6–346.0 items/m^2	Maharana et al. (2020)
Karnataka (Devbagh, Karwar, Kasarkod)	21.0–155.3 items/m^2	Maharana et al. (2020)
Goa (Vagator, Calangute, Colva)	17.0–95.6 items/m^2	Maharana et al. (2020)
South Andaman Islands	414.35 items/kg	Patchaiyappan et al. (2020)
Tuticorin and Vembar Coral islands, Tamil Nadu	50.0–103.8 items/kg	Patterson et al. (2020)
Tuticorin, Tamil Nadu	24.45–235.12 items/kg	Sathish et al. (2020)
Kanyakumari, Tamil Nadu	6860 items/kg	Sundar et al. (2021)
Silver Beach, Cuddalore, Tamil Nadu	204 items/kg	Vidyasakar et al. (2020)
Off Puducherry Coast	244.66 $\pm$ 15.022 particles/kg	Present study

intense fishing might exhibit higher concentrations of plastic pollutants. Furthermore, the availability and diversity of marine life in the chosen locations were considered as the interaction between marine organisms and microplastics, which is a key aspect of understanding the ecological implications of plastic pollution. The study area (Fig. 1) that has been selected and site photographs are presented in Figs. 2 and 3 (S1a-d, S2a-c and S3a) as follows, *Site 1: Naveen's Reef* (*Depth:* 18 m)-It is the oldest fishermen-built reef in Puducherry, covering an area of 10 $\times$ 10 square meters. This dive site, which is usually covered with marine life, has varying visibility factors due to its closeness to the harbor mouth, making it only accessible to SCUBA divers a few times a year (December–August). *Site 2: Dannys Eel Garden*(*Depth:* 18 m)—This site has a rocky bottom, believed to be the ballast of a ship, and houses some fantastic marine life, including a lot of juveniles, varieties of crabs, and other invertebrates. The shallowest of the reef ranges from 16 m, and the deepest is 18 m. An area of 10 $\times$ 30 m, the area is home to the widest variety of moray eel. *Site 3: Pirates Reef* (*Depth:* 24 m)—This site has a rocky bottom and huge rock structures throughout the site. The site is 8.24 nm from the shore of Puducherry. Ghost nets and other fishing gear are strewn all over this area.

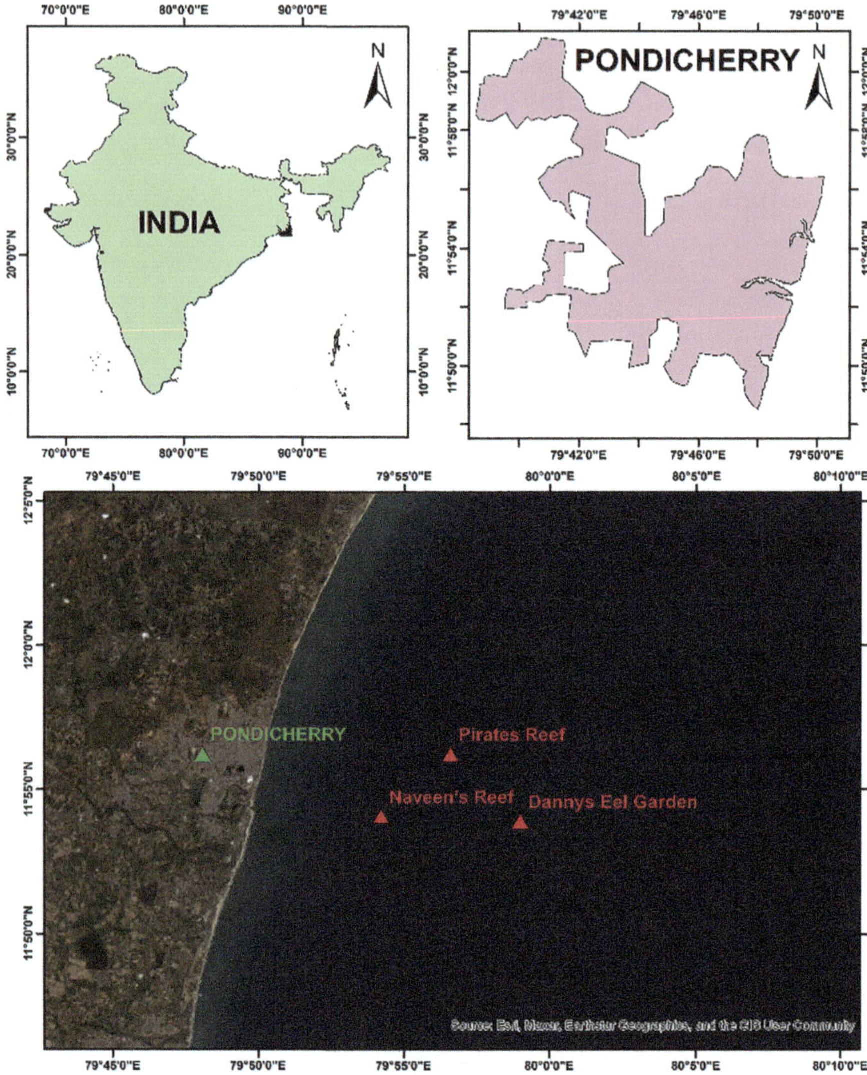

Fig. 1 Photographs of sample collection in Navvens reef

3 Materials and Methods

3.1 Collection of Sediment Sample for Microplastic Studies

The samples were collected from three different sampling sites, which allowed for a more representative sample of the overall characteristics of the coast. To ensure consistency, the samples were collected from a distance of 4.3 nm from the shoreline.

Fig. 2 Photographs of sample collection in Navvens reef

Fig. 3 Photograph sample collection locations

The samples were taken from a depth of 18–24 m. Sterile containers were used to collect the samples. This helped to ensure that the samples were not contaminated with any foreign substances that could impact the results of the analysis. Because it is hard to transport a container full of air to deep waters, the sampling containers were filled with water at the surface as the samples were gathered at a depth of 18–24 m. Once the sampling depth was achieved, sediment was collected from each location

(S1, S2 and S3). The collected samples were dried in a hot air oven at 50 °C overnight and stored in the laboratory for further analysis.

3.2 Density Separation

All samples underwent density separation utilising a Sodium chloride (NaCl) solution. A solution with a density of 1.52 g/mL was achieved by employing a 1:1 ratio of distilled water to NaCl, as outlined by Karami et al. (2017). The NaCl was chosen for its non-hazardous nature and high-density characteristics, facilitating the separation of less dense microplastic particles from denser inorganic counterparts such as sand, as highlighted by Bergmann et al. (2015). The setup involved a density separator comprising 200 mL filter funnels, 100–125 mL conical flasks, retort stands, rubber tubes, and hinged and binder clips. To prevent airborne contamination, aluminium foil covered the filter funnels and conical flasks. Prior to usage, NaCl underwent filtration through a Whatman GF/C glass microfibre filter membrane (pore size: 1.2 μm) to ensure purity. Each sample received 100 mL of NaCl, poured into the filter funnels and allowed to settle for one hour. Following the settling of denser materials, rubber tubes were unclipped gradually to discharge settled unwanted particles, specifically sand. The resulting supernatant was collected in clean conical flasks. Subsequently, the supernatant in the conical flasks underwent further filtration through a Whatman GF/C glass microfibre filter membrane (pore size: 1 μm). Each filter was individually stored in glass Petri dishes for subsequent observation and processing (Reed et al., 2018).

3.3 Microplastic Analysis

The visual identification of microplastics on filter membranes was carried out by using an Olympus CX41 digital electronic microscope. Microplastics were analysed and categorized into five different types (Alomar et al., 2017) fragments (Andrady, 2011), foam (Andrady, 2017), pellets (Ballent et al., 2016), film and line/fibres (Barnes, 2002). The significant microplastics found in filters were photographed. The number of microplastics found on the filter membranes was counted and represented as items per individual sample (Karthik et al., 2018) and a Scanning Electron Microscope (SEM) was performed for size distribution study.

3.4 Raman Spectroscopy for Polymer Identification

Raman spectroscopy was used to identify polymers using a representative sample of 50 particles. The spectra were scanned for a range of 200–3500 cm^{-1} in 30 s. The raw spectra were corrected for baseline and noise using Originlabs software, and the corrected spectra were matched with the Openspecy database to identify polymers. Because plastics in the environment undergo various types of degradation, only a few particles have an exact spectrum match to the pure polymers (Dowarah & Devipriya, 2019).

3.5 *Statistical Analysis*

The statistical analysis of variance (ANOVA) was used to investigate the potential differences between microplastic concentrations and seasonal variations. All the Statistical analysis was performed using JASP (0.18.1.0).

4 Results

4.1 *Abundance of Microplastics in Bottom Sediments*

The microplastic abundance in the coastal bottom sediment at three different sites for two years (2021–2023) was noticed with distinct patterns among the seasons and locations. The highest microplastic abundance was observed during the post-monsoon season in all the Sites (S1-9.5 $\pm$ 7.835 particle kg^{-1}; S2- 10.75 $\pm$ 8.923 particle kg^{-1} and 9.75 $\pm$ 8.067 particle kg^{-1}), while the lowest were observed during the pre-monsoon season in S1 (5.25 $\pm$ 4.774 particle kg^{-1}) and summer in S2 and S3 (S2- 6 $\pm$ 5.456 particle kg-1 and S3- 5.5 $\pm$ 4.294 particle kg-1) (Fig. 2). However, the highest abundance was observed in the year 2022–23 in S2 with an abundance of 12 $\pm$ 10.469 particle kg^{-1} during pre-monsoon. Similarly, the lowest (6 $\pm$ 6.044 particle kg^{-1}) was observed at S3 during the pre-monsoon season. Overall, Site 2 consistently displayed higher microplastic abundances than sites 1 and 3 (Fig. 3).

The size of the microplastics isolated from the collected sediments have ranged from 60 to 800 μm. In order to prevent the loss of particles during the process 1 μm glass fiber filter paper was used (Fig. 4). Four distinct morphologies were identified for the screened microplastics i.e., fragment, film, pellet and fibre (Fig. 5). The most common shape is fibres (63%), followed by fragment (33%), film (3%) and pellet (1%) (Fig. 6) and the microplastic anaylsed through SEM is presented in Figure as follows, A- Fragments- Site 3; B- Fibre- Site 2; C- Fiber-Site 2 and C- Film-Site 1.

At site 1, fibres were noticed as the dominant with a total of 155 particles, followed by fragments of 71 particles. In 2021–2022, the highest abundance of fibres (24 particles) was observed, whereas, in the subsequent year (2022–2023), a decrease in fibre abundance (22 particles) was observed, accompanied by an increase in fragments (9 particles). The summer season of both years (2021–2023) showed a constant decline in the number of fibres. There was a decline in the number of fragments during

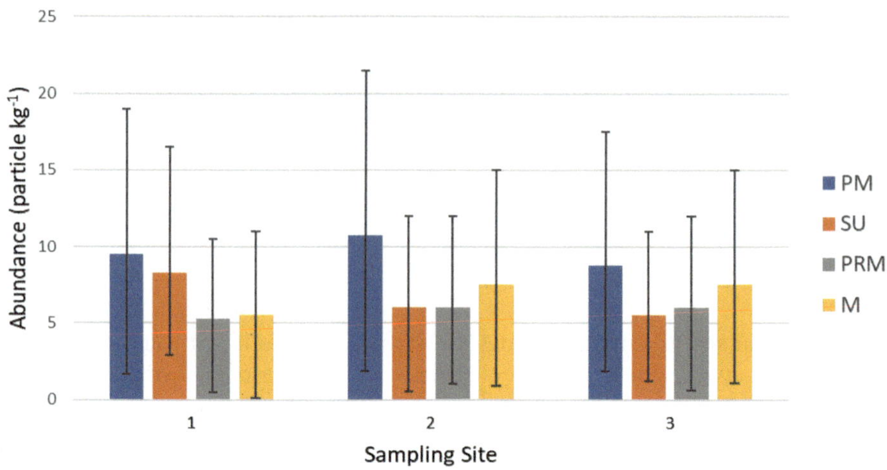

Fig.4 Abundance of microplastics in 2021–2022

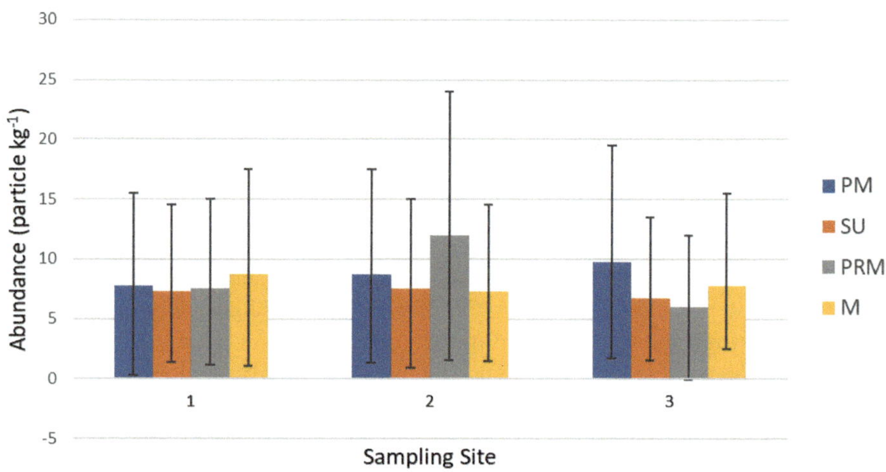

Fig. 5 Abundance of microplastics in 2022–2023Size and shape

the second year (2022–2023). Whereas, the fibre abundance was found to increase during the pre-monsoon season. Pellets and film were noticeably less (Fig. 7) and the microplastic particles were observed through microscopic is presented in Fig. 7 as follows, A- Fragment; B, D and E- Fibre; C- Film and F- Pellet and Film.

The analysis of the composition of microplastic in sediments at site 2 over the periods of 2021–2022 and 2022–2023 revealed that the post-monsoon season (2021–2022) exhibited the minimum presence of fibres (26 particles), with no pellets and a minimal presence of film (1 particle) and fragments (16 particles). Whereas the pre-monsoon season showed the minimum fibres (12 particles) and fragments (12

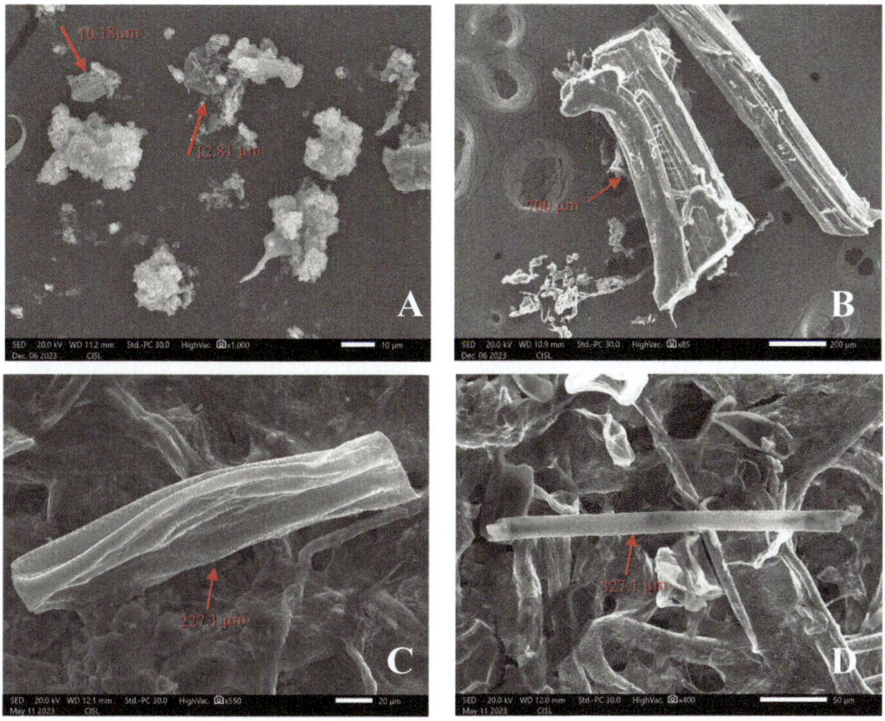

Fig. 6 SEM images of different sizes of microplastics observed in samples

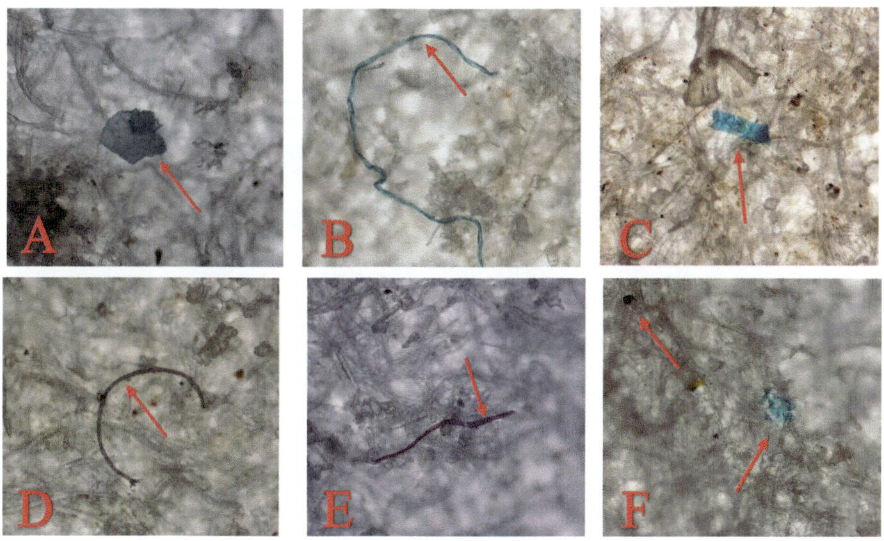

Fig. 7 Different morphology observed under the microscope

particles). The summer and monsoon seasons showed intermediate values for fibre and fragment categories (Fig. 8). In 2022–2023, there was a notable increase in fibre count of 28 particles during post-monsoon season. Further, the pre-monsoon season (2022–2023) was recorded with maximum counts of fibres (32 particles) and fragments (14 particles). The analysis of the composition of microplastic in the sediment samples collected at site 3 during the periods of 2021–2022 and 2022–2023 revealed the different morphologies. The minimum occurrence of fibres was observed during the summer seasons of both years, with 14 particles in 2021–2022 and (16 particles) in 2022–2023 (Fig. 9). The maximum fibre content was observed during the post-monsoon season of 2022–2023 (21 particles). Pellets consistently showed minimal presence, maintaining a constant count of zero in all the seasons for both years (2021–2023). Films were found to have displayed variability, as the highest was observed during the summer season of 2022–2023 (3 particles), while the maximum fragments were found during the post-monsoon season (2022–2023) of 18 particles. The pre-monsoon season of 2021–2022 showed the lowest fragment count (6 particles).

Among the variations in microplastic composition, site 2 stands out with the highest particle count of 263 (171 fibres, 3 pellets, 3 films and 86 fragments), indicating a potential reservoir. In contrast, site 3 exhibited the lowest particle count of 137, with a notable prevalence of fibres and fragments (Fig. 10). The recorded data showed a consistent pattern across all sites, with fibres being the most prevalent type of microplastic, followed by fragments, while pellets and films exhibit comparatively lower numbers.

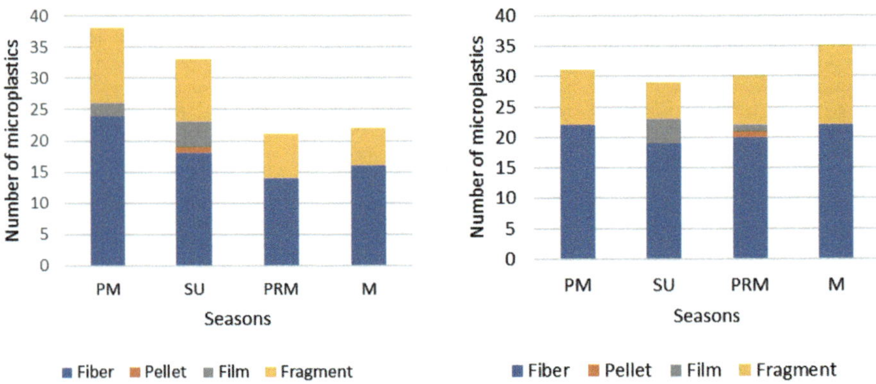

Fig. 8 Composition of fibre, pellet, film, and fragment in bottom sediment of site-1

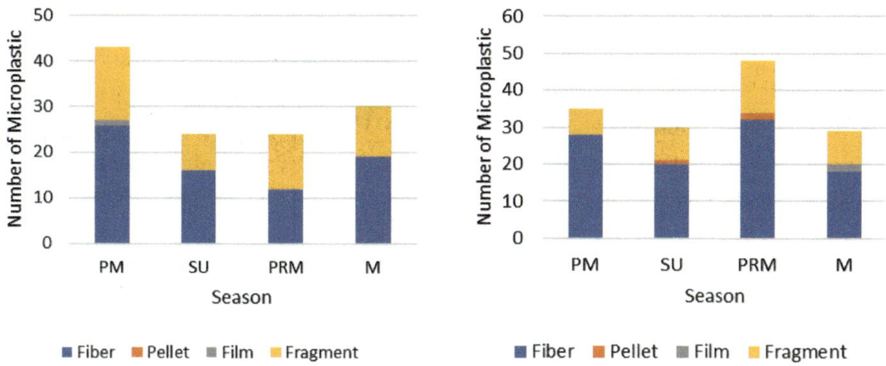

Fig. 9 Composition of fibre, pellet, film and fragment in bottom sediment of site-2

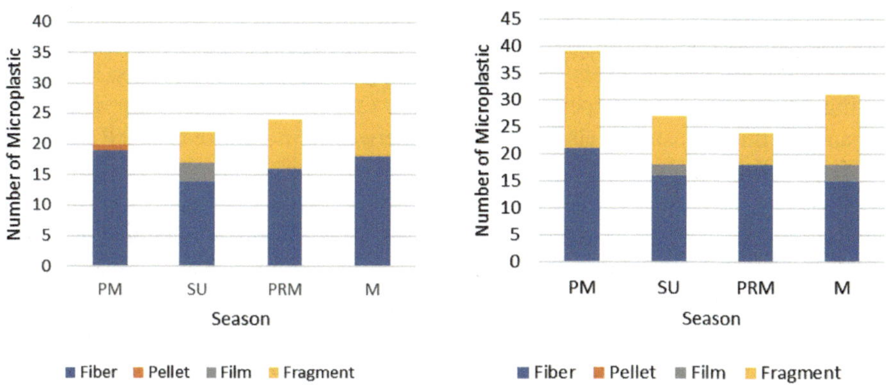

Fig. 10 Composition of fibre, pellet, films and fragment in bottom sediment at site-3

4.2 Polymer Composition

The extracted microplastics from the Puducherry coast bottom sediments showed six distinct polymers such as Polyacrylamide (Nylon), Polyester, PVC, PET-G, PVA, and PP, with Nylon accounting for 54%, Polypropylene (PP) for 7%, Polyester for 13%, Polyvinyl chloride (PVC) for 11%, Polyethylene terephthalate glycol (PET-G) for 10% and Polyvinyl alcohol (PVA) for 5% (Fig. 11). Raman spectroscopy was employed to identify the polymers (Fig. 12).

4.3 Statistical Analysis

The statistical analysis of variance performed to explore the relationship between seasons and microplastic levels in sediments during the study period (2021–2022)

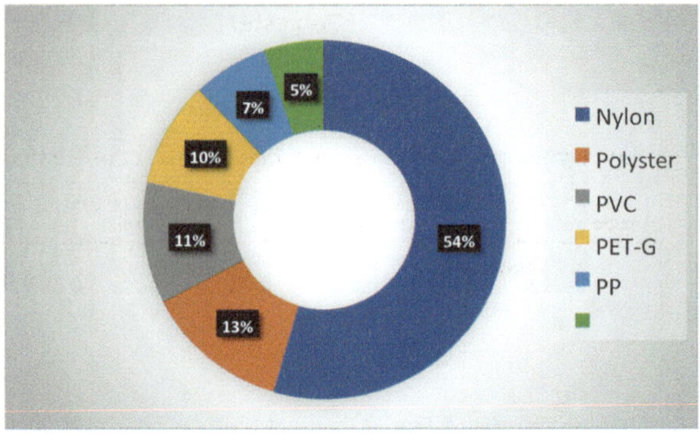

Fig. 11 Percentage of the polymer composition of microplastics in bottom sediment

Table 2, revealed significant variations among specific seasonal pairs. Specifically, the post-monsoon to summer, post-monsoon to pre-monsoon, and post- monsoon to monsoon comparisons exhibited statistically significant differences ($P < 0.05$), indicating that microplastic levels varied significantly between these seasons. However, during the subsequent study period (2022–2023) (Table 3), no statistically significant variance in microplastic levels was observed between different seasons ($P > 0.05$).

5 Discussion

The coastal region is the ultimate repository of all kinds of contaminants, arising out of human activities, and the accumulation of pollutants particularly, microplastics in the coastal belt is alarming and creates concern over the world. The disposal of plastic debris in riverbanks, deltas, coastlines, and in the ocean is rapidly increasing, making plastic pollution in the rivers and oceans is a growing environmental threat. Plastic pollution poses significant risks to ecosystems, human health, and aquatic life. Despite its impact, our understanding of microplastic pollution in marine environments remains limited, sparking curiosity among researchers to explore this area further. The microplastic abundance in the coastal bottom sediment samples at three sites for two years (2021–2023) were studied. In general, the highest microplastic abundance occurred during post-monsoon (S1- 9.5 ± 7.835 particle kg-1; S2- 10.75 ± 8.923 particle kg^{-1} and 9.75 ± 8.067 particle kg^{-1}), while the lowest was observed during pre-monsoon in S1 (5.25 ± 4.774 particle kg^{-1}) and summer in S2 and S3 (S2- 6 ± 5.456 particle kg^{-1} and S3- 5.5 ± 4.294 particle kg^{-1}). A similar study was carried out by Sundar et al. (2021) from the Kanyakumari coast and found microplastic abundance of 1.8 particles per gram in all sediment samples. The highest abundance was observed in the year 2022–2023 in S2 with an abundance of (12 ± 10.469 particle kg^{-1}) during pre-monsoon. Similarly, the lowest was observed at S3 during pre-monsoon with an abundance of (6 ± 6.044 particle kg^{-1}). The sudden variation in microplastic concentration might be due to variations in

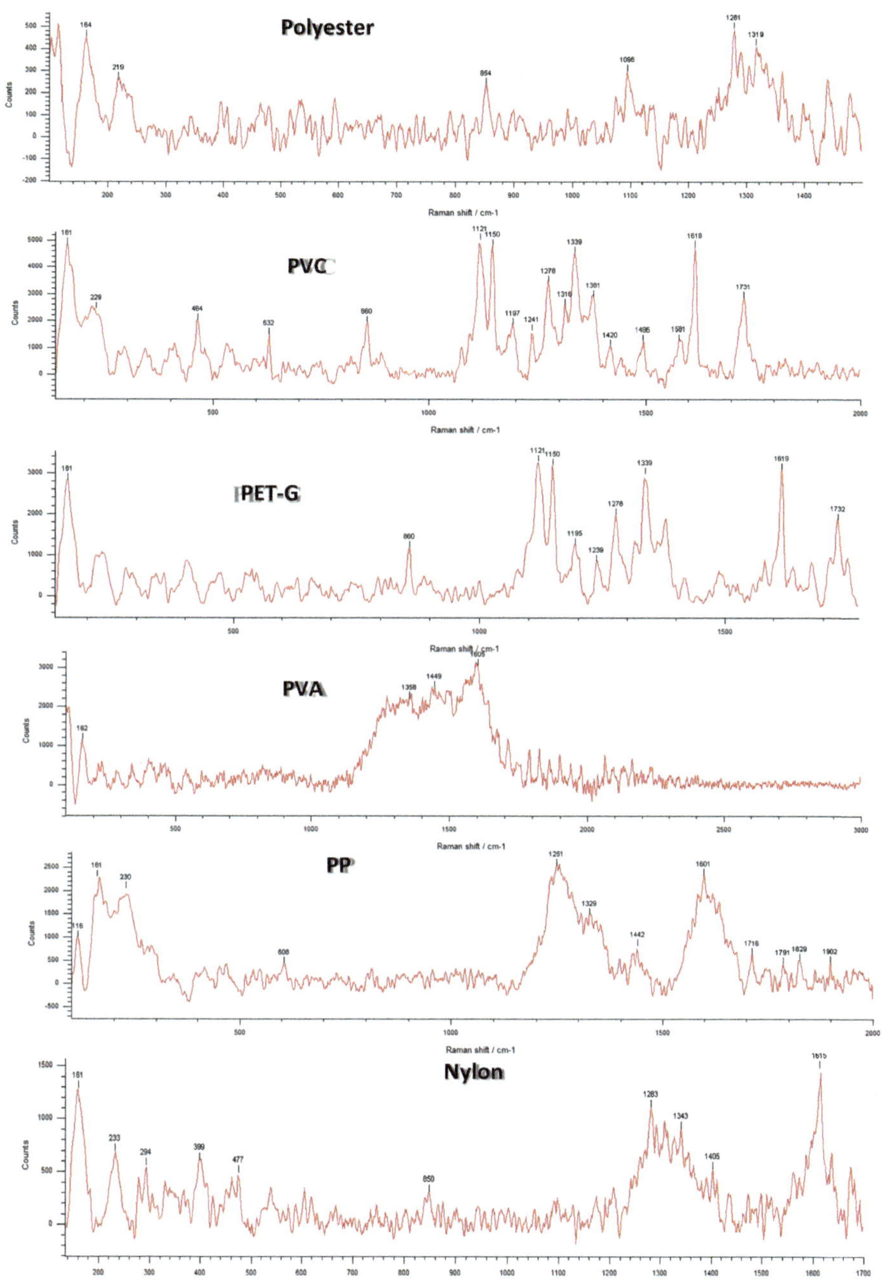

Fig. 12 Raman spectroscopy for Identification of polymer of the microplastics

Table 2 Statistical analysis of microplastics between seasons during 2021–2022

S. No	Seasons	*F* value	Degree of fraction	*P* value	Variance
1	PM/S	9.007	3.552	0.0464	$P < 0.05$
2	PM/PRM	38.09	2.711	0.01145	$P < 0.05$
3	PM/M	10.23	3.931	0.03377	$P < 0.05$
4	S/M	0.0538	3.793	0.8284	$P > 0.05$
5	S/PRM	0.8929	2.347	0.4315	$P > 0.05$
6	PRM/M	2.315	2.552	0.248	$P > 0.05$

Table 3 Analysis of variance (ANOVA) of microplastics between seasons during 2022–2023

S. No	Seasons	*F* value	Degree of fraction	*P* value	Variance
1	PM/S	6.564	2.571	0.0969	$P > 0.05$
2	PM/PRM	0.017	2.406	0.9052	$P > 0.05$
3	PM/M	1.316	3.741	0.3194	$P > 0.05$
4	S/M	2.314	2.914	0.2274	$P > 0.05$
5	S/PRM	0.538	2.06	0.5373	$P > 0.05$
6	PRM/M	0.098	2.238	0.7802	$P > 0.05$

anthropogenic activities or climate changes. Microplastic concentrations decreased exponentially with increasing depth in the top 5 m due the buoyant nature of the plastics and environmental factors (Reisser et al., 2020; Kooi et al., 2016). A recent study conducted around Puducherry examined the spatial distribution and characteristics of microplastics in Oussudu Lake. It found that microplastic levels were higher near the lake's center, with samples from deeper areas containing a larger concentration of microfibers and fragments, likely from fishing equipment and local runoff (Vanapalli et al., 2021). This study also provides insights into microplastic sources, attributing them to anthropogenic activities such as fishing and urban runoff, making it relevant for comparing patterns of microplastic distribution across coastal environments with diverse human impacts. Studying the microplastic's abundance in an ever-changing environment is still a challenge. The concentration and distribution of microplastics in the seabed of Pondicherry's coastal region are primarily shaped by anthropogenic activities and environmental factors, including ocean currents and wave dynamics. Studies suggest that human activities, such as urban runoff, waste disposal, and tourism, introduce a variety of microplastics (fibers, fragments, pellets, and films) into the marine environment. The coastal currents and wave actions further influence how these microplastics are distributed. Seasonal changes in monsoon-driven currents and tidal movements can lead to the redistribution of these particles across different depths and locations, affecting where they settle. During periods of stronger currents, lighter fragments and films are more likely to be suspended in the water column or transported to different areas, while denser pellets and fibers might settle closer to the coast, accumulating more heavily on the seabed. Seasonal variations also significantly affect the microplastic load, as stormier weather during monsoons may lead to higher inputs of terrestrial debris, altering the balance of microplastic types found across seasons. This seasonal discrepancy indicates that while anthropogenic sources contribute the majority of microplastics, the changing current and wave patterns redistribute these particles, creating a variable landscape of pollution on the seabed throughout the year.

Fibers and films, being lightweight and structurally flexible, are more susceptible to these forces, often remaining suspended in the water column or moving further offshore, while heavier particles like pellets tend to settle in areas of reduced hydrodynamic activity. Conversely, during the dry season, when currents and wave forces are lower, microplastics from local sources accumulate closer to shore as they are less likely to be dispersed by tidal or current actions. The difference in settling rates and transport tendencies of microplastic forms, coupled with seasonal shifts in oceanographic conditions, helps explain the significant seasonal variation in their concentrations. This understanding emphasizes the role of environmental factors in the spatial and temporal variability of microplastic pollution, underscoring the need for targeted mitigation strategies that account for seasonal oceanographic patterns. The study reported that the abundance of microplastics was found at an average count of 731 particles/kg dry weight of sediment in Chilika Lake and also reported that microplastic concentration was higher near the fishing harbor. Ranjani et al. (2022) reported similar observation on the coast of Chennai that the concentration of microplastics ranged from 84 to 498 items/kg, in the south-west monsoon whereas in the north-east monsoon, it was also observed in the range from 76 to 720 items/kg. In the south-west monsoon, the highest microplastic concentration was found in Adyar Beach and the lowest in Cooum, on the other hand, during the north-east monsoon, the highest was obtained in Ennore and the lowest in Cooum. However, the overall average values of microplastics obtained in this study area during south-west (247.4 items/kg) and north-east (302.7 items/kg) monsoons are lesser than the average microplastic values reported along the Indian coast (12.22–439 items/kg). The average microplastic abundance in the sediments were around or above 200 particles per kilogram in Deep Bay, 237 particles kg^{-1}; Tsing Yi, 198 particles kg^{-1}; Victoria Harbor, 263 particles kg^{-1} and Tolo Harbor, 279 particles kg^{-1} showed no significant variation in the sediment microplastics (Tsang et al., 2017) compared with the abundance of microplastics in different seas reported in the previous study, the level of available microplastic pollution in the coastal waters of Hong Kong is lower than the present study. Another study found that the mean abundance of microplastics in sediments ranged from 25 to 47,897 n/m^2 for beaches and 15–3320 n/kg for subtidal sediments. Despite the fact that there were few data points on microplastic mean abundance in subtidal zones, the results suggested a narrow distribution. Suspected hot spots, such as ports, have high concentrations (340–3320 n/kg) and relatively low abundances (15–145 n/kg) in coastal and continental shelf sediments (Noren, 2007; Claessens et al., 2011 and Shim et al., 2018).

References

Abidli, S., Toumi, H., Lahbib, Y., & Trigui El Menif, N. (2017). The first evaluation of microplastics in sediments from the complex lagoon-channel of Bizerte (Northern Tunisia). *Water, Air and Soil Pollution, 228*, 1–10.

Alomar, C., Sureda, A., Capó, X., Guijarro, B., Tejada, S., & Deudero, S. (2017). Microplastic ingestion by Mullus surmuletus Linnaeus, 1758 fish and its potential for causing oxidative stress. *Environmental Research, 159*, 135–142.

Andrady, A. L. (1994). Assessment of environmental biodegradation of synthetic polymers. *Journal of Macromolecular Science, Part c: Polymer Reviews, 34*(1), 25–76.

Andrady, A. L. (2011). Microplastics in the marine environment. *Marine Pollution Bulletin, 62*(8), 1596–1605.

Andrady, A. L. (2017). The plastic in microplastics: A review. *Marine Pollution Bulletin, 119*(1), 12–22.

Andrady, A. L., & Koongolla, B. (2022). Degradation and fragmentation of microplastics. *Plastics and the Ocean: Origin, Characterization, Fate, and Impacts*, 227–268.

Ashwini, S. K., & Varghese, G. K. (2020). Environmental forensic analysis of the microplastic pollution at "Nattika" Beach, Kerala Coast, India. *Environmental Forensics, 21*(1), 21–36.

Ballent, A., Corcoran, P. L., Madden, O., Helm, P. A., & Longstaffe, F. J. (2016). Sources and sinks of microplastics in Canadian Lake Ontario nearshore, tributary and beach sediments. *Marine Pollution Bulletin, 110*(1), 383–395.

Barnes, D. K. (2002). Invasions by marine life on plastic debris. *Nature, 416*(6883), 808–809.

Barnes, D. K., Galgani, F., Thompson, R. C., & Barlaz, M. (2009). Accumulation and fragmentation of plastic debris in global environments. *Philosophical Transactions of the Royal Society b: Biological Sciences, 364*(1526), 1985–1998.

Begum, A. N., Kumar, S. S., Krishnan, K. G., & Srinivasa, M. (2020). Distribution of microplastics in estuarine sediments along Tamil Nadu Coast, Bay of Bengal, India. *International Journal of Scientific Research & Engineering Trends, 6*(2), 814–821.

Bergmann, M., Gutow, L., & Klages, M. (2015). *Marine anthropogenic litter* (Vol. 447). Springer Nature.

Blumenröder, J., Sechet, P., Kakkonen, J. E., & Hartl, M. G. (2017). Microplastic contamination of intertidal sediments of Scapa Flow, Orkney: A first assessment. *Marine Pollution Bulletin, 124*(1), 112–120.

Claessens, M., De Meester, S., Van Landuyt, L., De Clerck, K., & Janssen, C. R. (2011). Occurrence and distribution of microplastics in marine sediments along the Belgian coast. *Marine Pollution Bulletin, 62*(10), 2199–2204.

Derraik, J. G. (2002). The pollution of the marine environment by plastic debris: A review. *Marine Pollution Bulletin, 44*(9), 842–852.

Di, M., & Wang, J. (2018). Microplastics in surface waters and sediments of the Three Gorges Reservoir, China. *Science of the Total Environment, 616*, 1620–1627.

Dilkes-Hoffman, L. S., Pratt, S., Lant, P. A., & Laycock, B. (2019). *The role of biodegradable plastic in solving plastic solid waste accumulation* (pp. 469–505). William Andrew Publishing.

Dimassi, S. N., Hahladakis, J. N., Yahia, M. N. D., Ahmad, M. I., Sayadi, S., & Al-Ghouti, M. A. (2022). Degradation-fragmentation of marine plastic waste and their environmental implications: A critical review. *Arabian Journal of Chemistry, 15*(11), Article 104262.

Dowarah, K., & Devipriya, S. P. (2019). Microplastic prevalence in the beaches of Puducherry, India and its correlation with fishing and tourism/recreational activities. *Marine Pollution Bulletin, 148*, 123–133.

Europe Plastic. (2019). *Plastics—The facts: An analysis of European plastics production, demand and waste data.* Plastics Europe, Brussels. https://www.plasticseurope.org/download_file/force/2367/181

Geyer, R., Jambeck, J. R., & Law, K. L. (2017). Production, use, and fate of all plastics ever made. *Science Advances, 3*(7), Article e1700782.

Goswami, P., Vinithkumar, N. V., & Dharani, G. (2020). First evidence of microplastics bioaccumulation by marine organisms in the Port Blair Bay. *Andaman Islands. Marine Pollution Bulletin, 155*, Article 111163.

Horton, A. A., Walton, A., Spurgeon, D. J., Lahive, E., & Svendsen, C. (2017). Microplastics in freshwater and terrestrial environments: Evaluating the current understanding to identify the knowledge gaps and future research priorities. *Science of the Total Environment, 586*, 127–141.

Jeyasanta, K. I., Patterson, J., Grimsditch, G., & Edward, J. P. (2020a). Occurrence and characteristics of microplastics in the coral reef, sea grass and near shore habitats of Rameswaram island, India. *Marine Pollution Bulletin, 160*, Article 111674.

Jeyasanta, K. I., Sathish, N., Patterson, J., & Edward, J. P. (2020b). Macro-, meso-and microplastic debris in the beaches of Tuticorin district, Southeast coast of India. *Marine Pollution Bulletin, 154*, Article 111055.

Karami, A., Golieskardi, A., Choo, C. K., Romano, N., Ho, Y. B., & Salamatinia, B. (2017). A high-performance protocol for extraction of microplastics in fish. *Science of the Total Environment, 578*, 485–494.

Karthik, R., Robin, R. S., Purvaja, R., Ganguly, D., Anandavelu, I., Raghuraman, R., Hariharan, G., Ramakrishna, A., & Ramesh, R. (2018). Microplastics along the beaches of Southeast Coast of India. *Science of the Total Environment, 645*, 1388–1399.

Klemchuk, P. P. (1990). Degradable plastics: A critical review. *Polymer Degradation and Stability, 27*(2), 183–202.

Kooi, M., Reisser, J., Slat, B., Ferrari, F. F., Schmid, M. S., Cunsolo, S., Brambini, R., Noble, K., Sirks, L. A., Linders, T. E., & Schoeneich-Argent, R. I. (2016). The effect of particle properties on the depth profile of buoyant plastics in the ocean. *Scientific Reports, 6*(1), 33882.

Krishnakumar, S., Srinivasalu, S., Saravanan, P., Vidyasakar, A., & Magesh, N. S. (2018). A preliminary study on coastal debris in Nallathanni island, Gulf of Mannar Biosphere Reserve, Southeast coast of India. *Marine Pollution Bulletin, 131*, 547–551.

Kumar, A., Sivakumar, R., Reddy, Y. S. R., Raja, B., Nishanth, T., & Revanth, V. (2016). Preliminary study on marine debris pollution along Marina beach, Chennai, India. *Regional Studies in Marine Science, 5*, 35–40.

Laglbauer, B. J., Franco-Santos, R. M., Andreu-Cazenave, M., Brunelli, L., Papadatou, M., Palatinus, A., Grego, M., & Deprez, T. (2014). Macrodebris and microplastics from beaches in Slovenia. *Marine Pollution Bulletin, 89*(1–2), 356–366.

Maharana, D., Saha, M., Dar, J. Y., Rathore, C., Sreepada, R. A., Xu, X. R., Koongolla, J. B., & Li, H. X. (2020). Assessment of micro and macroplastics along the west coast of India: Abundance, distribution, polymer type and toxicity. *Chemosphere, 246*, Article 125708.

Marghany, M., & Mansor, S. (2016). Introduction to coastal and marine environment concepts and significances. In M. Marghany (Ed.), *Applied studies of coastal and marine environments* (p. 438). IntechOpen.

Meijer, L. J., Van Emmerik, T., Van Der Ent, R., Schmidt, C., & Lebreton, L. (2021). More than 1000 rivers account for 80% of global riverine plastic emissions into the ocean. *Science Advances, 7*(18), 5803.

Mugilarasan, M., Venkatachalapathy, R., Sharmila, N., & Gurumoorthi, K. (2017). Occurrence of microplastic resin pellets from Chennai and Tinnakkara island: Towards the establishment of background level for plastic pollution. *Indian Journal of Geo Marine Sciences, 45*(6), 1210–1212.

Naji, A., Esmaili, Z., & Khan, F. R. (2017a). Plastic debris and microplastics along the beaches of the Strait of Hormuz. *Persian Gulf. Marine Pollution Bulletin, 114*(2), 1057–1062.

Naji, A., Esmaili, Z., Mason, S. A., & Dick Vethaak, A. (2017b). The occurrence of microplastic contamination in littoral sediments of the Persian Gulf. *Iran. Environmental Science and Pollution Research, 24*, 20459–20468.

Nigam, R. A. I. I. V. (1982). Plastic pellets on the Caranzalem beach sands, Goa. *India. Mahasagar, 15*(2), 125–127.

Noren, F. (2007). Small plastic particles in coastal Swedish waters. *Kimo Sweden, 11*, 1–11.

Patchaiyappan, A., Ahmed, S. Z., Dowarah, K., Jayakumar, S., & Devipriya, S. P. (2020). Occurrence, distribution and composition of microplastics in the sediments of South Andaman beaches. *Marine Pollution Bulletin, 156*, Article 111227.

Patterson, J., Jeyasanta, K. I., Sathish, N., Edward, J. P., & Booth, A. M. (2020). Microplastic and heavy metal distributions in an Indian coral reef ecosystem. *Science of the Total Environment, 744*, Article 140706.

Pavithran, V. A. (2021). Study on microplastic pollution in the coastal seawaters of selected regions along the northern coast of Kerala, southwest coast of India. *Journal of Sea Research, 173*, Article 102060.

Peng, G., Xu, P., Zhu, B., Bai, M., & Li, D. (2018). Microplastics in freshwater river sediments in Shanghai, China: A case study of risk assessment in mega-cities. *Environmental Pollution, 234*, 448–456.

Ramesh, S., & Nagalakshmi, R. (2022). Micro plastic contaminant in marine environment in Chennai coast. In *AIP conference proceedings* (Vol. 2516, Issue 1), *AIP conference proceedings* (Vol. 2516, p. 060001). https://doi.org/10.1063/5.0108892

Ranjani, M., Veerasingam, S., Venkatachalapathy, R., Jinoj, T. P. S., Guganathan, L., Mugilarasan, M., & Vethamony, P. (2022). Seasonal variation, polymer hazard risk and controlling factors of microplastics in beach sediments along the southeast Coast of India. *Environmental Pollution, 305*, Article 119315.

Reddy, M. S., Basha, S., Adimurthy, S., & Ramachandraiah, G. (2006). Description of the small plastics fragments in marine sediments along the Alang-Sosiya ship-breaking yard, India. *Estuarine, Coastal and Shelf Science, 68*(3–4), 656–660.

Reddy, M. S., Basha, S., Kumar, V. S., Joshi, H. V., & Ramachandraiah, G. (2004). Distribution, enrichment and accumulation of heavy metals in coastal sediments of Alang–Sosiya ship scrapping yard. *India. Marine Pollution Bulletin, 48*(11–12), 1055–1059.

Reed, S., Clark, M., Thompson, R., & Hughes, K. A. (2018). Microplastics in marine sediments near Rothera research station, Antarctica. *Marine Pollution Bulletin, 133*, 460–463.

Reisser, J., Slat, B., Noble, K., Du Plessis, K., Epp, M., Proietti, M., De Sonneville, J., Becker, Robin, R. S., Karthik, R., Purvaja, R., Ganguly, D., Anandavelu, I., Mugilarasan, M., & Ramesh, R. (2020). Holistic assessment of microplastics in various coastal environmental matrices, southwest coast of India. *Science of the Total Environment, 703*, 134947.

Rodrigues, M. O., Abrantes, N., Gonçalves, F. J. M., Nogueira, H., Marques, J. C., & Gonçalves, A. M. M. (2018). Spatial and temporal distribution of microplastics in water and sediments of a freshwater system (Antuã River, Portugal). *Science of the Total Environment, 633*, 1549–1559.

Sathish, M. N., Jeyasanta, K. I., & Patterson, J. (2020). Monitoring of microplastics in the clam Donax cuneatus and its habitat in Tuticorin coast of Gulf of Mannar (GoM). *India. Environmental Pollution, 266*, Article 115219.

Sathish, N., Jeyasanta, K. I., & Patterson, J. (2019). Abundance, characteristics and surface degradation features of microplastics in beach sediments of five coastal areas in Tamil Nadu, India. *Marine Pollution Bulletin, 142*, 112–118.

Shen, M., Song, B., Zeng, G., Zhang, Y., Huang, W., Wen, X., & Tang, W. (2020). Are biodegradable plastics a promising solution to solve the global plastic pollution? *Environmental Pollution, 263*, Article 114469.

Shim, W. J., Hong, S. H., & Eo, S. (2018). Marine microplastics: Abundance, distribution, and composition. In E. Y. Zeng (Ed.) *Microplastic contamination in aquatic environments* (pp. 1–26). Elsevier.

Singare, P. U. (2012a). Quantification study of non-biodegradable solid waste materials accumulated in the mangroves of Mahim Creek, Mumbai. *Marine Science, 2*(1), 1–5.

Singare, P. U. (2012b). Study of some major non-biodegradable solid wastes along Thane Creek of Mumbai. *World Environment, 2*(3), 24–30.

Singare, P. U., Mishra, R. M., & Trivedi, M. P. (2012a). Sediment contamination due to toxic heavy metals in Mithi River of Mumbai. *Advances in Analytical Chemistry, 2*(3), 14–24.

Singare, P. U., Trivedi, M. P., & Mishra, R. M. (2012b). Sediment heavy metal contaminants in Vasai Creek of Mumbai: Pollution impacts. *American Journal of Chemistry, 2*(3), 171–180.

Singh, S., Chakma, S., Alawa, B., Kalyanasundaram, M., & Diwan, V. (2023). Identification, characterization, and implications of microplastics in soil–A case study of Bhopal, central India. *Journal of Hazardous Materials Advances, 9*, Article 100225.

Su, L., Xue, Y., Li, L., Yang, D., Kolandhasamy, P., Li, D., & Shi, H. (2016). Microplastics in Taihu lake, China. *Environmental Pollution, 216*, 711–719.

Sundar, S., Chokkalingam, L., Roy, P. D., & Usha, T. (2021). Estimation of microplastics in sediments at the southernmost coast of India (Kanyakumari). *Environmental Science and Pollution Research, 28*, 18495–18500.

Thompson, R. C., Olsen, Y., Mitchell, R. P., Davis, A., Rowland, S. J., John, A. W. and Russell, Tiwari, M., Rathod, T. D., Ajmal, P. Y., Bhangare, R. C. & Sahu, S. K. (2019). Distribution and characterization of microplastics in beach sand from three different Indian coastal environments. *Marine Pollution Bulletin, 140*, 262–273.

Tsang, Y. Y., Mak, C. W., Liebich, C., Lam, S. W., Sze, E. T., & Chan, K. M. (2017). Microplastic pollution in the marine waters and sediments of Hong Kong. *Marine Pollution Bulletin, 115*(1–2), 20–28.

Van Cauwenberghe, L., Devriese, L., Galgani, F., Robbens, J., & Janssen, C. R. (2015). Microplastics in sediments: A review of techniques, occurrence and effects. *Marine Environmental Research, 111*, 5–17.

Vanapalli, K. R., Dubey, B. K., Sarmah, A. K., & Bhattacharya, J. (2021). Assessment of microplastic pollution in the aquatic ecosystems–An indian perspective. *Case Studies in Chemical and Environmental Engineering, 3*, Article 100071.

Veerasingam, S., Saha, M., Suneel, V., Vethamony, P., Rodrigues, A. C., Bhattacharyya, S., & Naik, B. G. (2016). Characteristics, seasonal distribution and surface degradation features of microplastic pellets along the Goa coast, India. *Chemosphere, 159*, 496–505.

Velmurugan, P. M., Vijayaprabhakaran, K., & Devika, P. T. (2023). Baseline study on identification, characterization, distribution and abundance of microplastics in surface water from Ennore to Kovalam along the east coast of India. *Physics and Chemistry of the Earth, Parts A/B/C, 130*, Article 103391.

Vidyasakar, A., Krishnakumar, S., Kasilingam, K., Neelavannan, K., Bharathi, V. A., Godson, P. S., Prabha, K., & Magesh, N. S. (2020). Characterization and distribution of microplastics and plastic debris along Silver Beach. *Southern India. Marine Pollution Bulletin, 158*, Article 111421.

Vidyasakar, A., Neelavannan, K., Krishnakumar, S., Prabaharan, G., Priyanka, T. S. A., Magesh, N. S., Godson, P. S., & Srinivasalu, S. (2018). Macrodebris and microplastic distribution in the beaches of Rameswaram Coral Island, Gulf of Mannar, Southeast coast of India: A first report. *Marine Pollution Bulletin, 137*, 610–616.

Zhao, S., Ward, J. E., Danley, M., & Mincer, T. J. (2018). Field-based evidence for microplastic in marine aggregates and mussels: Implications for trophic transfer. *Environmental Science and Technology, 52*(19), 11038–11048.

Occurrence of Microplastics in Coastal Waters of Kerala

Harsha Umesh and R. Mohanraj

1 Introduction

The quantum of plastic debris generated by human activity and its accumulation in the environment is one of the biggest threats to ecological sustainability and human health. Plastic use has continued to grow across the globe for the last 7 decades, and today, it's pivotal to modern life, mainly due to its low manufacturing cost, durability, non-degradable nature, and thermal and electrical insulation properties. Therefore, plastic usage has become inevitable in almost all sectors, including domestic appliances, clothing, packaging, storage, transportation, medical equipment, and construction (GESAMP, 2015). Plastic usage in the initial stages had some positive benefits, like the replacement of glass and metal containers, which in turn minimized the ill effects of pollution due to metal extraction and processing; over the period, the accumulation of plastic debris in the environment has overshadowed its positive effects. According to Organisation for Economic Co-operation and Development (OCED) the annual plastic production in 2020 is about 435 million tons and more than 80% of the plastics produced are thermoplastics. Thermoplastics are the products of polymerization of monomers to high-molecular-weight chains known as a thermoplastic polymers (Yang et al., 2011). It includes polyethylene terephthalate (PET), high, low, and linear low-density polyethylene (HDPE, LDPE, and LLDPE), polyvinyl chloride (PVC), polypropylene (PP), and polystyrene (PS). These polymers, upon physical treatment such as melting, extrusion, and pelletization are converted to desired products. Besides physical treatment, chemical additives such as antioxidants, plasticizers, clarifiers, bisphenol A-based polycarbonate, copolymer, pigments, and flame retardants are added to improve their performance and to facilitate their manufacture (Hermabessiere et al., 2017; Wei et al., 2014; Yang et al., 2011). Thus, the final

H. Umesh · R. Mohanraj (✉)
Department of Environmental Science and Management, Bharathidasan University, Tiruchirappalli, India
e-mail: mohan@bdu.ac.in

P. Thambidurai and A. K. Dikshit (eds.), *Coastal Environments of India*,
https://doi.org/10.1007/978-3-031-97341-3_14

plastic products make up a complex chemical composition. Global Plastic Outlook has reported that 367 million metric tons of plastic waste generated all over the world in 2020, a number which has more than doubled since 2000. Of this, only 9% was recycled, almost 72% was landfilled or disposed of in the natural environment, and 19% was incinerated. Unscientific and improper management of plastic poses serious hazards to the natural ecosystem and humans mainly due to its longevity in degradation. Plastic waste can persist for hundreds or even thousands of years depending on the properties of the plastics as well as the surrounding environmental conditions (Geyer et al., 2017; Jambeck et al., 2015; Plastics–the Facts, 2019). The environmental half-life of plastic debris is directly related to its polymer type and surrounding environmental conditions and it ranges from days to centuries (Ward et al., 2019). Over a period, certain physical, chemical, and biological processes in the surrounding environment shall slowly break down plastic wastes into numerous smaller fragments called microplastics (Plastics–the Facts, 2019). Photodegradation of plastics is the most important process which involves free radical-mediated reactions initiated by solar irradiation (Liu et al., 2019; Yang et al., 2019; Zhang et al., 2021). Extensive deterioration of plastics makes them brittle and powdery particles. Apart from the natural microplastic formation, there are certain primary sources for microplastics like the use of cosmetics, clothing, wastewater treatment, and biomedical applications. Microplastics (MPs) can be defined as plastic materials of micro size often between the range of 1 µm–5 mm dimension (Browne et al., 2010; Carson et al., 2011; Qiu et al., 2016; Van Cauwenberghe et al., 2013). In the aquatic environment, secondary microplastics are more abundant than primary ones.

Nowadays, microplastics are ubiquitous in nature and bioavailable for injection by marine organisms, soil organisms, and plants growing on microplastic-contaminated soils. The presence of microplastics is reported in marine organisms such as seabirds, fish, sand hoppers, sea turtles, crustaceans, and mussels, many studies have detected microplastics in invertebrates such as collembolan, oligochaeta (e.g. earthworms) and isopods (Chae & An, 2018; Rillig et al., 2017). The MPs have penetrated agroecosystems through wastewater irrigation and applications of biosolids; studies across the world quantified the presence of MP polymers like PS, PE, and PP in maize, rice and (Bosker et al., 2019; Urbina et al., 2020; Wang et al., 2020). Among the various compartments in the environment, the marine environment has captured scientific concern due to the occurrence of large quantities of plastic debris (Baker, 2011). In the past few decades, between 1950 and 2020, about 65 million tons of plastics have accumulated in the oceans, and most of it occurs on shorelines. Estimates show that each year 1.15–2.41 million tons of plastics are released into the oceans through river-fed estuaries (Lebreton et al., 2017; Schmidt et al., 2017). More particularly, coastal water bodies (e.g., bays and estuaries) and shorelines contain comparatively higher amounts of plastic debris due to multiple sources of pollution. According to OECD (2023), the annual discharge of macro plastics will tend to increase by 50% between 2020 and 2040 to 30 Mt, and correspondingly microplastic leakage will also continue to grow. The plastic accumulation in aquatic environments previously estimated at about 152 Mt in 2020, is expected to reach 300 Mt by 2040. The microplastics occur in various shapes, sizes, and densities. Low-density polymers used in single-use containers, such as polyethylene and polypropylene normally float in the shore waters, while denser polymers such as polyethylene terephthalate, polycarbonate, and polyvinyl chloride are likely to sink (Browne et al., 2010; Kane & Clare, 2019; Sharma & Chatterjee, 2017; Woodall et al., 2014). Recent reports also ascertain that MPs act as a carrier for other pollutants and cause synergistic effects in the food chain. Such complex organic pollutants may include polychlorinated biphenyls, plasticizers and heavy metals, which eventually can cause adverse mutagenic, teratogenic and carcinogenic effects in both animals and humans. Globally, about 15.9% of MP debris is contributed by Southeast Asia (Ajith et al., 2020), and beaches are the intermediate reservoirs and contribute to the MPs in coastal waters (Fok et al., 2017). In India, the most common anthropogenic sources of plastic debris in the shorelines include fishing, domestic sewage, tourist activities, recreation, religious, port, and industrial activities (Anburaj et al., 2018; Karthik et al., 2018; Robin et al., 2020; Tiwari et al., 2019; Vidyasakar et al., 2020). In this context, the present study attempts to experimentally analyze the microplastics in the coastal water samples collected from the southwestern coast /Malabar coast of Kerala.

2 Methodology

2.1 Study Area

Beach water samples were collected at 40 sites along the Arabian Sea coastline covering five districts of Kerala, including Kollam, Alappuzha, Ernakulam, Thrissur, Malappuram, and Kozhikode (Fig. 1) for the analytical experiments to study the microplastic contamination. Although the major sources of plastic debris in these districts are common, the type of plastics, the intensity of the source, and the anthropogenic activities may significantly vary. Upon observations, it was noted that fishing, domestic sewage, tourism, religious rituals, port and industrial activities were the most common sources.

2.2 Experimental Methods

Water samples (unfiltered) were collected from 40 locations in a 2-L pre-cleaned polyethylene container, and a standard procedure was followed for sampling the seawater. The basic water quality parameters, including pH, Dissolved Oxygen, Temperature, Electrical Conductivity, and Salinity, were recorded at the sampling site using respective electrodes and sensors. Microplastic analysis has been carried out using the most common wet hydrogen peroxidation technique. An amount of 20 ml of aqueous 0.05 M Fe (II) solution was added to the sample, followed by 20 ml

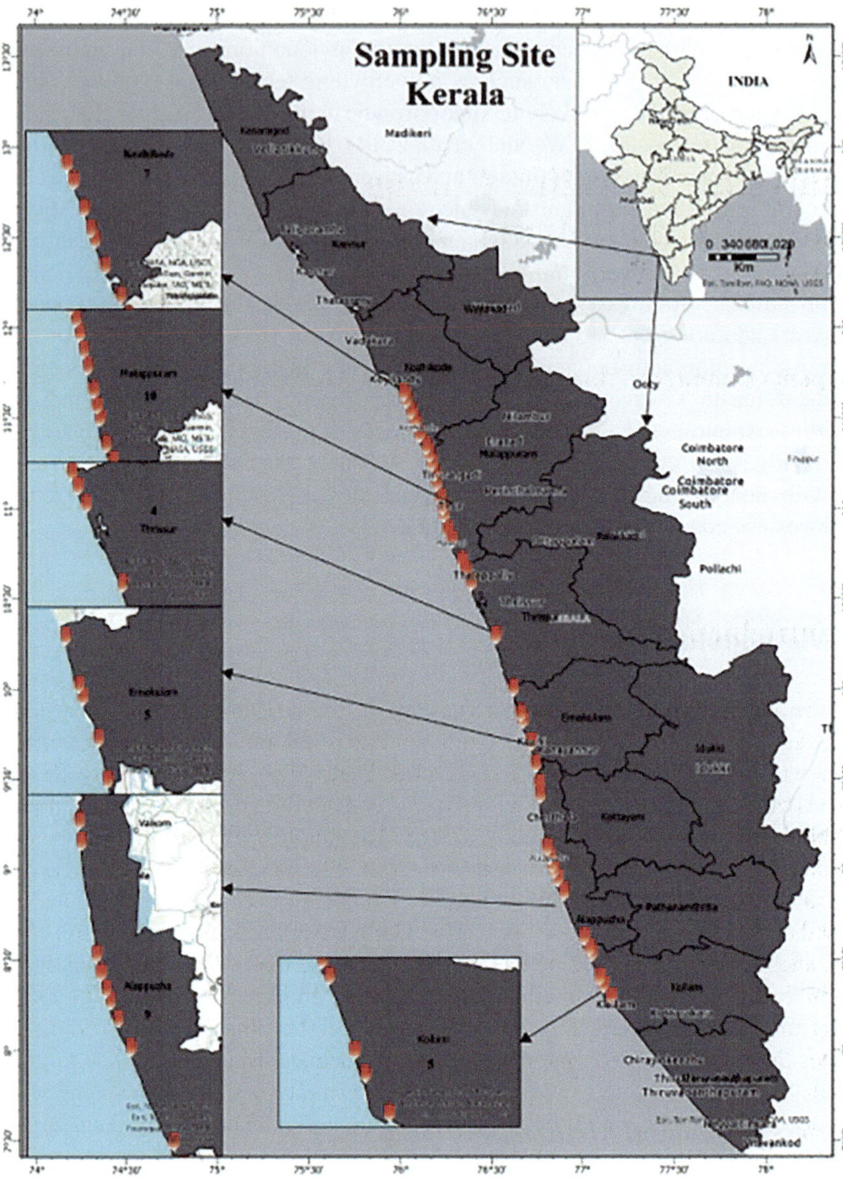

Fig. 1 Sampling sites in the study area—Kerala coastal region

of 30% hydrogen peroxide. It was left undisturbed for 5 min and then heated to 75 °C on a hotplate until gas bubbles appeared. Further, 20 ml of 30% hydrogen peroxide was added if the natural organic material was evident. This procedure was repeated until no more biological debris was visible. The sample was filtered using a vacuum filtration device, with glass microfiber filter paper. Subsequently, the sample was placed into the Buchner funnel, and the vacuum was switched on. The microplastics in the filter sheets were counted under a microscope at a magnification of 40X (Marine Debris Program, 2015).

2.3 FTIR (Fourier-Transform Infrared Spectroscopy) Analysis

Fourier-transform infrared (FTIR) spectroscopy is widely used to characterize microplastics (Ojeda et al., 2009). It produces individual band patterns by specific infrared spectra for different plastic types (Hidalgo-Ruz et al., 2012). In a FTIR analysis the samples are subjected to irradiation with infrared light (wave number range 400–4000 cm^{-1} for Mid-IR) enabling the identification of fundamental vibrational modes (Hidalgo-Ruz et al., 2012). The molecular compounds present in the sample absorb IR radiation which is measured in transmission or reflection mode. Fourier-transform infrared (FTIR) spectroscopy is extensively used for micro plastic characterization; it generates unique spectral finger prints for distinct polymer types (Ojeda et al., 2009). FTIR spectroscopy is applicable to solid, liquid and gaseous samples, referring its potential in environmental assessment (Veerasingam et al., 2020). In the present study, water samples are collected using sterile containers to avoid contamination. Samples are filtered through membranes with a pore size of 0.2–5 μm. Filtered micro plastic residues are dried and subjected for FTIR analysis.

3 Results and Discussion

The recorded concentrations of the MPs and the physico-chemical parameters of all forty samples have been summarized in Table 1. The pH values of all marine water samples were observed to be in the range of 7.8–8.3, and the values are within the permissible limits prescribed by the Central Pollution Control Board, India (CPCB, 2007). The temperature of water samples was observed to be in the range of 25.5–36 °C. Among these forty samples, the highest temperature was observed in the following sites: S9, S10, S11, S12, S13, and S14 (Vadi beach at Kollam, Sakthikulangara harbour, Puthenthura, Alappad coastal, Parayakadavu coast and Azheekal respectively). The dissolved oxygen was in the range of 0.3–1.2 mg/L, and among all samples, the lowest amount was observed in sites S1 and S4 (Puthuvype Beach and Cherai Beach, respectively). The total suspended solid content in all marine water samples was observed to be in the range of 230–13,642 mg/L. The highest concentration was observed in site S27 (Koottayi), and the lowest concentration was observed in site S6 (Maruvakad). The total organic carbon content in the samples ranged from 0.56 to 1.9 mg/L. The concentrations of other water quality parameters are shown in Table 1. The occurrence of microplastics in an aquatic system is likely to influence the water quality in terms of physico-chemical characteristics. More notably, the MPs can elevate the levels of suspended solids and, in turn, increase the BOD levels (da Costa et al., 2023). This could be potentially attributed to the adherence of microorganisms to the microplastics, leading to the enhancement of organic material and eventually depleting the dissolved oxygen in the water. The hydrophobic nature of microplastics and their long half-life generally favors biofilm formation of microbes including bacteria, fungi, algae and protozoa (Di Pippo et al., 2020; Guan et al., 2020; Michels et al., 2018; Verhagen et al., 2011) The macronutrients present such as nitrogen and phosphorus are directly correlated with biofilm production, while salinity has a negative relationship with the growth rate of biofilms (Yang et al., 2019). The availability of dissolved oxygen is positively correlated with the abundance of autotrophs in biofilms (Artham et al., 2009).

3.1 *Quantification of Microplastics*

Microplastics were detected in all samples. A few microscopic images of the microplastics observed in the samples are shown in Fig. 2. Fibers were the most abundant particle type, comprising 90% of all particles, with a very low proportion of fragments and pellets. The mean number of microplastic particles recorded in the marine water samples of each district are as follows; Kollam-69, Alappuzha-92, Ernakulam, 91, Thrissur 39, Malappuram 123 and Kozhikode 131 with the abundance of microplastic pollution across the coastal area in the following decreasing order of Kozhikode > Malappuram > Alappuzha > Ernakulam > Kollam > Thrissur (Fig. 3). The increased pollution in the Kozhikode coastal area could be due to anthropogenic factors. The high levels of tourism activities persistent in the region could have contributed to the increased microplastic pollution. In Malappuram, the fishing activities and the debris from fishing nets could be attributed to the higher count of microplastics. Whereas comparatively lower fishing and commercial activities in Thrissur could have resulted in lower microplastic counts, as seen in Fig. 4. The higher percentage of fibers denotes the disintegrated portion of the primary microplastics. The highest number of microplastics was observed at site number 37 in Calicut and the lowest number was observed at site number 8 at Thaikal beach, as seen in Fig. 4.

3.2 *Micro-plastic Distribution Based on Color*

Microplastics were also characterized based on the colours. The majority of the microplastics obtained were blue in colour (39%), followed by black (37%), red

Table 1 Physico-chemical parameters of water samples

Site No.	pH	Temp. (°C)	DO (mg/l)	EC (mS/cm)	Total solids (mg/l)	TSS (mg/l)	TDS (mg/l)	Salinity (%)	Alkalinity (mg/l)
1	8.1	25.5	0.3	28	37,406	1312	36,094	26	170
2	8.1	25.5	0.8	27	35,706	864	34,842	29	174
3	8.1	25.5	1.2	28	41,424	4962	36,462	26	175
4	8.1	27.5	0.3	29	39,966	6116	33,850	26	175
5	8.3	27.5	1	29	36,996	438	36,558	26	180
6	8.2	27.5	0.7	28	36,050	230	35,820	28	156
7	8.3	29.5	0.4	29	44,868	8794	36,074	31	100
8	8.3	29.5	0.4	29	34,614	2972	31,642	25	105
9	8.1	36	0.3	29	38,562	2314	36,248	26	150
10	7.8	36	0.5	28	43,652	6570	37,082	28	154
11	8.2	36	0.4	29	34,762	250	34,512	30	150
12	8.2	36	0.7	29	45,954	8582	37,372	28	164
13	8.2	36	1	29	49,336	11,580	37,756	29	180
14	8.3	36	1.2	28	42,318	4136	38,182	28	175
15	8.1	27	0.9	26	35,996	3172	32,824	24	175
16	8.1	27	0.8	30	46,676	8436	38,240	28	170
17	8.2	27	0.9	29	34,136	302	33,834	28	152
18	8.3	27.5	1	29	38,384	3686	34,698	28	168
19	8.2	27.5	1.1	29	46,068	9854	36,484	27	134
20	8.3	27.5	1.2	29	40,712	4956	35,756	28	150
21	8.2	32.9	1.1	29	39,350	2516	36,834	30	200
22	8.2	32.9	1.2	29	38,376	540	37,836	28	234
23	8.2	29.6	1.1	30	47,398	12,876	34,522	28	234
24	8.2	28.3	1.2	29	38,240	2052	36,188	27	233
25	8.1	28.3	1.1	30	43,304	6560	36,744	27	200
26	8.2	28.3	1.1	29	36,934	3098	33,836	33	230
27	8.2	34.4	0.9	30	50,066	13,642	36,424	30	232
28	8.3	34.4	1	29	39,904	3902	36,002	26	200
29	8.1	34.4	1.1	30	42,264	7822	34,442	28	200
30	8.3	35.2	1.1	30	39,924	3896	36,028	27	186
31	8.2	35.2	1.1	28	38,208	1022	37,186	30	174
32	8.1	33	0.9	29	40,790	5068	35,722	29	198
33	8.2	33	1	29	35,898	3042	32,856	29	168
34	8.1	33	1.1	29	46,450	9590	36,860	32	250
35	8.1	33	1	28	35,448	850	34,598	30	245

(continued)

Table 1 (continued)

Site No.	pH	Temp. (°C)	DO (mg/l)	EC (mS/cm)	Total solids (mg/l)	TSS (mg/l)	TDS (mg/l)	Salinity (%)	Alkalinity (mg/l)
36	8.2	27	1	30	36,348	1298	35,050	29	228
37	8.3	27	1.1	30	37,642	1612	36,030	29	244
38	8.2	27	0.9	29	40,718	4250	36,468	28	250
39	8.1	27	1	28	46,638	9590	37,048	27	234
40	8.2	27	1.1	28	35,618	1830	33,788	29	250

Site No.	TOC (mg/l)	Ca (mg/l)	Mg (mg/l)	Total hardness (mg/l)	Acidity	Cl^- (mg/l)	Na (mg/l)	K (mg/l)
1	0.96	1209	11,860	13,069	29.6	15,540	**65**	9
2	1.23	1530	10,927	12,457	22.2	14,342	68	9
3	1.08	1322	11,787	13,109	14.8	13,562	59	8
4	0.63	1360	11,840	13,200	22.2	18,340	67	9
5	0.56	964	13,123	14,087	37	19,143	62	8
6	1.65	1389	14,191	15,580	29.6	18,210	54	8
7	0.9	1740	11,260	13,000	37	17,100	60	8
8	0.6	1168	11,278	12,446	29.6	17,441.4	63	9
9	0.78	1290	12,110	13,400	22.2	18,880	54	10
10	1.5	1009	13,551	14,560	29.6	15,321	56	8
11	1.09	1456	14,044	15,500	37	16,442	62	9
12	0.99	986	14,295	15,281	14.8	16,661	43	7
13	0.87	1568	11,730	13,298	14.8	15,089	53	8
14	1.11	1010	11,150	12,160	22.2	17,366	47	8
15	1.2	976	12,558	13,534	29.6	14,340.3	66	9
16	1.65	1097	12,054	13,151	29.6	14,747.2	40	8
17	1.7	1135	12,545	13,680	14.8	13,987	68	10
18	1.03	990	12,330	13,320	22.2	15,461.3	62	10
19	0.94	976	12,914	13,890	37	14,437	65	9
20	1.05	1874	13,927	13,890	29.6	15,693	63	11
21	1.35	1985	12,404	14,389	22.2	16,389	61	8
22	1.91	1150	15,010	16,160	14.8	15,442	54	9
23	1.11	1013	14,115	15,128	22.2	14,889	46	7
24	0.65	1630	11,253	12,883	29.6	13,786	59	10
25	1.35	1050	13,990	15,040	29.6	15,909	61	10
26	1.86	1281	12,254	13,535	14.8	15,909.96	55	11
27	1.45	980	12,905	13,885	22.2	14,362.8	55	9
28	1.26	1070	14,570	15,640	37	13,441.2	63	10

(continued)

Table 1 (continued)

Site No.	TOC (mg/l)	Ca (mg/l)	Mg (mg/l)	Total hardness (mg/l)	Acidity	Cl^- (mg/l)	Na (mg/l)	K (mg/l)
29	1.88	1340	12,660	14,000	22.2	14,630	60	8
30	1.38	1132	11,012	12,144	29.6	13,782.96	54	11
31	1.33	1150	13,709	14,859	37	16,563	53	8
32	1.09	1289	14,850	16,139	37	18,464	61	9
33	1.22	990	12,410	13,400	29.6	19,363.4	66	11
34	1.1	971	12,038	13,009	22.2	20,844.6	57	10
35	1.31	1340	13,630	14,970	14.8	20,062	56	10
36	1.26	938	14,024	14,962	22.2	21,257.4	62	9
37	1.03	970	13,030	14,000	29.6	20,346	54	8
38	1.29	984	12,848	13,832	22.2	22,290.96	66	11
39	1.34	1005	16,865	17,870	29.6	21,463	62	10
40	1.26	1205	11,450	12,655	25.6	14,658	54	11

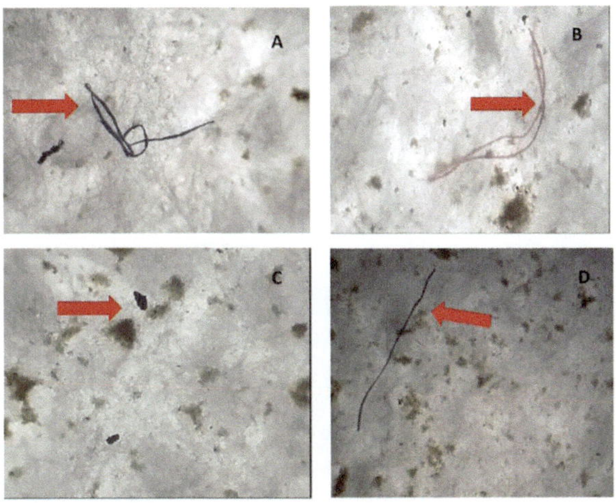

Fig. 2 Micro-plastics observed under the microscope at 40Xmagnification

(17%), grey (4%), and green (3%). The blue colours could be sourced from the microplastics generated from fishing gears like nets, which are mostly in blue shades of colour (Figs. 5 and 6). All the plastics observed are predominantly secondary microplastics which could have degraded from the human discarded wastes and wastes from fishing activities.

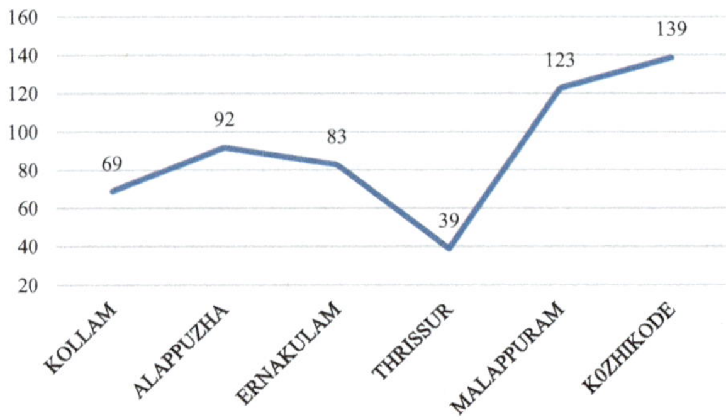

Fig. 3 Micro-plastic distribution district-wise

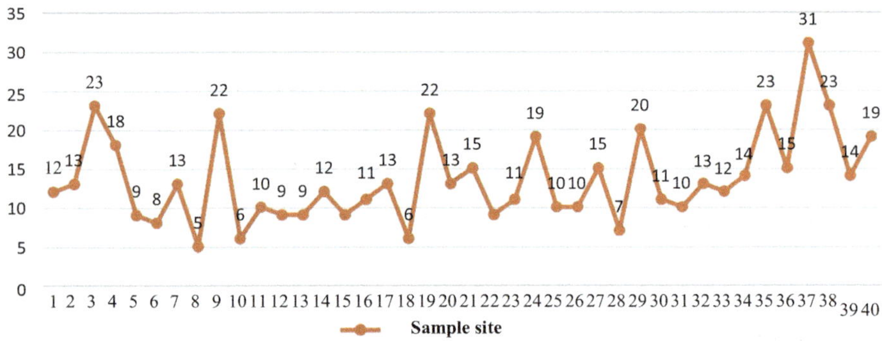

Fig. 4 Micro-plastic distribution site-wise

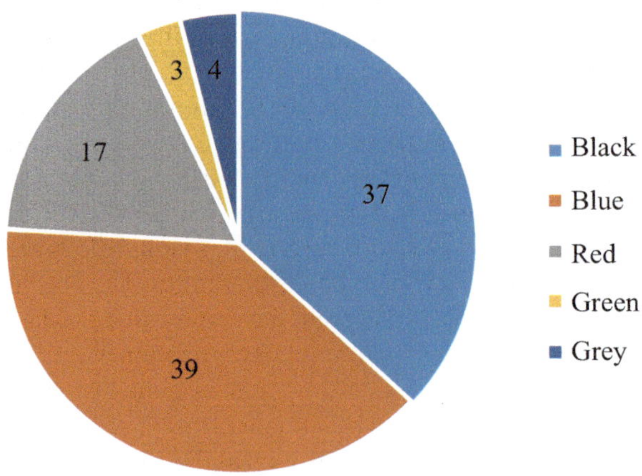

Fig. 5 Total microplastic colour distribution in study area

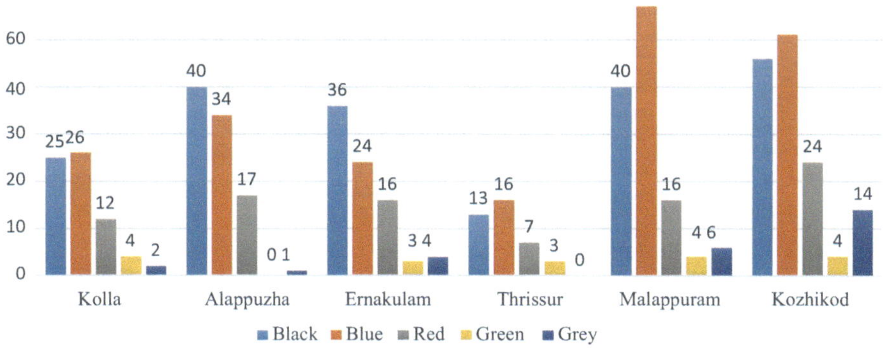

Fig. 6 Microplastic distribution in district-wise

3.3 *FTIR Analysis*

The FTIR analysis, the vibrations at 2922 and 2337 cm^{-1}, are attributed to asymmetric and symmetric C–H stretching vibrations, and it confirms the presence of polypropylene (PP), which is in agreement with the values obtained by Kaczmarek et al. (2005) at 2952, 2918, and 2840 cm^{-1}, or 2950, 2917, and 2838 cm^{-1} (Castillo et al., 2016) is shown in Table 2. The Nylon compound (polyamide) agrees with the published spectra by Jung et al. (2018). The vibration at 3623 cm^{-1} represented

Table 2 FTIR functional group

Frequency range (cm^{-m})	Functional group
3854	O-Hstretchingvibration
3587.12	Phenols
3373–3422	BondedN-H/C-H/O-Hstretchingamines
2918.2–2954	C-H
2500–3300	Carboxylacid
2322.8–2138.1	C-N
2047.30	Silicon compounds
1733.59	Ketones
1405–1445	Alkanes
1421–1415	C-O/C-Hbending
1382–1036	C-O
1215–1325	Alkylketone
1020–1220	Alkylamine
1026	Vibration of C-Oinalcoholhydroxyl group
469	Alkylhalides

the stretching of N–H bonds. Additional vibrations, which were observed at 2914, 2336, 1456 and 1374 cm^{-1} agree with literature values at 2932, 2858, 1634, 1538 1464, and 1372 cm^{-1} and mild vibration at 3100 cm^{-1} is attributed to NH Fermi resonance at 3086 cm^{-1} (Jung et al., 2018; Lee et al., 2008). The vibrations at 1008 and 1014 cm^{-1} confirm the presence of PVP. The existence of polypropylene, nylon, and PVP was confirmed, and it is obvious that the fluoropolymer discovered by FTIR is PP, according to the FTIR examination of MPs collected in samples. The FTIR Interferogram of two samples from each district is shown in Fig. 7.

4 Conclusion

A microplastic availability study was carried out on coastal water of southwestern Kerala through water samples from forty coastal sites and determined the distribution of microplastics, along with some of the physicochemical properties of water associated with them. Although no significant relationship between the physicochemical properties of water and the microplastics was observed in the present study, earlier studies have ascertained their role of nutrients, pH, salinity and dissolved oxygen in microplastic biofilm growth rate. The results of the study reveal that microplastic pollution originates from land-based activities. The highest abundance of microplastics was recorded in the Kozhikode district, which is due to the tourism and intensive pilgrimage activity. A vast number of multi-coloured small–sized fibers were observed in all the water samples, and the major polymer constituent was a blue colour. Further, microplastics undergo degradation, which makes them more suitable substrates for the absorption of toxic pollutants and, thus, more harmful to the marine biota. This study provides the baseline information on microplastic pollution in coastal areas, and this data could be useful for planning strategies to control plastic pollution in coastal areas. Further studies are required on an ongoing basis to understand the impact of microplastic ingestion by wildlife and humans. The management protocol must include strategies to reduce the influx of plastics into the coastal environment and measures to mitigate the long-term hazards of microplastics on the Kerala coast.

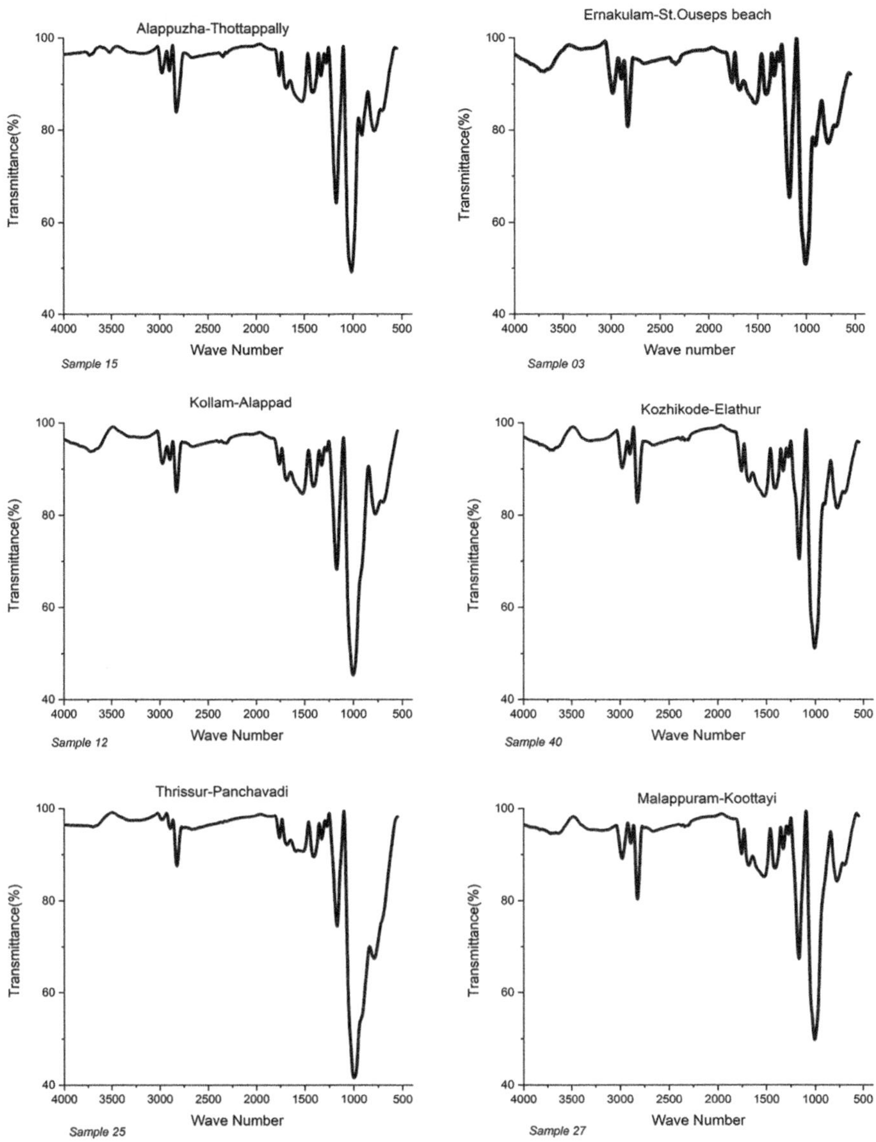

Fig. 7 FTIR Interferogram of water samples at Kollam, Alapuzha, Ernakulam, Thrissur, Malppuram, Kozhikode

Acknowledgements Authors acknowledge Rashtriya Uchchatar Shiksha Abhiyan (RUSA 2.0, Biological Sciences), scheme of MHRD, Government of India for the funding support.

Declarations The authors declare no competing interests.

References

Ajith, N., Arumugam, S., Parthasarathy, S., Manupoori, S., & Janakiraman, S. (2020). Global distribution of microplastics and its impact on marine environment—a review. *Environmental Science and Pollution Research, 27*(21), 25970–25986. https://doi.org/10.1007/s11356-020-09015-5

Artham, T., Sudhakar, M., Venkatesan, R., Nair, C. M., Murty, K., & Doble, M. (2009). Biofouling and stability of synthetic polymers in sea water. *International Biodeterioration & Biodegradation, 63*(7), 884–890.

Baker, A. (2011). *Proceedings of the second research workshop on microplastic marine debris.* NOAA Technical Memorandum, November, 54.

Bosker, T., Bouwman, L. J., Brun, N. R., Behrens, P., & Vijver, M. G. (2019). Microplastics accumulate on pores in seed capsule and delay germination and root growth of the terrestrial vascular plant *Lepidium sativum. Chemosphere, 226*, 774–781. https://doi.org/10.1016/j.chemosphere.2019.03.163

Browne, M. A., Galloway, T. S., & Thompson, R. C. (2010). Spatial patterns of plastic debris along estuarine shorelines. *Environmental Science and Technology, 44*(9), 3404–3409. https://doi.org/10.1021/es903784e

Carson, H. S., Colbert, S. L., Kaylor, M. J., & McDermid, K. J. (2011). Small plastic debris changes water movement and heat transfer through beach sediments. *Marine Pollution Bulletin, 62*(8), 1708–1713. https://doi.org/10.1016/j.marpolbul.2011.05.032

Castillo, A., Al-Maslamani, D. I., & Obbard, J. (2016). Prevalence of microplastics in the marine waters of Qatar. *Marine Pollution Bulletin, 111*. https://doi.org/10.1016/j.marpolbul.2016.06.108

Chae, Y., & An, Y. J. (2018). Current research trends on plastic pollution and ecological impacts on the soil ecosystem: A review. *Environmental Pollution, 240*, 387–395. https://doi.org/10.1016/j.envpol.2018.05.008

CPCB. (2007). *Assessment of fugitive emissions & in cement manufacturing industries.* Central Pollution Control Board (Ministry of Environment & Forests, Govt. of India) Parivesh Bhawan, East Arjun Nagar Delhi.

da Costa, J. P., Chamkha, M., Ksibi, M., & Sayadi, S. (2023). Effects of microplastics' physical and chemical properties on aquatic organisms: State-of-the-art and future research trends. *TrAC Trends in Analytical Chemistry*, 117192.

Di Pippo, F., Venezia, C., Sighicelli, M., Pietrelli, L., Di Vito, S., Nuglio, S., & Rossetti, S. (2020). Microplastic-associated biofilms in lentic Italian ecosystems. *Water Research, 187*, Article 116429.

Fok, L., Cheung, P. K., Tang, G., & Li, W. C. (2017). Size distribution of stranded small plastic debris on the coast of Guangdong, South China. *Environmental Pollution, 220*, 407–412. https://doi.org/10.1016/j.envpol.2016.09.079

GESAMP. (2015). Report of the forty-second session of GESAMP. *Journal Series GESAMP Reports and Studies, 92*(September), 79.

Geyer, R., Jambeck, J. R., & Law, K. L. (2017). Production, use, and fate of all plastics ever made. *Science Advances, 3*(7), 25–29. https://doi.org/10.1126/sciadv.1700782

Guan, J., Qi, K., Wang, J., Wang, W., Wang, Z., Lu, N., & Qu, J. (2020). Microplastics as an emerging anthropogenic vector of trace metals in freshwater: Significance of biofilms and comparison with natural substrates. *Water Research, 184*, Article 116205.

Hermabessiere, L., Dehaut, A., Paul-Pont, I., Lacroix, C., Jezequel, R., Soudant, P., & Duflos, G. (2017). Occurrence and effects of plastic additives on marine environments and organisms: A review. *Chemosphere, 182*, 781–793. https://doi.org/10.1016/j.chemosphere.2017.05.096

Hidalgo-Ruz, V., Gutow, L., Thompson, R. C., & Thiel, M. (2012). Microplastics in the marine environment: A review of the methods used for identification and quantification. *Environmental Science & Technology, 46*(6), 3060–3075. https://doi.org/10.1021/es2031505

Jambeck, J. R., Roland, G., Wilcox, C., Siegler, T. R., Perryman, M., Andrady, A., Narayan, R., & Law, K. L. (2015). Marine pollution. Plastic waste inputs from land into the ocean. *Science, 347*(6223), 764–768.

Jung, M. R., Horgen, F. D., Orski, S. V., Rodriguez C., V., Beers, K. L., Balazs, G. H., Jones, T. T., Work, T. M., Brignac, K. C., Royer, S. J., Hyrenbach, K. D., Jensen, B. A., & Lynch, J. M. (2018). Validation of ATR FT-IR to identify polymers of plastic marine debris, including those ingested by marine organisms. *Marine Pollution Bulletin, 127*, 704–716. https://doi.org/10.1016/j.marpolbul.2017.12.061

Kaczmarek, H., OlŁdak, D., Malanowski, P., & Chaberska, H. (2005). Effect of short wavelength UV-irradiation on ageing of polypropylene/ cellulose compositions. *Polymer Degradation and Stability, 88*(2), 189–198. https://doi.org/10.1016/j.polymdegradstab.2004.04.017

Kane, I., & Clare, M. (2019). Dispersion, accumulation and the ultimate fate of microplastics in deep-marine environments: A review and future directions. https://doi.org/10.31223/osf.io/ahjes

Karthik, R., Robin, R. S., Purvaja, R., Ganguly, D., Anandavelu, I., Raghuraman, R., Hariharan, G., Ramakrishna, A., & Ramesh, R. (2018). Microplastics along the beaches of southeast coast of India. *Science of the Total Environment, 645*, 1388–1399. https://doi.org/10.1016/j.scitotenv.2018.07.242

Lebreton, L. C. M., Van Der Zwet, J., Damsteeg, J. W., Slat, B., Andrady, A., & Reisser, J. (2017). River plastic emissions to the world's oceans. *Nature Communications, 8*, 1–10. https://doi.org/10.1038/ncomms15611

Lee, J. W., Nam, J. H., Kim, Y. H., Lee, K. H., & Lee, D. H. (2008). Bacterial communities in the initial stage of marine biofilm formation on artificial surfaces. *Journal of Microbiology, 46*(2), 174–182. https://doi.org/10.1007/s12275-008-0032-3

Liu, K., Wang, X., Fang, T., Xu, P., Zhu, L., & Li, D. (2019). Source and potential risk assessment of suspended atmospheric microplastics in Shanghai. *Science of the Total Environment, 675*, 462–471. https://doi.org/10.1016/j.scitotenv.2019.04.110

Marine Debris Program, N. (2015). Laboratory methods for the analysis of microplastics in the marine environment: Recommendations for quantifying synthetic particles in waters and sediments. July.

Michels, J., Stippkugel, A., Lenz, M., Wirtz, K., & Engel, A. (2018). Rapid aggregation of biofilm-covered microplastics with marine biogenic particles. *Proceedings of the Royal Society B, 285*(1885), 20181203.

OECD. (2023). *Towards eliminating plastic pollution by 2040 a policy scenario analysis*. November.

Ojeda, T., Dalmolin, E., Forte, M. M., Jacques, R., Bento, F., & Camargo, F. (2009). Abiotic and biotic degradation of oxo-biodegradable polyethylenes. *Polymer Degradation and Stability, 94*, 965–970. https://doi.org/10.1016/j.polymdegradstab.2009.03.011

Plastics – the Facts 2019. (2019).

Qiu, Q., Tan, Z., Wang, J., Peng, J., Li, M., & Zhan, Z. (2016). Extraction, enumeration and identification methods for monitoring microplastics in the environment. *Estuarine, Coastal and Shelf Science, 176*, 102–109. https://doi.org/10.1016/j.ecss.2016.04.012

Rillig, M. C., Ziersch, L., & Hempel, S. (2017). Microplastic transport in soil by earthworms. *Scientific Reports, 7*(1), 1–6. https://doi.org/10.1038/s41598-017-01594-7

Robin, R. S., Karthik, R., Purvaja, R., Ganguly, D., Anandavelu, I., Mugilarasan, M., & Ramesh, R. (2020). Holistic assessment of microplastics in various coastal environmental matrices, south-west coast of India. *Science of the Total Environment, 703*, Article 134947. https://doi.org/10.1016/j.scitotenv.2019.134947

Schmidt, C., Krauth, T., & Wagner, S. (2017). Export of plastic debris by Rivers into the Sea. *Environmental Science and Technology, 51*(21), 12246–12253. https://doi.org/10.1021/acs.est.7b02368

Sharma, S., & Chatterjee, S. (2017). Microplastic pollution, a threat to marine ecosystem and human health: A short review. *Environmental Science and Pollution Research, 24*(27), 21530–21547. https://doi.org/10.1007/s11356-017-9910-8

Tiwari, M., Rathod, T. D., Ajmal, P. Y., Bhangare, R. C., & Sahu, S. K. (2019). Distribution and characterization of microplastics in beach sand from three different Indian coastal environments. *Marine Pollution Bulletin, 140*, 262–273. https://doi.org/10.1016/j.marpolbul.2019.01.055

Urbina, M. A., Correa, F., Aburto, F., & Ferrio, J. P. (2020). Adsorption of polyethylene microbeads and physiological effects on hydroponic maize. *Science of the Total Environment, 741*, Article 140216. https://doi.org/10.1016/j.scitotenv.2020.140216

Van Cauwenberghe, L., Claessens, M., Vandegehuchte, M. B., Mees, J., & Janssen, C. R. (2013). Assessment of marine debris on the Belgian Continental Shelf. *Marine Pollution Bulletin, 73*(1), 161–169. https://doi.org/10.1016/j.marpolbul.2013.05.026

Veerasingam, S., Ranjani, M., Venkatachalapathy, R., Bagaev, A., Mukhanov, V., Litvinyuk, D., Verzhevskaia, L., Guganathan, L., & Vethamony, P. (2020). Microplastics in different environmental compartments in India: Analytical methods, distribution, associated contaminants and research needs. *TrAC - Trends in Analytical Chemistry, 133*, Article 116071. https://doi.org/10.1016/j.trac.2020.116071

Verhagen, P., De Gelder, L., Hoefman, S., De Vos, P., & Boon, N. (2011). Planktonic versus biofilm catabolic communities: Importance of the biofilm for species selection and pesticide degradation. *Applied and Environmental Microbiology, 77*(14), 4728–4735.

Vidyasakar, A., Neelavannan, K., Krishnakumar, S., Prabaharan, G., Priyanka, T. S. A., Magesh, N. S., Godson, P. S., & Srinivasalu, S. (2018). Macrodebris and microplastic distribution in the beaches of Rameswaram Coral Island, Gulf of Mannar, Southeast coast of India: A first report. *Marine Pollution Bulletin, 137*, 610–616. https://doi.org/10.1016/j.marpolbul.2018.11.007

Vidyasakar, A., Krishnakumar, S., Kasilingam, K., Neelavannan, K., Bharathi, V. A., Godson, P. S., Prabha, K., & Magesh, N. S. (2020). Characterization and distribution of microplastics and plastic debris along Silver Beach, Southern India. *Marine Pollution Bulletin, 158*, Article 111421. https://doi.org/10.1016/j.marpolbul.2020.111421

Wang, S., Chen, H., Zhou, X., Tian, Y., Lin, C., Wang, W., Zhou, K., Zhang, Y., & Lin, H. (2020). Microplastic abundance, distribution and composition in the mid-west Pacific Ocean. *Environmental Pollution, 264*, Article 114125. https://doi.org/10.1016/j.envpol.2020.114125

Ward, J. E., Zhao, S., Holohan, B. A., Mladinich, K. M., Griffin, T. W., Wozniak, J., & Shumway, S. E. (2019). Selective ingestion and egestion of plastic particles by the blue mussel (*Mytilus edulis*) and Eastern Oyster (*Crassostrea virginica*): Implications for using bivalves as bioindicators of microplastic pollution. *Environmental Science and Technology, 53*(15), 8776–8784. https://doi.org/10.1021/acs.est.9b02073

Wei, X. F., Bao, R. Y., Cao, Z. Q., Yang, W., Xie, B. H., & Yang, M. B. (2014). Stereocomplex crystallite network in asymmetric PLLA/PDLA blends: Formation, structure, and confining effect on the crystallization rate of homocrystallites. *Macromolecules, 47*(4), 1439–1448. https://doi.org/10.1021/ma402653a

Woodall, L. C., Sanchez-Vidal, A., Canals, M., Paterson, G. L. J., Coppock, R., Sleight, V., Calafat, A., Rogers, A. D., Narayanaswamy, B. E., & Thompson, R. C. (2014). The deep sea is a major sink for microplastic debris. *Royal Society Open Science, 1*(4). https://doi.org/10.1098/rsos.140317

Yang, C. Z., Yaniger, S. I., Jordan, V. C., Klein, D. J., & Bittner, G. D. (2011). Most plastic products release estrogenic chemicals: A potential health problem that can be solved. *Environmental Health Perspectives, 119*(7), 989–996. https://doi.org/10.1289/ehp.1003220

Yang, Y., Liu, G., Song, W., Ye, C., Lin, H., Li, Z., & Liu, W. (2019). Plastics in the marine environment are reservoirs for antibiotic and metal resistance genes. *Environment International, 123*(1), 79–86. https://doi.org/10.1016/j.envint.2018.11.061

Zhang, K., Hamidian, A. H., Tubić, A., Zhang, Y., Fang, J. K. H., Wu, C., & Lam, P. K. S. (2021). Understanding plastic degradation and microplastic formation in the environment: A review. *Environmental Pollution, 274*. https://doi.org/10.1016/j.envpol.2021.116554

GPSR Compliance

The European Union's (EU) General Product Safety Regulation (GPSR) is a set of rules that requires consumer products to be safe and our obligations to ensure this.

If you have any concerns about our products, you can contact us on ProductSafety@springernature.com

In case Publisher is established outside the EU, the EU authorized representative is:

Springer Nature Customer Service Center GmbH
Europaplatz 3
69115 Heidelberg, Germany

Batch number: 10584688

Printed by Printforce, the Netherlands